CAMBRIDGE ASTROPHYSICS SERIES

Globular clusters

GLOBULAR CLUSTERS

*Based on the proceedings of a
Nato Advanced Study Institute held at
the Institute of Astronomy
University of Cambridge, August 1978*

EDITED BY

D. HANES & B. MADORE

CAMBRIDGE UNIVERSITY PRESS

Cambridge

London   New York   New Rochelle
Melbourne   Sydney

Published by the Press Syndicate of the University of Cambridge
The Pitt Building, Trumpington Street, Cambridge CB2 1RP
32 East 57th Street, New York, NY 10022, USA
296 Beaconsfield Parade, Middle Park, Melbourne 3206, Australia

© Cambridge University Press 1980

First published 1980

Printed in Great Britain at the University Press, Cambridge

British Library cataloguing in publication data

Nato Advanced Study Institute on Globular Clusters,
University of Cambridge, 1978
Globular clusters. – (Cambridge astrophysics series).

1. Stars – Globular clusters – Congresses
I. Title   II. Hanes, D   III. Madore, B
IV. Series
523.8'55   QB853   79–41472

ISBN 0 521 22861 1

CONTENTS

	List of contributors	vii
	Preface	viii
	Introductory remarks	ix
1	Milestones in globular cluster research: a guide to the literature	
	D. A. Hanes	1
2	Terminology and fundamental data of globular clusters	
	B. F. Madore	21
3	Stellar evolution and globular clusters	
	V. Castellani	65
4	Abundance anomalies in globular clusters	
	R. P. Kraft	87
5	Populations in globular clusters	
	K. C. Freeman	103
6	The correlation of cyanogen, calcium and the heavy elements on the giant branch of Omega Centauri	
	J. Norris	113
7	Nucleosynthesis in evolved globular clusters	
	I. Iben, Jr	125
8	Infrared observations of red giants in globular clusters	
	S. E. Persson and J. A. Frogel	143
9	Infrared observations of globular clusters in M31 and a comparison with galactic globulars and elliptical galaxies	
	J. A. Frogel, S. E. Persson and J. G. Cohen	159

Contents

10	Globular clusters and galaxy evolution *S. van den Bergh*	175
11	The dwarf spheroidal galaxies *R. Zinn*	191
12	Globular clusters as extragalactic distance indicators *D. A. Hanes*	213
13	Luminosity distributions and density profiles *I. R. King*	249
14	On the dynamics of globular clusters *J. E. Gunn*	271
15	Dynamical theory of binaries in clusters *D. C. Heggie*	281
16	Some thoughts concerning the origin of globular clusters *J. E. Gunn*	301
17	Globular clusters as survivors *S. M. Fall*	309
18	X-ray burst sources in globular clusters and the galactic bulge *W. H. G. Lewin*	315
19	X-rays from globular clusters: steady emission *J. G. Jernigan*	351
20	Summary of contemporary research on globular clusters *K. C. Freeman*	361
	Glossary	369
	Author index	371
	Object index	379
	Subject index	384

CONTRIBUTORS

V. Castellani
Laboratorio Astrofisica Spaziale, Casella Postale 67, 00044 Frascati, Italy

J. G. Cohen
Kitt Peak National Observatory,† 950 North Cherry Avenue, PO Box 26732, Tucson, Arizona 85726, USA

S. M. Fall
Institute of Astronomy, University of Cambridge, Madingley Road, Cambridge CB3 0HA, UK

K. C. Freeman
Mount Stromlo Observatory, Australian National University, Canberra, ACT, Australia

J. A. Frogel
Cerro Tololo Inter-American Observatory,† Casilla 63-D, La Serena, Chile

J. E. Gunn
California Institute of Technology, Pasadena, California 91125, USA

D. A. Hanes
Institute of Astronomy, University of Cambridge, Madingley Road, Cambridge CB3 0HA, UK

D. C. Heggie
University of Edinburgh, Department of Mathematics, King's Buildings, Mayfield Road, Edinburgh EH9 3JZ, UK

I. Iben, Jr
Department of Astronomy, University of Illinois at Urbana-Champaign, Urbana, Illinois 61801, USA

J. G. Jernigan, Jr
Space Research Center, Massachusetts Institute of Technology, Room 37-607, Cambridge, Massachusetts 02139, USA

I. R. King
Department of Astronomy, University of California, Berkeley, California 94720, USA

R. P. Kraft
Lick Observatory, Board of Studies in Astronomy and Astrophysics, University of California, Santa Cruz, California 95064, USA

W. H. G. Lewin
Space Research Center, Massachusetts Institute of Technology, Room 37-627, Cambridge, Massachusetts 02139, USA

B. F. Madore
Department of Astronomy, University of Toronto, Toronto, Ontario, Canada

J. Norris
Mount Stromlo Observatory, Australian National University, Canberra, ACT, Australia

S. E. Persson
Hale Observatories, 813 Santa Barbara St, Pasadena, California 91101, USA

S. van den Bergh
Dominion Astrophysical Observatory, 5071 W. Saanich Road, Victoria BC V8X 3X3, Canada

R. J. Zinn
Hale Observatories, 813 Santa Barbara St, Pasadena, California 91101, USA

† Supported by the National Science Foundation under contract No. AST 74-04128.

PREFACE

This book was first conceived of as the members of the Institute of Astronomy, Cambridge, prepared in 1978 to host a NATO-sponsored Advanced Study Institute on the topic of 'Globular Clusters'. It was realized then that no general volume in the area existed and that a long-standing need could be met, at a time moreover when the recent exciting discoveries chronicled in this book made the lack more acutely felt.

Happily, those scientists who were to present review papers at the Institute shared our views and our enthusiasm and agreed to prepare the chapters included here; to them we express our thanks. We have together sought to create a volume which provides a comprehensive overview of globular cluster research as it stands at present. Dr Ivan King and Dr Robert Kraft, already committed to extensive reviews elsewhere, graciously permitted the editors to prepare written summaries from taped versions of their presentations and were kind enough to review (and improve upon) our efforts.

We would like, too, to thank the secretarial and technical staff of the Institute of Astronomy, Cambridge, and of the David Dunlap Observatory and the University of Toronto, Canada, for their untiring efforts in preparing the manuscripts and the figures; they are too many to name individually, but know of our gratitude. Nor would the book have reached fruition without the original impetus given the project by Martin Rees, who encouraged our efforts at every stage.

Finally, we gratefully acknowledge the generous sponsorship afforded the Advanced Study Institute by NATO, and thank the Institute of Astronomy, Cambridge, for hosting the meeting which sparked the preparation of this book.

University of Cambridge *Dave Hanes*
Institute of Astronomy *Barry Madore*

INTRODUCTORY REMARKS

The history of the study of globular clusters is a long one, dating at least as far back as A. Ihle's discovery of M22 in A.D. 1665. A striking feature, though, is the rapidly increasing *diversity* of globular cluster research, with new techniques being applied and new terminology – that of X-ray astronomy, for example – entering the lexicon for the first time in the last decade. This book would have had a very different appearance ten years ago, and it is of course impossible to predict what fresh discoveries may come in the near future.

Harlow Shapley, writing in 1949, had this to say: 'To one who has worked long with the globular clusters it seems as if practically nothing has yet been satisfactorily solved. The problems ahead are numerous, interesting, significant, and some of them for the present impossible.' A few of the puzzles which so perplexed him have been (more or less) solved, but by and large his remarks could be repeated today without an eyebrow being raised in surprised response. That of course is what makes globular cluster research such a rich area of study, as we hope to chronicle in this volume.

D.A.H.
B.F.M.

1
Milestones in globular cluster research: a guide to the literature

DAVID A. HANES

1.1 Introduction

It is my intention to present in this chapter a guide to the most recent and comprehensive review papers available in various areas of globular cluster research. In this way, I hope to provide a signpost to the literature so that the reader to whom some particular area is unfamiliar can easily be referred to a logical starting point for subsequent study. It is *not* my intention to present a critical appraisal of all the literature, which would in any event demand much more space than is available. That sort of critical assessment will be found in the other chapters of this book.

In a number of areas I have gone beyond my stated aims and also identified the classical papers in the literature, often half a century or more old. Needless to say, the more recent reviews will lead the interested reader to these papers eventually, but their significance then and now is such that they warrant special mention. On the other side of the coin, in some areas I have not found comprehensive recent reviews and I have simply identified one or more pertinent papers in the recent literature as exemplifying the ongoing research in those areas.

It seems to me that this kind of review can proceed in one of two logical ways. The first of these is the strictly chronological, and is exemplified by Section B of Sawyer's (1947) excellent *Bibliography of Individual Globular Clusters*, wherein she presents a year-by-year summary of research into the globular clusters in our own galaxy for the years 1679–1946. The second approach, that of subdividing the literature into broadly distinct areas, has an obvious and appealing simplicity for the reader who presumably knows where his interests lie. I shall adopt that topical approach, preserving the chronology where possible within a particular subtopic.

The subdivision into subtopics is a matter of personal preference and, while I have cross-referenced closely similar topics as much as possible, this chapter must certainly still reflect my own subjective views. In particular, the division into sections 1.4 (Globular clusters as stellar aggregates) and

1.5 (Globular clusters star-by-star) is not clearcut. I can only apologize in advance for any ambiguity and suggest that these sections be considered more or less together. Nor can this chapter as a whole hope to have any real degree of completeness; indeed, it is the recognition of the impossibility of such an aim that has led me to write this chapter at all – the review papers suggested here will 'cascade' the reader through the literature more thoroughly than any single tabulation could ever achieve.

1.2 General topics

1.2.1 Recent review papers

The fact that globular cluster research has been especially active in recent years is reflected not only in the appearance of this book but also in the appearance in 1979 in a single volume of *Annual Reviews of Astronomy and Astrophysics* of no less than three review papers dedicated to the subject. Harris & Racine (1979) consider the integrated properties of globular clusters, with especial reference to the properties of the galaxies to which they belong. They treat globular clusters as probes of galactic structure and evolutionary history, and test their uniformity within disparate systems. Kraft (1979) considers the question of abundances within the globular clusters and satellite subsystems of our own galaxy; he first reviews methods of abundance determinations and thereafter addresses the question of the chemical non-homogeneity within and between clusters and halo stars. Finally, Freeman & Norris (1979) present a review in a chapter entitled simply 'Globular clusters'. I regret that I have not yet seen that chapter in preprint form†, but I am confident that it will maintain the high standards we have come to expect both from the authors and from the series of *Annual Reviews*. These three papers together make Volume 17 an auspicious place from which to begin a survey of globular cluster research.

1.2.2 Ongoing research

The interested reader new to some area of astronomical research could hardly do better than begin his study with a perusal of the most recent edition of *Reports on Astronomy*, published triennially under the auspices of the International Astronomical Union. Summaries of research in the preceding three-year period are presented in a variety of categories, with

† Note added in proof: that chapter, anticipated in a prospectus, did not in fact appear in Volume 17.

the report of IAU Commission 37 (*Star Clusters and Associations*) being of particular interest here. In addition to providing lucid descriptions of the most recent advances in the area, nicely bringing together the research efforts of many individuals and groups, the reports incorporate short descriptions of new catalogues and collections of data, as well as tabular summaries, object by object, of recent studies carried out in specific clusters. Attention should also be paid to the reports of IAU Commission 27 (*Variable Stars*) and IAU Commission 33 (*Structure and Dynamics of the Galactic System*).

These IAU reports have the benefit of a three-year compilation time and give unhurried overviews of the directions various lines of research are leading. For a more immediate taste of the vigour of cluster research, an exploration of a recent edition of *Astronomy and Astrophysics Abstracts*, with its six-monthly summary of every article published in the area, is invaluable.

1.2.3 Historically important reviews

Shapley's (1930) book *Star Clusters* has a deserved reputation as the classic early work, reviewing as it does not only the observational data available to that time but also chronicling our developing awareness of the importance of globular clusters in the understanding of the structure and history of our own galaxy. Sawyer (1947) presents a summary, year-by-year and object-by-object, of cluster research through more than two and a half centuries, and compiles all available data. (A supplement, Sawyer-Hogg (1963), was subsequently published.) Shapley (1949) surveyed *A Half Century of Globular Clusters*; interestingly, he recorded on the third-last page his puzzlement over the low luminosity of the putative globular clusters in M31, a comment which anticipated Baade's revision (in 1952) of the extragalactic distance scale.

A decade later, Sawyer-Hogg's (1959a) beautifully complete review of associations and of open and galactic globular clusters appeared. In the same year, the proceedings of an AAS-sponsored symposium The Differences Among Globular Clusters were published in the *Astronomical Journal*, featuring reports on: the areas of difference among globular clusters (Sawyer-Hogg 1959b); the photometry of clusters in our own galaxy and in extragalactic systems (Kron & Mayall 1959); the integrated spectra of globular clusters (Morgan 1959); a comparison of globular clusters in our own galaxy and in the Magellanic Clouds (Thackeray 1959); and the colours, magnitudes and metallicities of stars in globular clusters (Arp 1959). Some interesting discussion sessions were included, with the moderator (Allan Sandage) concluding that 'the problem of interpretation appears to be more complicated than we thought only a few years ago'.

Baade (1963), in a posthumous volume edited by C. Payne-Gaposchkin, reviewed in an informal and readable way the important part played by globular clusters in galactic studies as well as in their own right. Finally, Arp (1965) continued in this vein in a well-known review paper which was intended mainly as a summary of how globular clusters serve as indicators of galactic structure and of the evolutionary history of our own galaxy.

1.2.4 Matters of definition

The simple definition of a globular cluster is by no means straightforward. The alternative definitions which rely on geometrical properties on the one hand (the identification of a dynamical grouping) and the stellar content on the other (say, as evidenced by the morphology of the colour–magnitude diagram) may conflict – as, for example, in the young and intermediate-age clusters in the Magellanic Clouds which are structurally similar to conventional globular clusters. This problem was considered in the general discussion following Arp's (1959) report and was reconsidered by Arp (1965). Harris & Racine (1979) present the most recent assessment, preferring on balance to rely on photometric data to isolate the 'true' globular clusters. The various parameters which have been adopted over the years to describe real structural features or morphological features in the colour–magnitude diagram are tabulated in Madore (p. 21, this volume), and their sources are identified there.

The question of definition extends also to the classical identification of globular clusters as representative of a 'Population II' component of the galaxy. This identification, suggested by Baade (1944), had in fact been anticipated by Shapley (1915) in a study of the colour–magnitude diagram of M13 (and of other clusters shortly thereafter). An entertaining account of Baade's observations is to be found in Whitney (1971), while Shapley (1949), Struve & Zebergs (1962), and Baade (1963) provide historical reviews. Blaauw (1965) explains the physical, kinematic and compositional criteria distinguishing different populations, while in the same volume Arp (1965) re-emphasizes the importance of globular clusters in this understanding. The most recent and comprehensive reviews are those of van den Bergh (1975) and Tinsley (1978), while Freeman (p. 103, this volume) draws an interesting analogy when he suggests that the galactic globular cluster ω Cen may have stellar populations reminiscent of those within galaxies.

1.3 Data compilations

Needless to say, many of the review papers mentioned elsewhere in this chapter will contain tabular material summarizing observational data to date, or will contain references to such information. Here, however, I will simply list some of the better known compilations for quick reference; these and other sources are described in more detail in Madore (p. 21, this volume).

Observational (photometric, spectroscopic, positional, kinematic) data are to be found in the compilations of Sawyer-Hogg (1959a), Arp (1965), White (1970), Alcaino (1973), Kukarkin (1974), Woltjer (1975), Harris (1976), Davis Philip (1977) and Harris & Racine (1979). Structural parameters for galactic globular clusters are to be found in Peterson & King (1975). Variable stars in galactic globular clusters are identified and their properties summarized in Sawyer-Hogg's (1973) *Third Catalogue of Variable Stars in Globular Clusters* (with a fourth edition imminent). Finally, the properties of known globular clusters in external galaxies are elucidated in Harris & Racine (1979).

1.4 Globular clusters as stellar aggregates

1.4.1 Star counts and surface photometry

King (p. 249, this volume; and 1975) reviews the methods and problems in this fruitful area of research. Early functional representations of the stellar distribution within clusters (von Zeipel 1908) were replaced by the elegant semi-theoretical functions introduced in a series of classical papers by King (1962, 1966a, b). Compilations of star counts are to be found in King, Hedemann, Hodge & White (1968), and the structural parameters thereby derived are tabulated in Peterson & King (1975).

1.4.2 Integrated photometric properties

See Madore (p. 21, this volume) and Harris & Racine (1979) for the most recent summaries of available data.

1.4.3 Reddenings

See Madore (p. 21, this volume) and Harris & Racine (1979) for tabulated reddenings from the best available sources. Madore (p. 21) presents detailed descriptions of the techniques whereby reddenings are estimated.

1.4.4 Spectral types

Van den Bergh (1969, and p. 183, this volume) describes in some detail the interpretation of the observed frequency distributions of globular cluster integrated spectral types in our own galaxy and in M31; the data themselves are to be found in van den Bergh (1969), where various spectral features are used to define the line-strength parameter L employed by Frogel, Persson & Cohen (p. 165, this volume) and by van den Bergh (p. 183, this volume). The classical paper in the field is that of Mayall (1946), who first focused attention on the vast range of spectroscopic characteristics seen. Kukarkin (1974) provides a recent summary in his Table D.

1.4.5 Spectrum scans

Moderate resolution (~ 35 Å) spectrum scans were presented for 21 galactic globular clusters by van den Bergh & Henry (1962). This work was nicely complemented by Aller & Faulkner's (1964) study (at typical resolutions of 20–40 Å) of several southern globular clusters, including a few in the Magellanic Clouds. D. Hanes & J. Brodie (in preparation) have repeated these studies with modern spectrophotometric instrumentation at 12 Å resolution, while L. Searle (in preparation) is continuing his analysis of low resolution scans of globular clusters in M31. Searle's results are peripherally described in Racine, Oke & Searle's (1978) study of spectrum scans of globular clusters associated with the giant elliptical galaxy NGC 4486. Spinrad & Peimbert (1975) have explained the importance of including a spectrophotometric component corresponding to a globular cluster population in any recipe of stellar-population-synthetic modelling of galaxies.

1.4.6 Chemical composition

The question of the abundances in the stars within globular clusters and the galactic halo, and of the non-homogeneity of such abundances, is one which has recently been actively and enthusiastically explored; there has been a flood of information in this area, as noted by Kraft (p. 87, this volume). The most recent reviews are: those of Kraft (1979, and p. 87, this volume), who concentrates primarily on the question of abundance anomalies but who (in Kraft 1979) nicely reviews the techniques whereby abundances can be estimated; the chapter of Madore (p. 21, this volume), who likewise describes these techniques, and in tabular form summarizes the abundances deduced for the clusters within our own galaxy; the chapter of Persson & Frogel (p. 143, this volume), who give an account of infrared indices as abundance indicators for single stars within globular clusters; and the chapter of Frogel et al. (p. 159, this volume), who summarize methods of gauging the composition from measures of the integrated infrared light of a cluster.

Kraft's excellent and comprehensive reviews will lead the enquiring reader to a plethora of other equally fascinating reports; I will mention here a few of special interest. Mayall (1946) first established the great range in spectroscopic characteristics evidenced by galactic globular clusters; and ten years later Morgan (1956) demonstrated the existence of a correlation between the galactocentric distances and metal depletion of globular clusters, firmly establishing their importance in the history of the chemical evolution of the galaxy. This dependence was summarized by Arp (1965), while Plaut (1965) reaffirmed the importance of RR Lyrae stars in our understanding of galactic structure and the run of chemical abundances. Danziger (1970) elaborated on this when he described (in part) how the structure of the horizontal branch and investigations of stellar pulsation theory put constraints on the helium abundance within globular clusters; Castellani (p. 65, this volume) provides the latest word in this area. Meanwhile, Pagel (1970) reviewed methods of determining the chemical composition of the old stars and restated the importance of such studies in understanding the chemical evolution of the galaxy.

An exciting development was the discovery by Zinn (1973) of chemical inhomogeneities within a single cluster. As Kraft (p. 87, this volume) relates, this discovery, coupled with (among other evidence) Freeman & Rodgers' (1975) further demonstration of inhomogeneities on the horizontal branch of ω Cen, sparked a flurry of activity. Another manifestation of the complexity of the abundance distribution within globular clusters is the so-called second-parameter effect, whereby clusters of the same mean metallicity have very different colour–magnitude diagrams (see fig. 4.3 of Kraft, p. 93, this volume).

The unravelling of the physical effects underlying these observations is far from complete, as Kraft describes. Several recent symposia have dedicated large amounts of time to these questions; I would mention in particular; the reviews by Bell (1976) and Spinrad (1976); the published version of Renzini's (1977) lecture series on the evolution of Population II stars; Castellani's (1977) excellent review on the enrichment history in our own galaxy; the paper by Butler, Bell, Dickens & Epps (1978) on RR Lyrae stars in ω Cen; and finally Gustafsson's (1979) review of the techniques of model stellar atmospheres in estimating abundances.

1.4.7 Mass functions

The most recent and explanatory review in this area is that of Freeman (1977), who summarizes data on globular clusters in the galaxy and in the Magellanic Clouds, as well as detailing the problems involved in these studies (such as segregation by mass, etc.). See King (p. 265, this volume) for a discussion of the implications of adopting mass functions of various kinds in modelling globular clusters as they are actually observed.

1.4.8 Mass/luminosity ratios, masses

King (p. 265, this volume, and 1975) discusses the theoretical and observational problems here, while the best recent review is that of Illingworth (1975). Tabulations of available determinations are to be found in Illingworth (1976) and in Illingworth & King (1976).

1.4.9 Cluster dynamics

Just as the detailed investigation of abundances within globular clusters was sparked by the development of modern astronomical instrumentation, so too the study of cluster dynamics has been greatly forwarded by the evolution of the modern computer. The outstanding recent review in the area is that of Lightman & Shapiro (1978), who in a comprehensive article summarize the observational attributes of globular clusters and then review their postulated dynamical history – from the early stages of violent relaxation through quasi-steady evolution and mass segregation to core collapse, the death of globular clusters, and beyond. This is a comprehensive and impressive review.

In more specific areas, several of the chapters in this volume are of great interest: Gunn (p. 271) reports upon his investigations into the observed dynamics of a real cluster, and concludes that the observational evidence strongly favours anisotropic velocity distributions; Fall (p. 309) reviews the physical processes which lead to the disruption of clusters within galaxies; and Heggie (p. 281) assesses critically the role played by binary stars within globular clusters. Looking back several years, we find several good reviews in earlier publications. The first of these is Michie's (1964) early but still pithy description of the problem and its history, with a survey of the methods of attack. An engaging semi-popular account of cluster dynamics is to be found in King (1971), while the same year produced not only Bouvier's (1971) straightforward review of the basic questions but also

Aarseth's (1971) and Hénon's (1971) careful descriptions of the alternative approaches of N-body calculations (Aarseth) and Monte Carlo methods (Hénon). More reviews were to come as the pace of research quickened: Hénon (1973) contrasted the results of Monte Carlo studies, fluid-dynamical approaches and the exact N-body integrations; Aarseth (1973) presented an introduction to numerical studies of small self-gravitating stellar systems, and pointed up the importance of binaries; Freeman (1975) in a massive review paper, described in detail the complex subject of stellar dynamics in stellar systems of all sizes and descriptions; and Aarseth & Lecar (1975) neatly summarized progress to that time, noting with interest as they did that the year marked only the fifteenth anniversary of the first N-body calculation in stellar dynamics.

Finally, the first half of the volume *Dynamics of Stellar Systems*, which appeared in 1975, was dedicated to the study of spherical systems: useful chapters include those of Spitzer (1975), King (1975), Wielen (1975), Heggie (1975) and Hénon (1975), with these last two dealing with binaries in clusters (Heggie) and with the perplexing problem of what happens to a cluster after the (apparently inevitable) stage of core collapse (Hénon).

Two less-sweeping contributions are of note: Cudworth (1976) presented an example of the kind of careful proper motion study so needed in galactic globular clusters; and Tremaine (1976) suggested the concept of dynamical friction creating galaxy nuclei from agglomerated globular clusters, a subject considered further in van den Bergh's chapter (p. 187, this volume).

1.4.10 Gas in globular clusters

Renzini (1977, 1979) reviews the evidence for mass loss in the stellar-evolutionary explanation of the structure of the horizontal branch. However, it is puzzling that thorough searches for gas within globular clusters have so far done little but set stringent upper limits on the observed quantities, whether optical (Hesser & Shawl 1977) or radio (Kerr & Knapp 1972). Cohen (1976) has found evidence for mass loss from single globular cluster giants, while the presence of a planetary nebula in M15 (Peimbert 1977) and dark patches (dust?) in the fields of certain globular clusters (Sawyer-Hogg 1959b) bespeak a non-zero amount of interstellar matter. Faulkner & Freeman (1977) and VandenBerg (1978) have modelled expected gas flows in efforts to understand the surprisingly low upper limits.

1.4.11 X-rays from globular clusters

A question perhaps not entirely unrelated to the presence of gas in clusters is the identification and understanding of the X-ray sources within them. The initial discovery has been described (along with a description of the satellite instrumentation used in X-ray astronomy) by Lightman (1976), and an informal and interesting history is to be found in Clark (1977). The published sessions on X-ray astronomy at the Eighth Texas Symposium on Relativistic Astrophysics include two reviews of great interest: Ostriker (1977), on the observed properties of X-ray sources and the theoretical constraints set thereby; and Liller (1977), on the optical nature of the clusters known to be X-ray sources. Lewin's (1977) useful review there has been brought even more completely up-to-date in this volume (p. 315). Elsewhere, Grindlay (1977) reviews the situation as it stood in 1976, summarizing the data and models then available; but the frantic pace of research in this area soon renders such reports out of date. The most recent reviews are to be found in Lewin (p. 315, this volume) and Jernigan (p. 351, this volume). It must be remembered that in so young an area new discoveries may change the perspectives wholesale, and no review can expect to be completely correct for long.

1.4.12 Black holes within globular clusters

Black holes within globular clusters have been proposed as explaining both the X-ray sources (as, for example, in Bahcall & Ostriker 1976) and also the sometimes peculiar surface brightness distributions (Newell, DaCosta & Norris 1976). The former of these is reviewed by Lightman & Shapiro (1978) and by Lewin (p. 335, this volume), with the latter author offering observational evidence that such may in fact be the case, at least for low-mass black holes. The other suggestion, that a massive black hole lurks within M15, is less compelling, as King (p. 265, this volume) explains.

1.5 Globular clusters star-by-star

1.5.1 Photometry and spectroscopy of individual stars

No compilations of such data exist except in the contexts already discussed: those of the colour–magnitude diagrams of clusters and in the studies of metallicities or kinematics of stars within clusters – see, for example, section 1.3 and section 1.4.6. The close scrutiny of single stars in determinations of metallicities in clusters and in the galactic halo is well described by Kraft (1979, and p. 87, this volume), by Madore (p. 44, this volume), by Persson & Frogel (p. 143, this volume), by Norris (p. 113, this volume), by Searle & Zinn (1978), and in many more of the references of section 1.4.6.

1.5.2 Variable stars within globular clusters

Variable stars within globular clusters have a long history, being first found by Solon Bailey in 1895 during his time at the Arequipa, Peru, station of the Harvard Observatory; the history is well told in Shapley (1949). The most recent catalogue (Sawyer-Hogg 1973) contains more than two thousand variables in 108 clusters, with more than 90 per cent of these being RR Lyrae pulsators. The importance of such stars as distance indicators has been semi-popularly explained by Kraft (1959), while a more rigorous review is that of Preston (1964) who outlined the application of RR Lyrae stars in studies of galactic structure. Plaut (1965) reaffirmed this aspect and discussed the RR Lyraes in the context of globular clusters: their presence (or not) as a function of cluster metallicity, and the like. Christy (1966) neatly reviewed the pulsation theory needed for an understanding of such stars.

An important volume appeared in 1972, honouring the lifelong work of Helen Sawyer-Hogg. Especially important within that volume were: the review by Sawyer-Hogg (1972), summarizing the present state of variable star statistics; the review by Rosino (1972), summarizing observations of RR Lyrae stars in particular; the review by van Agt (1972), on variable stars in the dwarf spheroidals; and the review by Graham (1972) on the RR Lyrae stars in the Magellanic Clouds. The same year also produced Tsesevich's (1972) review of the study of the change of periods of RR Lyrae stars, an area long of interest to Sawyer-Hogg and her co-workers.

Shortly thereafter, Iben (1974a) reviewed the theoretical situation for the variable stars of both Populations I and II; and more recently still Rosino (1978) has summarized all the varieties of variable stars in globular clusters and discussed what we can and should learn through their detailed study. The most recent and comprehensive review though is that of Castellani (p. 65, this volume) who explains in considerable detail how the structure of the horizontal branch and the pulsation properties of the RR Lyrae stars themselves yield astrophysical insights into the compositions and masses of the cluster stars.

1.5.3 Evolution of single stars within globular clusters

Clearly this topic cannot be entirely divorced from that of section 1.5.2, and indeed the review papers overlap to a large extent. Here the most prolific – and fortunately very readable! – reviewer is Iben. A semi-popular introduction to the subject is to be found in Iben (1970); while in Iben (1972a) he first informally reviews the importance of clusters in understanding the chemical enrichment history of the galaxy and then in Iben (1972b) elegantly surveys the available theory and evidence. An equally important contribution in the same volume is that of Larsson-Leander (1972) who reviews the observational attributes of colour–magnitude and HR diagrams in the determination of stellar ages.

Iben (1974b), in a sequel to Iben (1967), reviews the post-main-sequence evolution of single stars (of high mass as well as of low) and describes the expected horizontal branch morphology and pulsation properties of evolved globular cluster stars. Castellani (1977) summarizes the dependence of the morphology of the colour–magnitude diagram upon age, chemical composition, and the like; and similar summaries were the objectives of Iben (1977) and Renzini (1977), with Iben strongly emphasizing the need for a good physical understanding (rather than simply a models-and-mathematics overview), while Renzini considered in addition questions of mass loss and the second-parameter problem.

A new set of theoretical isochrones were described by Demarque & McClure (1977), who fitted them to observed colour–magnitude diagrams to conclude that the metal-richest clusters in the galaxy were significantly younger than the most metal-poor. Sweigart (1978) reviews some recent stellar evolution calculations by various authors with a view to considering the systematic differences, if any. And finally the most recent review is that of Castellani (p. 65, this volume), who explains how constraints on stellar masses, compositions and ages are set by the observed disposition of stars in the colour–magnitude diagram and in other parameters. Iben's review (p. 125, this volume) deals rather more specifically with possible explanations of the second-parameter effect, but does provide a concise summary of standard evolution theory.

1.6 Globular clusters in our galaxy

1.6.1 Distances of globulars in the galaxy

The methods of distance determination are nicely summarized in Madore (p. 30, this volume). Tabulations of the best values are available in Madore, in Harris & Racine (1979), and in Harris (1976); the sources of each value are identified in those reviews.

1.6.2 Globular clusters as probes of galactic structure

The realization, evidenced by the distribution of globular clusters, that the galactic centre was considerably removed from the solar neighbourhood was certainly one of the most important discoveries in modern astronomy. The original announcement to this effect was contained in Shapley's (1918) publication, elegantly entitled 'Remarks on the arrangement of the Sidereal universe', and in subsequent papers. The history is well told by Shapley (1930), by Kienle (1971), by Whitney (1971), and by Berendzen, Hart & Seeley (1976). Accurate determinations of the distance to the galactic centre through study of the distribution of globular clusters have since been attempted by Fernie (1962), by Arp (1965), and most recently by Harris (1976). Van den Bergh (1968) has reviewed the early attempts, and in a slightly different context has re-emphasized the importance of the single globular cluster NGC 6522 in Baade's (1951) determination of the distance to the centre of the galaxy through a study of the RR Lyrae stars in the nuclear bulge of the galaxy.

1.6.3 The kinematics of the globular clusters in the galaxy

The most recent review is again that of Harris & Racine (1979). The earliest study (van Hoerner 1955) led to the conclusion that cluster orbits are very elongated, a point of view repeated by Arp (1965). The velocities used in most such studies are the classical values of Mayall (1946) and Kinman (1959), despite recent demonstrations that many of the measured values may need considerable modification (J. Hesser, private communication). In addition to orbital eccentricities, the rotation, if any, of the cluster system has long been a question of interest. Arp (1965), for example, concluded with Kinman (1959) that the cluster system rotates at ~ 110 km s^{-1} (in an absolute sense) with respect to the galactic centre; but Arp could not decide if the rotation was differential. He did confirm Kinman's conclusion that the subsystem of metal-richest clusters is somewhat flatter than that of the metal-poorest clusters. A more recent treatment is that of Hartwick & Sargent (1978), who indeed find differential rotation, with the inner, metal-richer subsystem rotating more rapidly. However, the (rotational?) flattening of the metal-rich subsystem reported by Arp (1965) was not detected by Harris (1976) in a careful study.

1.6.4 The run of metallicity in the galaxy

Yet again the summary of Harris & Racine (1979) represents the logical starting point for further study in this area. The early discovery by Morgan (1956), summarized in Arp (1965), of a run of cluster metallicity with galactocentric distance prompted thoughts on the enrichment history of the galaxy during the formative collapse phases; this kind of speculation is reviewed by Spinrad (1976) and by Audouze & Tinsley (1976). Searle (1977) has envisaged globular clusters experiencing random numbers of enrichment events during the collapse phase, and has observationally pursued the subject as explained in Searle & Zinn (1978). As Harris & Racine (1979) explain, it remains contentious whether or not the observed abundance gradient levels off at large galactocentric distances or continues to fall.

1.6.5 New clusters: the completeness of the known sample

The level of completeness of the known sample is not easily judged. Bailey suggested in 1915 (as reported by Arp (1965)) that all galactic globular clusters were known; Edmondson (1935) suggested a total cluster population near two hundred; while de Kort (1941) proposed a total population near two thousand. Arp (1965) concluded that the galactic sample was about 94 per cent complete, but Woltjer (1975) concluded that significant numbers (~ 50) might still be hidden beyond the galactic centre, a conclusion endorsed by Harris (1976) and Oort (1977). Harris & Racine (1979) and Madore (p. 26, this volume) present identifications and positional information for as complete a sample as is now available.

1.6.6 Mass of the galaxy

The observed velocities of globular clusters have been used by Hartwick & Sargent (1978) to deduce a mass of $(3-10) \times 10^{11}$ $M_\odot$ for the galaxy, within a radius of 60 kpc. See Harris & Racine (1979) for a discussion.

1.7 Globular clusters in galaxies

1.7.1 Properties of globular clusters in disparate systems

The most recent reviews in the area are Harris & Racine (1979), Hanes (p. 233, this volume), van den Bergh (p. 183, this volume), and Frogel *et al.* (p. 159, this volume). Globular clusters are found to be remarkably similar in:

(i) photometric properties (Hanes, p. 233; Frogel *et al.*, p. 159; but see van den Bergh, p. 184);

(ii) the range of metallicities observed (Frogel *et al.*, p. 165); and

(iii) observed luminosity functions (Hanes, p. 229; Harris & Racine 1979).

In addition, the total cluster population seems to scale roughly as the mass of the parent galaxy (Jaschek 1957; Harris & Racine 1979; Hanes, p. 232), though strong exceptions to this rule are observed (Hanes, p. 232).

Van den Bergh (1972, 1974) emphasizes how globular clusters may differ from galaxy to galaxy, interpreting this in terms of different evolutionary histories; Harris & Racine on the other hand remark 'we may emphasize the impressive similarities displayed by systems of globular clusters in a vast diversity of galaxies', a conclusion shared by Hanes (p. 233).

1.7.2 The globular clusters in the Magellanic Clouds

The puzzling nature of the dynamically evolved but otherwise youthful rich clusters in the Magellanic Clouds has had an interesting history. Following Thackeray's (1959) early comments, it was Gascoigne (1961) who first strongly focused attention upon them; and he reviewed the history of their investigation in Gascoigne (1964), identifying at that time three different kinds of rich cluster within the Clouds. Van den Bergh (1968) presented a lucid historical review, and the peculiarities were summarized again by Thackeray (1971) and by Gascoigne (1971). Spectroscopic analyses were presented by Andrews & Lloyd Evans (1971), who concluded that the clusters reflected the same kind of compositeness and correlation of spectral type and colour with colour–magnitude diagram morphology as are seen in the moderately metal-deficient clusters in our own galaxy.

Freeman (1974) stressed that the blue young clusters in the Clouds were *dynamically* genuine globular clusters, but otherwise unlike any seen in our own galaxy, and asked why they continued to form in that environment but not locally. Van den Bergh (1975) nicely summarized the observational data on the clusters in the Magellanic Clouds. Freeman (1977) discussed the mass functions within various clusters; interestingly, the slope varies from place to place. Finally, Harris & Racine (1979) present the most complete summary to date.

1.7.3 Clusters and dwarf spheroidal galaxies

The most recent and comprehensive review is that of Zinn (p. 191, this volume); other, earlier reviews are those of van den Bergh (1968), van Agt (1972, dealing with the variable stars in particular), Hodge (1971) and van den Bergh (1975). Zinn (p. 191) especially considers the role played by the dwarf spheroidals in aiding our understanding of the second-parameter problem.

1.7.4 Other extragalactic systems

The best review is that of Harris & Racine (1979); see also Hanes (p. 213, this volume) and Frogel et al. (p. 159, this volume).

1.7.5 Globular clusters as extragalactic distance indicators

See Hanes (p. 213, this volume) for a complete review.

1.7.6 The origin of globular clusters

Van den Bergh (p. 175, this volume) summarizes and considers various pieces of evidence afforded us by globular clusters in an assessment of several suggested galaxy formation scenarios. Gunn (p. 301, this volume) deals more rigorously with the formation of a globular cluster proper.

1.8 Conclusions

The simplest conclusion is that the term 'globular cluster research' covers a vast amount of territory, and that a short summary can do nothing beyond simply touching upon the highlights. I cannot pretend to have achieved more than that, but it is still my hope that this survey will be of some help – though perhaps only to the new arrivals in the field, as they first familiarize themselves with the profound and exciting results in the science of globular cluster astronomy.

References

Aarseth, S. J. (1971). *Astrophys. Space Sci.* **14**, 20 and 118.
Aarseth, S. J. (1973). *Vistas Astron.* **15**, 13.
Aarseth, S. J. & Lecar, M. (1975). *Ann. Rev. Astron. Astrophys.* **13**, 1
Alcaino, G. (1973). *Atlas of Galactic Globular Clusters with Colour–Magnitude Diagrams.* Universidad Católica de Chile, Santiago.
Aller, L. H. & Faulkner, D. J. (1964). In *The Galaxy and the Magellanic Clouds*, IAU Symposium no. 20, ed. F. J. Kerr & A. W. Rodgers, p. 358. Australian Academy of Science.
Andrews, P. J. & Lloyd Evans, T. (1971). In *The Magellanic Clouds*, ed. A. B. Müller, p. 88. Dordrecht: D. Reidel.
Arp, H. C. (1959). *Astron. J.* **64**, 441.
Arp, H. C. (1965). In *Galactic Structure*, ed. A. Blaauw & M. Schmidt, p. 401. University of Chicago Press.
Audouze, J. & Tinsley, B. M. (1976). *Ann. Rev. Astron. Astrophys.* **14**, 43.
Baade, W. (1944). *Astrophys. J.* **100**, 137.
Baade, W. (1951). *Publ. Univ. Mich. Obs.* **10**, 7.
Baade, W. (1963). In *Evolution of Stars and Galaxies*, ed. C. Payne-Gaposchkin. Harvard University Press.
Bahcall, J. N. & Ostriker, J. P. (1976). *Nature*, **262**, 37.
Bell, R. (1976). In *Abundance Effects in Classification*, IAU Symposium no. 72, ed. B. Hauck & P. C. Keenan, p. 49. Dordrecht: D. Reidel.
Berendzen, R., Hart, R. & Seeley, D. (1976). *Man Discovers the Galaxies.* New York: Science History Publications.
Blaauw, A. (1965). In *Galactic Structure*, ed. A. Blaauw & M. Schmidt, p. 435. University of Chicago Press.
Bouvier, P. (1971). In *Structure and Evolution of the Galaxy*, ed. L. N. Mavridis, p. 250. Dordrecht: D. Reidel.
Butler, D., Bell, R. A., Dickens, R. J. & Epps, E. (1978). In *The HR Diagram*, IAU Symposium no. 80, ed. A. G. Davis Philip & D. S. Hayes, p. 183. Dordrecht: D. Reidel.
Castellani, V. (1977). In *Chemical and Dynamical Evolution of Our Galaxy*. IAU Colloquium no. 45, ed. E. Basinska-Grezesik & M. Mayor, p. 133. Geneva Observatory Publ.
Christy, R. F. (1966). *Ann. Rev. Astron. Astrophys.* **4**, 353.
Clark, G. W. (1977). *Sci. Amer.* **237**, 42.
Cohen, J. G. (1976). *Astrophys. J. Lett.* **203**, L127.
Cudworth, K. (1976). *Astron. J.* **81**, 975.
Danziger, I. J. (1970). *Ann. Rev. Astron. Astrophys.* **8**, 161.
Davis Philip, A. G. (1977). *Vistas Astron.* **21**, 407.
De Kort, J. (1941). *Bull. astron. Inst. Netherlands*, **9**, 189.
Demarque, P. & McClure, R. D. (1977). In *The Evolution of Galaxies and Stellar Populations*, ed. B. M. Tinsley & R. B. Larson, p. 199. Yale University Observatory Press.
Edmondson, F. K. (1935). *Astron. J.* **45**, 1.
Faulkner, D. A. & Freeman, K. C. (1977). *Astrophys. J.* **211**, 77.
Fernie, J. D. (1962). *Astron. J.* **67**, 769.
Freeman, K. C. (1974). In *Research Programmes for Large Telescopes*, ed. A. Reiz, p. 177. ESO/SRC/CERN.
Freeman, K. C. (1975). In *Galaxies and the Universe*, ed. A. Sandage, M. Sandage & J. Kristian, p. 409. University of Chicago Press.
Freeman, K. C. (1977). In *The Evolution of Galaxies and Stellar Populations*, ed. B. M. Tinsley & R. B. Larson, p. 133. Yale University Observatory Press.

Freeman, K. C. & Rodgers, A. W. (1975). *Astrophys J. Lett.* **201**, L71.
Gascoigne, S. C. B. (1961). In *Problems of Extragalactic Research*, IAU Symposium no. 15, ed. G. C. McVittie, p. 49. New York: Macmillan.
Gascoigne, S. C. B. (1964). In *The Galaxy and the Magellanic Clouds*, IAU Symposium no. 20, ed. F. J. Kerr & A. W. Rodgers, p. 354. Australian Academy of Science.
Gascoigne, S. C. B. (1971). In *The Magellanic Clouds*, ed. A. B. Müller, p. 25. Dordrecht. D. Reidel.
Graham, J. A. (1972). In *Variable Stars in Globular Clusters and in Related Stellar Systems*, IAU Colloquium no. 21, ed. J. D. Fernie, p. 120. Dordrecht: D. Reidel.
Grindlay, J. E. (1977). *Highlights of Astronomy*, **4**, 111.
Gustafsson, B. (1979). In *Stars and Star Systems*, ed. B. E. Westerlund, p. 135. Dordrecht: D. Reidel.
Harris, W. E. (1976). *Astron. J.* **81**, 1095.
Harris, W. E. & Racine, R. (1979). *Ann. Rev. Astron. Astrophys.* **17**, 241.
Hartwick, F. D. A. & Sargent, W. L. W. (1978). *Astrophys J.* **221**, 512.
Heggie, D. C. (1975). In *Dynamics of Stellar Systems*, IAU Symposium no. 69, ed. A. Hayli, p. 73. Dordrecht: D. Reidel.
Hénon, M. H. (1971). *Astrophys. Space Sci.* **14**, 151.
Hénon, M. H. (1973). In *Dynamics of Stellar Systems*, p. 181. Saas Fee, Switzerland.
Hénon, M. H. (1976). In *Dynamics of Stellar Systems*, IAU Symposium no. 69, ed. A. Hayli, p. 133. Dordrecht: D. Reidel.
Hesser, J. E. & Shawl, S. J. (1977). *Astrophys. J. Lett.* **217**, L143.
Hodge, P. W. (1971). *Ann. Rev. Astron. Astrophys.* **9**, 35.
Iben, I., Jr (1967). *Ann. Rev. Astron. Astrophys.* **5**, 571.
Iben, I., Jr (1970). *New Frontiers in Astronomy*, p. 113. San Francisco: W. H. Freeman & Co.
Iben, I., Jr (1972a). In *Stellar Ages*, IAU Colloquium no. 17, ed. G. Cayrel de Strobel & A. M. Delplace, ch. 1. Observatoire de Paris-Meudon.
Iben, I., Jr (1972b). In *Stellar Ages*, IAU Colloquium no. 17, ed. G. Cayrel de Strobel & A. M. Delplace, ch. 11. Observatoire de Paris-Meudon.
Iben, I., Jr (1974a). In *Stellar Instability and Evolution*, IAU Symposium no. 59, ed. P. Ledoux, A. Noels & A. W. Rodgers, p. 3. Dordrecht: D. Reidel.
Iben, I., Jr (1974b). *Ann. Rev. Astron. Astrophys.* **12**, 215.
Iben, I., Jr (1977). In *Advanced Stages in Stellar Evolution*, 7th Course of the Swiss Society of Astronomy and Astrophysics, Saas-Fee, ed. P. Bouvier & A. Maeder, p. 1. Geneva Observatory Publ.
Illingworth, G. (1975). In *Dynamics of Stellar Systems*, IAU Symposium no. 69, ed. A. Hayli, p. 151. Dordrecht: D. Reidel.
Illingworth, G. (1976). *Astrophys. J.* **204**, 73.
Illingworth, G. & King, I. R. (1976). *Publ. astron. Soc. Pacific*, **88**, 607.
Jaschek, C. O. R. (1957). *Z. Astrophys.* **44**, 23.
Kerr, F. J. & Knapp, G. (1972). *Astron. J.* **77**, 573.
Kienle, H. (1971). In *Structure and Evolution of the Galaxy*, ed. L. N. Mavridis, p. 1. Dordrecht: D. Reidel.
King, I. R. (1962). *Astron. J.* **67**, 471.
King, I. R. (1966a). *Astron. J.* **71**, 64.
King, I. R. (1966b). *Astron. J.* **71**, 276.
King, I. R. (1971). *Sky and Telescope*, **41**, 139.
King, I. R. (1975). In *Dynamics of Stellar Systems*, IAU Symposium no. 69, ed. A. Hayli, p. 99. Dordrecht: D. Reidel.
King, I. R., Hedemann, E., Hodge, S. M. & White, R. E. (1968). *Astron. J.* **73**, 456.
Kinman, T. D. (1959). *Mon. Not. Roy. astron. Soc.* **119**, 157.

Kraft, R. P. (1959). *Sci. Amer.* **201**, 48.
Kraft, R. P. (1979). *Ann. Rev. Astron. Astrophys.* **17**, 309.
Kron, G. E. & Mayall, N. U. (1959). *Astron. J.* **64**, 428.
Kukarkin, B. V. (1974). *The Globular Star Clusters*. Moscow: Sternbergh State Astron. Inst. Nauka.
Larsson-Leander, G. (1972). In *Stellar Ages*, IAU Colloquium no. 17, ed. G. Cayrel de Strobel & A. M. Delplace, ch. 2. Observatoire de Paris-Meudon.
Lewin, W. H. G. (1977). In *Proceedings of the 8th Texas Symposium on Relativistic Astrophysics*, ed. M. Papagiannis, *Ann. N.Y. Acad. Sci.* **302**, 210.
Lightman, A. P. (1976). *Sky and Telescope*, **52**, 243.
Lightman, A. P. & Shapiro, S. L. (1978). *Rev. Mod. Phys.* **50**, 437.
Liller, W. (1977). In *Proceedings of the 8th Texas Symposium on Relativistic Astrophysics*, ed. M. Papagiannis, *Ann. N.Y. Acad. Sci.* **302**, 248.
Mayall, N. U. (1946). *Astrophys. J.* **104**, 290.
Michie, R. W. (1964). *Ann. Rev. Astron. Astrophys.* **2**, 49.
Morgan, W. W. (1956). *Publ. astron. Soc. Pacific*, **68**, 509.
Morgan, W. W. (1959). *Astron. J.* **64**, 432.
Newell, E. B., DaCosta, G. S. & Norris, J. (1976). *Astrophys J. Lett.* **208**, L55.
Oort, J. H. (1977). *Astrophys. J. Lett.* **218**, L97.
Ostriker, J. P. (1977). In *Proceedings of the 8th Texas Symposium on Relativistic Astrophysics*, ed. M. Papagiannis, *Ann. N.Y. Acad. Sci.* **302**, 229.
Pagel, B. (1970). *Vistas Astron.* **12**, 313.
Peimbert, M. (1977). In *Planetary Nebulae*, IAU Symposium no. 76, ed. Y. Terzian, p. 215. Dordrecht: D. Reidel.
Peterson, C. J. & King, I. R. (1975). *Astron. J.* **80**, 427.
Plaut, L. (1965). In *Galactic Structure*, ed. A. Blaauw & M. Schmidt, p. 267. University of Chicago Press.
Preston, G. W. (1964). *Ann. Rev. Astron. Astrophys.* **2**, 23.
Racine, R., Oke, J. B. & Searle, L. (1978). *Astrophys. J.* **223**, 82.
Renzini, A. (1977). In *Advanced Stages in Stellar Evolution*, 7th Course of the Swiss Society of Astronomy and Astrophysics, Saas-Fee, ed. P. Bouvier & A. Maeder, p. 149. Geneva Observatory Publ.
Renzini, A. (1979). In *Stars and Star Systems*, ed. B. E. Westerlund, p. 155. Dordrecht: D. Reidel.
Rosino, L. (1972). In *Variable Stars in Globular Clusters and in Related Systems*, IAU Colloquium no. 21, ed. J. D. Fernie, p. 51. Dordrecht: D. Reidel.
Rosino, L. (1978). *Vistas Astron.* **22**, 39.
Sawyer, H. B. (1947). *Publ. David Dunlap Obs.* **1**, no. 20, 381.
Sawyer-Hogg, H. B. (1959a). *Handbuch der Physik*, ed. S. Flügge, p. 129. Berlin: Springer-Verlag.
Sawyer-Hogg, H. B. (1959b). *Astron. J.* **64**, 425.
Sawyer-Hogg, H. B. (1963). *Publ. David Dunlap Obs.* **2**, no. 12, 335.
Sawyer-Hogg, H. B. (1972). In *Variable Stars in Globular Clusters and in Related Systems*, IAU Colloquium no. 21, ed. J. D. Fernie, p. 3. Dordrecht: D. Reidel.
Sawyer-Hogg, H. B. (1973). *Publ. David Dunlap Obs.* **3**, no. 6.
Searle, L. (1977). In *The Evolution of Galaxies and Stellar Populations*, ed. B. M. Tinsley & R. B. Larson, p. 219. Yale University Observatory Press.
Searle, L. & Zinn, R. (1978). *Astrophys. J.* **225**, 357.
Shapley, H. (1915). *Mt Wilson Obs. Contr.* no. 116, 51.
Shapley, H. (1918). *Astrophys. J.* **48**, 154.
Shapley, H. (1930). *Star Clusters*. New York: McGraw-Hill.
Shapley, H. (1949). *Harvard Obs. Reprint* no. 320.

Spinrad, H. (1976). In *Abundance Effects in Classification*, IAU Symposium no. 72, ed. B. Hauck & P. C. Keenan, p. 183. Dordrecht: D. Reidel.
Spinrad, H. & Peimbert, M. (1975). In *Galaxies and the Universe*, ed. A. Sandage, M. Sandage & J. Kristian, p. 37. University of Chicago Press.
Spitzer, L. (1975). In *Dynamics of Stellar Systems*, IAU Symposium no. 69, ed. A. Hayli, p. 1. Dordrecht: D. Reidel.
Struve, O. & Zebergs, V. (1962). *Astronomy of the Twentieth Century*. New York: Macmillan.
Sweigart, A. V. (1978). In *The HR Diagram*, IAU Symposium no. 80, ed. A. G. Davis Philip & D. S. Hayes, p. 333. Dordrecht: D. Reidel.
Thackeray, A. D. (1959). *Astron. J.* **64**, 437.
Thackeray, A. D. (1971). In *The Magellanic Clouds*, ed. A. B. Müller, p. 3. Dordrecht: D. Reidel.
Tinsley, B. M. (1978). In *The HR Diagram*, IAU Symposium no. 80, ed. A. G. Davis Philip & D. S. Hayes, p. 247. Dordrecht: D. Reidel.
Tremaine, S. (1976). *Astrophys. J.* **203**, 345.
Tsesevich, V. (1972). *Vistas Astron.* **13**, 241.
Van Agt, S. (1972). In *Variable Stars in Globular Clusters and in Related Systems*, IAU Colloquium no. 21, ed. J. D. Fernie, p. 35. Dordrecht: D. Reidel.
VandenBerg, D. A. (1978). *Astrophys J.* **224**, 394.
van den Bergh, S. (1968). *J. Roy. astron. Soc. Canada*, **62**, 145.
van den Bergh, S. (1969). *Astrophys. J. Suppl.* **19**, 145.
van den Bergh, S. (1972). In *External Galaxies and Quasi-stellar Objects*, IAU Symposium no. 44, ed. D. S. Evans, p. 1. Dordrecht: D. Reidel.
van den Bergh, S. (1974). In *The Formation and Dynamics of Galaxies*, IAU Symposium no. 58, ed. J. R. Shakeshaft, p. 157. Dordrecht: D. Reidel.
van den Bergh, S. (1975). *Ann. Rev. Astron. Astrophys.* **13**, 217.
van den Bergh, S. & Henry, R. C. (1962). *Publ. David Dunlap Obs.* **2**, 279.
van Hoerner, S. (1955). *Z. Astrophys.* **32**, 255.
von Zeipel, H. (1908). *Ann. Obs. Paris Mem.* **25**, F.
White, R. E. (1970). *Astrophys. J. Suppl.* **19**, 343.
Whitney, C. A. (1971). *The Discovery of Our Galaxy*. New York: Knopf.
Wielen, R. (1975). In *Dynamics of Stellar Systems*, IAU Symposium no. 69, ed. A. Hayli, p. 119. Dordrecht: D. Reidel.
Woltjer, L. (1975). *Astron. Astrophys.* **42**, 109.
Zinn, R. (1973). *Astrophys. J.* **182**, 183.

2
Terminology and fundamental data on globular clusters

BARRY F. MADORE

2.1 Introduction

Recently several publications have provided comprehensive surveys of the fundamental properties of the galactic globular system. These have built on the earlier works of Sawyer-Hogg (1959) and Arp (1965), while following soon after the compilations of White (1970) and Alcaino (1973). Perhaps the most impressive of these new publications is the book *The Globular Star Clusters*, by B. V. Kukarkin (1974) (originally published in Russian with some explanatory English text, it is now available in complete translation), which attempts to consolidate in a homogeneous way a wide range of individual studies of globular clusters. Woltjer (1975) and Harris & Racine (1979) have also reviewed the properties of galactic globular clusters, the latter drawing heavily on the previous work of Harris (1976), with the nominal intent of using globular clusters as probes of the distances to, and the chemical evolution of, nearby galaxies. A less interpretive but still very useful cataloguing of observations with bibliographic references is given by Davis Philip (1977), and has the added advantage of being available on tape from the Strasbourg Stellar Data Centre. Finally, the ultimate reference source to globular cluster data is the reference card catalogue by Alter, Balasz & Ruprecht (1970).

This chapter will not attempt to supersede these excellent surveys; nor will it attempt to reconcile their differences. Globular cluster research is still very much an ongoing process, and in this chapter it is hoped that the interested researcher will find a guide both to the sources of basic data and to the terminology peculiar to this corner of astronomy.

The terminology used frequently in globular cluster studies divides naturally (but not exclusively) along two lines: terms related to the colour–magnitude diagram (or its theoretical counterpart, the temperature–luminosity diagram); and terms related to the apparent structure of the clusters. This is essentially a division between stellar content and stellar dynamics. Needless to say, both aspects ultimately aim to shed light on the origin and evolution of globular star clusters, not only as individual systems but as attendant subsets of larger stellar systems, the galaxies.

2.2 The colour–magnitude diagram

In the interpretation of data on individual stars in globular clusters, the colour–magnitude (CM) diagram has been of paramount importance. The reason for this is the relative facility with which data on the cluster stars can be obtained and then compared directly with theory. Magnitudes and colours have long provided the basis of comparison with theoretical luminosities and temperatures: the distribution, within the CM diagram, of stars in a single globular cluster is a distinctive indication of the evolutionary status of these stars, and can be matched (with varying degrees of success) to theoretical stellar evolutionary models. Decades of progress in techniques on both the observational and theoretical fronts have not changed the importance of the CM diagram; it continues as

Fig. 2.1. Taken from Baade's original (1944) paper on stellar populations. The shaded area represents the young Population I, while the hatched area represents the Population II typical of globular clusters.

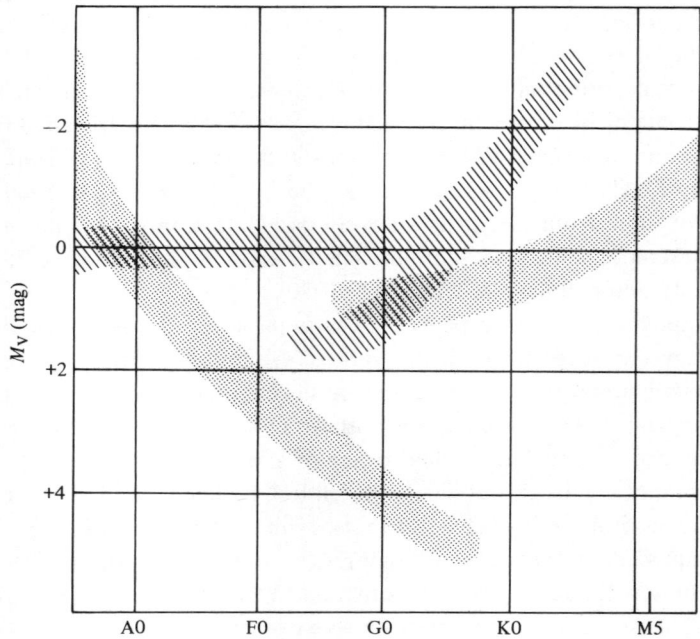

always to be the focal point in the study and interpretation of globular clusters.

Many of the terms introduced in this section are used in the remainder of the text. Also, because of their dependence upon the CM diagram for their definition, many of the terms are illustrated in figs. 2.2, 2.9 and 2.10 and the reader would be well advised to consult these.

2.2.1 Morphology of the CM diagram

Red giant branch

Historically, as shown by Baade (1944), the primary indication that globular clusters differed fundamentally from the near-by population of stars in open clusters came from the presence of a bright red giant branch (RGB) (fig. 2.1). It is now well established that this region is populated by low-mass stars ($M \sim 0.8\ M_\odot$) which are in an advanced stage of stellar evolution, burning hydrogen in a shell surrounding a gradually heating inert helium core. Evolution is toward higher luminosities and cooler temperatures. Typical absolute magnitudes and colours at the tip of the RGB are $M_V \sim -2.0$ and $(B-V)_o \sim 1.4$ mag (fig. 2.5), although these values are dependent on the exact chemical composition of the individual stars during this phase of their evolution.

Horizontal branch

The next most distinctive feature of the CM diagrams of most globular clusters is the horizontal branch (HB). As shown in fig. 2.6, it is found at $M_V \sim +0.5$ mag and covers a significant range in colour. Three components are distinguished:

(i) The red horizontal branch (RHB), which intersects the giant branch (GB) to the red.

(ii) The RR Lyrae gap, a region of pulsational instability (see below) in which stars are found but usually not plotted because of their light variability ($(B-V)_o \sim 0.15$–0.40 mag).

(iii) The blue horizontal branch (BHB), which starts at $(B-V)_o \sim 0.2$ mag and curves to lower visual luminosities and higher temperatures (i.e. bluer colours).

The strength of any one of these components varies from cluster to cluster, and is indicative of the detailed evolution beyond the GB hydrogen-shell burning, involving processes which include the helium-core flash and probably a phase of mass loss, and which bring stars 'raining down' along the HB.

Zero-age horizontal branch

In theoretical studies it is possible to isolate the HB and follow evolution away from it as additional sources of energy generation come into play. Models that have just started helium-core burning under non-degenerate conditions define the zero-age horizontal branch (ZAHB).

Double-shell sources

(i) *Above horizontal branch.* Models with double-shell sources (DSS) evolve off the ZAHB toward higher luminosities and cooler temperatures through a region above the horizontal branch (AHB; see Strom, Strom, Rood & Iben 1970). The DSS phase results from hydrogen and helium burning in two separate regions surrounding an inert carbon–oxygen core.

(ii) *Asymptotic giant branch.* As the evolution proceeds through the AHB and away from the HB, stars ascend to even higher luminosities and cooler temperatures while continuing to be in a phase of DSS burning. The resulting asymptotic giant branch (AGB) falls somewhat above the hydrogen-shell burning GB for higher temperatures but slowly converges on it with decreasing temperature.

Fig. 2.2. The approximate positions in the CM diagram of the principal morphological features in a typical globular cluster. See text for meaning of the abbreviations (Arp 1962).

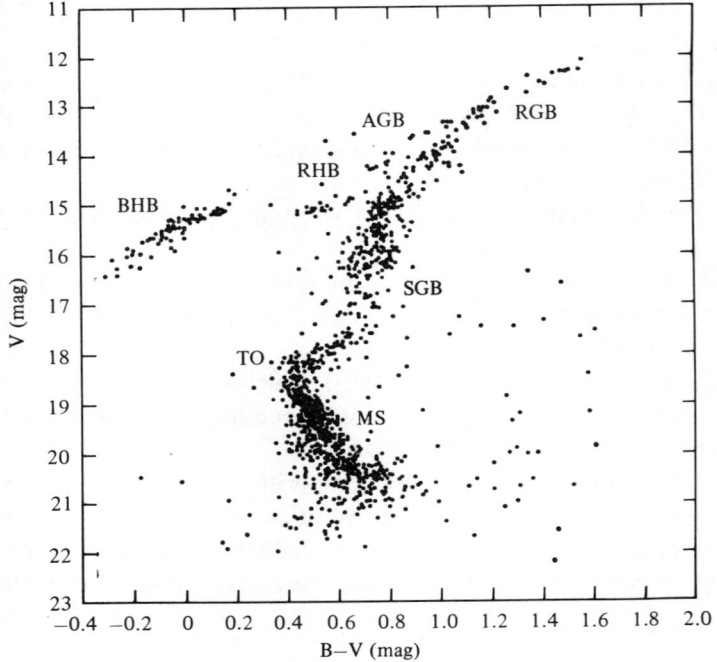

Zero-age main sequence or main sequence

Some four magnitudes fainter than the HB is found the densely populated main sequence (MS). Here stars have been burning hydrogen in their cores for the entire age of the globular cluster. Because of the relatively long time scale for evolution on the MS as compared to the more advanced stages, the position of the MS is practically the same as the original, or zero-age main sequence (ZAMS), except near the turnoff point (see below). Better understood (from a theoretical point of view) than the quickly-evolving brighter regions of the globular cluster colour–magnitude diagram, the position of the MS is found to be quite sensitive to the metal content of its stars (fig. 2.5). This will be important in the subsequent discussions of age and distance determinations, especially considering the increased difficulty in observing to so faint a magnitude level.

Turnoff point

At the high-mass (hot) end of the MS, the increasing pace of evolution draws stars away from the MS; alternative sources of energy are called upon at the end of hydrogen-core burning. This turnoff point (TO) can be well calibrated as a function of age (and metallicity) to provide a dating method for the few globular clusters that have accurate photometry to such faint levels.

Fig. 2.3. A comparison of observed luminosity functions for three globular clusters normalized at $M_V = \pm 2$ mag (Simoda & Kimura 1968).

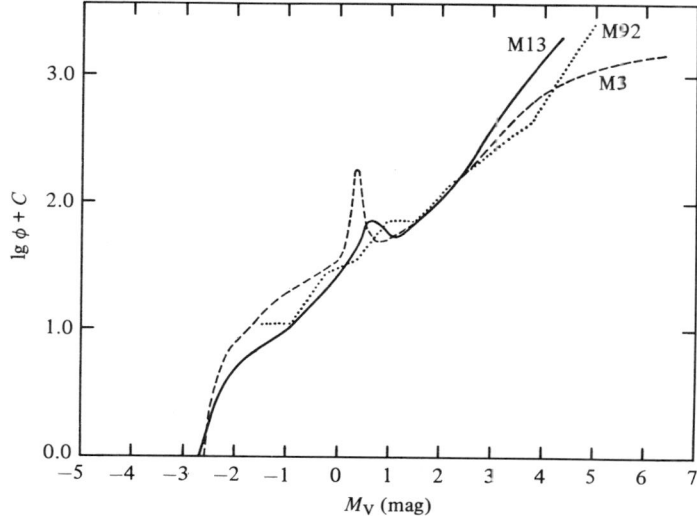

Table 2.1. *Integrated and related properties*

NGC/IC (1)	Name (2)	RA (1950) (3)	Dec. (4)	Long. (5)	Lat. (6)	Vel. (7)	Res. (8)	Mod. (9)	M_V^0 (10)
104	47Tuc	00 21.8	−72 21	305.9	−44.9	−18	3	13.46	−9.43
288		00 50.2	−26 52	149.7	−89.4	−47		14.70	−6.60
362		01 01.6	−71 07	301.5	−46.3	221	2	14.90	−8.32
1261		03 10.9	−55 25	270.6	−52.1	46		15.70	−7.32
	Pal 1	03 25.7	79 28	130.0	19.1	4		18.7:	
	AM-01	03 53.6	−49 45	258.0	−48.0			22.7:	−5.5:
0423−21	Eri	04 22.6	−21 18	251.1	−41.3	−44		19.8	−4.2:
	Pal 2	04 43.1	31 23	170.5	−9.0	−133			
1851		05 12.4	−40 05	244.5	−35.0	310	1	15.40	−8.10
1904	M79	05 22.2	−24 33	227.2	−29.3	198	1−2	15.65	−7.65
2298		06 47.2	−35 57	245.6	−16.0	64	1	15.80	−6.40
2419		07 34.8	39 00	180.4	25.3		1	19.94	−9.57
2808		09 10.9	−64 39	282.2	−11.3	101	2	15.52	−9.22
	Pal 3	10 03.0	00 18	240.3	41.9	22		20.0:	−5.3:
3201		10 15.5	−46 09	277.2	8.6	493	4	14.15	−7.40
	Pal 4	11 26.6	29 16	202.3	71.8	168		19.85	−5.65
4147		12 07.6	18 49	252.9	77.2	188	1−2	16.28	−6.02
4372		12 23.0	−72 24	301.0	−9.9	66	4	14.90	−7.10
4590	M68	12 36.8	−26 29	299.6	36.0	−116	3	15.01	−6.81
4833		12 56.0	−70 36	303.6	−8.0	204	3	14.90	−7.55
5024	M53	13 10.5	18 26	333.0	79.8	−112	1−2	16.34	−8.62
5053		13 13.9	17 57	335.6	78.9			16.00	−6.20
5139	ω Cen	13 23.8	−47 13	309.1	15.0	230	3−4	13.92	−10.27
5272	M3	13 39.9	28 38	42.2	78.7	−154	2−3	15.00	−8.65
5286		13 43.0	−51 07	311.6	10.6	45	2	15.61	−7.99
5466		14 03.2	28 46	42.2	73.6	−63		15.96	−6.86
5634		14 27.0	−05 45	342.2	49.3			16.90	−7.33
5694		14 36.7	−26 19	331.1	30.4	−187		17.80	−7.60
IC 4499		14 52.7	−82 02	307.4	−20.5			17.12	−6.52
5824		15 00.9	−32 53	332.6	22.1	−58	0	17.32	−8.32
	Pal 5	15 13.5	00 05	0.9	45.9			16.75	−5.00
5897		15 14.5	−20 50	342.9	30.3		3	15.60	−7.05
5904	M5	15 16.0	02 16	3.9	46.8	50	3	14.51	−8.76
5927		15 24.4	−50 29	326.6	4.9	−88	0	16.10	−7.77
5946		15 31.8	−50 30	327.3	4.2		1	16.7:	−7.05:
5986		15 42.8	−37 37	337.0	13.3	2	2	15.90	−8.78
1608+15	Pal 14	16 08.8	15 05	28.8	42.2	81		19.2:	
6093	M80	16 14.1	−22 52	352.7	19.4	19	1	15.28	−8.08
6101		16 20.0	−72 06	317.7	−15.8		3?	15.70	−6.40
6121	M4	16 20.6	−26 24	351.0	16.0	65	4	12.73	−6.80
6144		16 24.2	−25 56	351.9	15.7		1?	15.69	−6.57
6139		16 24.3	−38 44	342.4	6.9	20	0	16.93:	−7.75:
6171	M107	16 29.7	−12 57	3.4	23.0	−147	2	15.03	−6.90
6205	M13	16 39.9	36 33	59.0	40.9	−240		14.35	−8.49
6218	M12	16 44.6	−01 52	15.7	26.3	−16	3−4	14.30	−7.70
6229		16 45.6	47 37	73.6	40.3	−152		17.50	−8.07
6235		16 50.4	−22 06	358.9	13.5			16.31:	−6.16:
6254	M10	16 54.5	−04 02	15.1	23.1	69	3	14.05	−7.48
6256	TRZ 12	16 56.0	−37 00	347.8	3.4			16.6:	
	Pal 15	16 57.6	−00 28	18.9	24.3				
6266	M62	16 58.1	−30 03	353.6	7.3	−77	1	15.38	−8.78
6273	M19	16 59.5	−26 12	356.9	9.4	114	1−2	16.35	−9.20
6284		17 01.5	−24 41	358.4	9.9	22	1	15.89:	−6.94:
6287		17 02.1	−22 38	0.1	11.0		0	15.92:	−6.67:
6293		17 07.1	−26 30	357.6	7.8	−73	1−2	15.41:	−7.21:
6304		17 11.4	−29 24	355.8	5.4	−98	1	15.50	−7.08
6316		17 13.4	−28 05	357.2	5.8		0	16.99:	−7.99:
6325		17 15.0	−23 42	1.0	8.0		2	16.70	−6.00
6341	M92	17 15.6	43 11	68.4	34.9	−118		14.50	−7.98
6333	M9	17 16.2	−18 28	5.5	10.7	224		15.34:	−7.41:
6342		17 18.2	−19 32	4.9	9.7		0	17.5:	−7.60:
6356		17 20.7	−17 46	6.7	10.2	32	0	17.07	−8.67
6355		17 20.9	−26 19	359.6	5.4		0	16.6:	−7.00:
6352		17 21.6	−48 25	341.4	−7.2		1−2	14.47	−6.32
	TRZ 2	17 24.3	−30 46	356.3	2.3				
6366		17 25.1	−05 02	18.4	16.0			15.1:	−5.10:
6362		17 26.6	−67 01	325.5	−17.6	−18		14.65	−6.35
	TRZ 4	17 27.4	−31 33	356.0	1.3				
	HP 1	17 27.9	−29 57	357.4	2.1				
1730−33	Liller 1	17 30.1	−33 21	354.8	−0.2				
6380	Ton 1	17 32.0	−39 02	350.3	−3.6				
	TRZ 1	17 32.6	−30 26	357.6	1.0				

Fundamental data

(B−V) (11)	(U−B) (12)	E(B−V) (13)	R (kpc) (14)	[Fe/H] (15)	OG (16)	Var/per (17)	[m₁] (18)	Δ (19)	L (20)
0.89	0.37	0.04	8.2	−0.44		28/10			
0.66	0.09	0.03	12.3	−1.41		1/1			
0.76	0.14	0.04	10.2	−1.20	I	15/10			
0.70	0.14	0.02	16.1	−1.24		15/0			
		0.12	52.0:	−1.9:		0/0			
0.8:	0.5:	0.1:	300:						
0.8:	0.5:	0.09	83:	−2.2					
1.9:	1.8:	1.2:				0/0			
0.78	0.22	0.07	16.4	−1.29		10/0			
0.63	0.03	0.01	20.0	−1.58		7/3			
0.73	0.19	0.11	17.9	−1.41		2/0			
0.66	0.07	0.03	101.3	−2.00	II	36/0			4
0.93	0.29	0.22	11.5	−1.09		9/0			
		0.03	99.4:	−2.2:		1/0			
0.98	0.37	0.21	9.7	−1.26	I	88/84			
0.80		0.00	96.3	−2.4:		2/2			
0.60	0.07	0.02	20.2	−1.77	I:	16/15			2
0.97	0.32	0.45	7.8	−1.7:		2/0			
0.63	0.04	0.03	10.2	−2.04	II	42/38			0
0.96	0.28	0.38	7.6	−2.15	II	16/9			
0.64	0.10	0.05	18.1	−1.85	II	47/36	0.118	0.21	2
0.64	0.06	0.03	16.2	−2.09	II	11/10			
0.79	0.19	0.11	7.1	−1.6:P		179/159			
0.69	0.10	0.01	12.4	−1.57	I	212/186	0.153	0.20	4
0.87	0.29	0.27	7.5	−1.38		8/0			
0.71	0.04	0.05	15.3	−1.91	II	23/21			
0.67	0.12	0.07	17.5	−1.70	II	7/1			
0.69	0.07	0.08	26.1	−1.91		0/0			0
0.88	0.36	0.24	15.4	−1.0:		129/0			
0.75	0.15	0.14	17.1	−1.67	II	27/9			1
0.70		0.03	16.5	−1.24		5/5			
0.75	0.08	0.06	6.9	−1.45		7/7			
0.71	0.12	0.03	6.7	−1.25	I	97/92	0.154	0.23	6
1.31	0.84	0.55	5.0	−0.67		11/1			
1.24	0.54	0.56	5.3:	−1.5:		3/0			
0.90	0.30	0.27	4.5	−1.26		5/0			2
		0.03	60:	−1.6:					
0.85	0.24	0.21	3.2	−1.54		8/3	0.134	0:25	4
0.68	0.10	0.08	8.6	−1.8:		0/0			
1.03	0.44	0.35	7.0	−1.30	I	43/42			2
1.01	0.45	0.36	2.8	−0.9:		1/0			
1.39	0.73	0.68	3.0:	−1.27:					
1.14	0.55	0.37	4.3	−0.79	I	25/23			7
0.69	0.03	0.02	9.1	−1.42		11/7	0.168	0.28	3
0.82	0.21	0.19	5.1	−1.64		1/1	0.195		4
0.71	0.05	0.01	30.6	−1.44		22/15		0.20	6
1.04	0.40	0.38	2.8:	−1.2:	I	2/0			
0.92	0.23	0.26	5.5	−1.43		4/2	0.150	0.25	2
1.68	1.04					0/0			
		0.09:							
1.17	0.52	0.46	3.2	−1.14	I	89/74			6
1.00	0.37	0.38	2.4	−1.61		4/0		0.40	4
0.95	0.37	0.27	2.0:	−1.01:		6/0			6
1.20	0.65	0.36	1.7:	−0.39:		3/0			
0.98	0.29	0.34	2.0:	−1.86:		5/0			1
1.33	0.85	0.58	3.7	−0.37		21/0			12
1.27	0.66	0.48	3.5:	−0.44:					
1.54	0.88	0.80	2.5	−0.7:					
0.62	0.00	0.01	10.0	−2.12	II	15/13	0.990	0.09	0
0.94	0.30	0.36	2.7:	−1.81:	II	13/11			0
1.29	0.73	0.49	6.7:	−0.41					
1.11	0.60	0.28	8.6	−0.37		10/1	0.271	0.48	10
1.46	0.76	0.76	2.3:	−1.05:					
1.06	0.63	0.25	4.3	−0.06					
1.47	0.98	0.65	5.6:	−0.1:		2/0			
0.85	0.28	0.12	5.6	−0.9:	I	33/15			
						15/0			
						1/0			

Table 2.1 (cont).

NGC/IC (1)	Name (2)	RA (1950) (3)	Dec. (4)	Long. (5)	Lat. (6)	Vel. (7)	Res. (8)	Mod. (9)	M_V^0 (10)
6388		17 32.6	−44 43	345.5	−6.7	81	0	16.83	−9.98
	Ton 2	17 32.7	−38 31	350.8	−3.4				
6401		17 35.5	−23 53	3.5	4.0		0	16.7:	−7.20:
6397		17 36.8	−53 39	338.2	−12.0	11	4	12.30	−6.65
6402	M14	17 35.0	−03 14	21.3	14.8	−116	0	16.90	−9.34
	Pal 6	17 40.6	−26 12	2.1	1.8			18.1:	
6426		17 42.4	03 12	28.1	16.2			17.30	−6.10
	TRZ 5	17 45.0	−24 46	3.8	1.7				
6440		17 45.9	−20 21	7.7	3.8	−118	0	16.40	−6.75
6441		17 46.8	−37 02	353.5	−5.0	−70	0−1	16.50	−9.08
	TRZ 6	17 47.5	−31 16	358.6	−2.2				
6453		17 48.0	−34 37	355.7	−4.0			16.4:	−6.50:
6496		17 55.5	−44 14	348.1	−10.0		3?	15.0:	−5.80:
	TRZ 9	17 58.7	−26 52	3.6	−2.0				
6517		17 59.1	−08 57	19.2	6.8			18.10:	−7.80:
6522		18 00.4	−30 02	1.0	−3.9	−24	1	15.64	−7.04
6535		18 01.3	−00 18	27.2	10.4			16.35:	−5.75:
6528		18 01.6	−30 04	1.1	−4.2	114	1	16.40	−6.90
6539		18 02.1	−07 38	20.8	6.8			15.7:	−6.10:
6544		18 04.3	−25 01	5.8	−2.2	−12	2−3	15.35	−7.10
6541		18 04.4	−43 44	349.3	−11.2	−148	2−3	14.60	−7.96
6553		18 06.3	−25 56	5.3	−3.1	−27	0	16.40	−8.15
6558		18 07.0	−31 47	0.2	−6.0			16.1:	
IC 1276	Pal 7	18 08.0	−07 14	21.8	5.7			18.50	
	TRZ 11	18 09.6	−24 46	8.4	−2.2				
6569		18 10.4	−31 50	0.5	−6.7		0	16.47:	−7.77:
6584		18 14.6	−52 14	342.1	−16.4	160	1	16.17:	−6.99:
6624		18 20.5	−30 23	2.8	−7.9	69	0	15.45	−7.13
6626	M28	18 21.5	−24 53	7.8	−5.6	00	2	14.99:	−8.06:
6638		18 27.9	−25 32	7.9	−7.2	−14	0	15.66	−6.51:
6637	M69	18 28.1	−32 23	1.7	−10.3	74	2	15.60	−7.90
6642		18 28.8	−23 30	9.8	−6.4	−84	1−2	15.11	
6652		18 32.5	−33 02	1.5	−11.4	−124	1	16.23:	−7.32:
6656	M22	18 33.3	−23 58	9.9	−7.6	−144	4	13.55	−8.45
	Pal 8	18 38.5	−19 52	14.1	−6.8			18.4:	
6681	M70	18 40.0	−32 21	2.9	−12.5	198		15.40	−7.32
6712		18 50.3	−08 47	25.3	−4.3	−124	2	15.51	−7.30
6715	M54	18 51.9	−30 32	5.6	−14.1	122	0	17.11	−9.41
6717	Pal 9	18 52.1	−22 47	12.9	−10.9			16.55:	
6723		18 56.2	−36 42	0.1	−17.3	−3	2	14.80	−7.48
6749		19 02.5	01 42	36.1	−2.2				
6752		19 06.4	−60 04	336.5	−25.6	−39	4	13.20	−7.80
6760		19 08.6	00 57	36.1	−3.9		0	15.90	−6.80
	TRZ 7	19 14.4	−34 45	3.4	−20.0				
6779	M56	19 14.6	30 05	62.6	8.3	−145	2	15.60	−7.35
	Pal 10	19 16.0	18 28	52.4	2.7			18.6:	
1925−30	Arp 2	19 25.6	−30 27	8.6	−20.8				
6809	M55	19 36.9	−31 03	8.8	−23.3	169	4	14.40	−7.45
	Pal 11	19 42.6	−08 09	31.8	−15.6			16.40	
6838	M71	19 51.5	18 39	56.7	−4.6	80	3?	13.90	−5.60
6864	M75	20 03.2	−22 04	20.3	−25.8	−198	0	16.85	−8.30
6934		20 31.7	07 14	52.1	−18.9	−360	1	16.22	−7.34
6981	M72	20 50.7	−12 44	35.1	−32.7	−255		16.29	−6.94
7006		20 59.1	16 00	63.8	−19.4	−362		18.12	−7.52
7078	M15	21 27.6	11 57	65.0	−27.3	−109	3	15.26	−8.91
7089	M2	21 30.9	−01 03	53.4	−35.8	−5	2	15.45	−8.95
7099	M30	21 37.5	−23 25	27.2	−46.8	−175	3	14.60	−7.10
	Pal 12	21 47.3	−21 28	30.5	−47.6			16.46	−4.30
	Pal 13	23 04.2	12 28	87.1	−42.7			17.10	−2.60
7492		23 05.7	−15 54	53.3	−63.5			16.70	−5.20

(1) New General Catalogue or Index Catalogue number.
(2) Other catalogue numbers or names.
(3) and (4) Right ascension and declination for epoch (1950) taken from Kukarkin (1974).
(5) and (6) Galactic longitude and latitude from Kukarkin (1974).
(7) Radial velocity in km s^{-1} from Davis Philip (1977) and Hartwick & Sargent (1978).
(8) Visual resolution of the brightest stars on a scale of 0 to 4 in the sense of increasing resolution (van den Bergh 1967b).
(9) Apparent distance modulus. Mostly from the review of Harris & Racine (1979) except for some very uncertain cases.
(10), (11) and (12) Absolute integrated visual magnitude and apparent (B−V) and (U−B) colours (Harris & Racine 1979).

Fundamental data

(B−V) (11)	(U−B) (12)	E(B−V) (13)	R (kpc) (14)	[Fe/H] (15)	OG (16)	Var/per (17)	$[m_1]$ (18)	Δ (19)	L (20)
1.17	0.66	0.32	6.3	−0.48		9/0			
1.58	0.89	0.79	2.3:	−0.7:		2/0 3/0			
0.75	0.15	0.18	7.1	−1.83		3/3			
1.28	0.64	0.58 1.8:	4.4 6.1:	−1.28	I	77/40 0/0	0.125	0.30	4
1.03 2.77	0.34 2.1:	0.40 1.8:	9.6	−1.35:	II	13/11			
1.98	1.51	1.11	4.8	−0.28					14
1.28	0.83	0.45	1.9	−0.24		10/0			
1.28	0.65	0.67:	2.1:	−1.1:		0/0			
0.93	0.42	0.07	2.4:	−0.1:		0/0			
1.79	0.94	1.14	3.2:	−1.33					
1.22	0.67	0.50	2.6	−1.04		10/9			5
0.97	0.31	0.36	5.3:	−1.9:		1/0			
1.45	1.10	0.65	1.8	−0.43		0/0			10
1.89	1.10	1.22	6.9:	−1.2:		1/0			
1.36	0.67	0.63	4.4	−1.02					
0.77	0.14	0.13	3.0	−1.59		1/0			
1.62	1.24	0.79	3.3	−0.4:		18/4			10
1.09	0.48	0.40	1.0:	−0.9:		9/0			
1.74	1.0:	0.92	5.8	−0.8:		5/1			
1.34	0.63	0.63	1.6:	−0.54:		5/0			
0.79	0.17	0.11	7.3:	−1.40		1/0			
1.10	0.57	0.25	1.4	−0.34		4/0			11
1.10	0.45	0.33	3.1:	−1.08		18/10	0.164	0.26	
1.16	0.58	0.36	1.9:	−0.6:		3/0			
0.99	0.50	0.17	2.2	−0.47		8/2			13
1.08	0.50	0.36	3.2	−0.88:		2/0			
0.92	0.39	0.11	6.4:	−0.5:		0/0			
0.99	0.30	0.35	6.1	−1.69	II	32/27	0.124	0.31	4
1.19	0.69	0.30	22:						
0.71	0.14	0.07	2.9	−1.17		2/0			
1.16	0.56	0.35	4.0	−0.43	I	21/16			8
0.85	0.24	0.14	13.0	−1.55	I	80/37			6
0.93	0.37	0.18	7.5:						
0.75	0.24	0.03	2.7	−0.85	I	25/19			
1.76	0.74	0.96:							
0.65	0.07	0.03	6.0	−1.62		2/0			
1.66	1.00	0.91	6.3	−1.06		4/0			
		0.12:							
0.86	0.18	0.22 1.2:	9.7 7.9:	−1.79		12/4 1/0		0.21	1
		0.11:							
0.69	0.10	0.07	4.8	−1.78		6/5			2
		0.35:	6.7	−0.63		0/0			
1.13	0.54	0.28	7.6	−0.28		4/2	0.283	0.52	
0.87	0.28	0.17	11.7	−1.30		11/0		0.26	
0.74	0.19	0.12	12.0	−1.38		51/30		0.23	
0.72	0.13	0.03	12.9	−1.27	I	40/28			
0.74	0.17	0.13	32.1	−1.66	I	71/58		0.33	2
0.68	0.06	0.12	10.3	−2.01	II	111/68	0.118	0.14	−1
0.67	0.08	0.06	10.5	−1.53	II	21/21	0.142	0.23	2
0.58	0.03	0.01	7.6	−2.03		12/4	0.401	0.12	0
0.90	0.29	0.02	15.6	−1.55		3/0			
0.69	0.14	0.05	25.7	−2.03		4/4			
0.48	0.17	0.00	21.3	−2.0:		4/4			

(13) Adopted (B−V) colour excess from Harris & Racine (1979).
(14) Adopted distance in kiloparsec (Harris & Racine 1979).
(15) [Fe/H] metallicity as suggested by Harris & Racine (1979), but see also table 2.3.
(16) Oosterhoff Group.
(17) The ratio of the number of variable stars discovered, to the number with measured periods (Sawyer-Hogg 1973). No entry indicates that no search has been made; zeros indicate failure to detect.
(18) Strömgren metallicity parameter from Johnson & McNamara (1969).
(19) Δ, a photoelectric scanner index used to measure the continuum break at λ 4000 Å (van den Bergh & Henry 1962).
(20) Line-strength index of van den Bergh (1969).

Sub-giant branch

Linking the MS and the GB, the sub-giant branch (SGB) is the ascending path taken by stars as they leave the MS and begin hydrogen-shell burning as their primary source of energy.

Upper horizontal branch

It has been suggested by theoretical work (Demarque & Hirshfeld 1975) that an upper horizontal branch (UHB) may extend upward and blueward from the red end of the RHB. It has yet to be confirmed in any object, although a recent study of the dwarf galaxy in Draco (Stetson 1979) indicates that such a feature may in fact be present.

2.2.2 Luminosity functions

Most researchers have been interested in the morphology of the CM diagram for globular clusters. A much more exacting task is to define the luminosity function, ϕ, which must be an unbiased account of the total relative numbers of stars at specified magnitude levels. Luminosity functions are available for only a few well-studied systems: M3 (Sandage 1957); M5 (Simoda & Tanikawa 1972); M92 (Hartwick 1970, van den Bergh 1975); and M13 (Simoda & Kimura 1968). A comparison of three of these luminosity functions is shown in fig. 2.3.

2.3 Distances to globular clusters

In table 2.1 we present the distances to galactic globular clusters, as given by Harris & Racine (1979). For comparison the reader may wish to consult Kukarkin (1974) and Woltjer (1975), the differences between independent compilations often giving a better estimate of the true errors involved in such complicated determinations.

2.3.1 Reddening

In determining the distance to a globular cluster by using its apparent CM diagram, one correction is fundamental: the foreground reddening. Not too surprisingly, this correction is deeply interwoven into the determination of metallicity, and although we shall discuss these two properties in separate sections it is worth noting from the outset that the two – reddening and metallicity – are by no means easily disentangled in any practical sense (see Kukarkin & Kireeva 1976; Kukarkin 1974).

Fundamental data

Without faint-star photometry and a prior knowledge of the intrinsic colours of the stars in globular clusters, it is impossible to correct the CM diagram immediately for the effects of interstellar reddening. Several procedures have been developed; they are outlined below. Comprehensive listings of reddening determinations are given by Burstein & McDonald (1975), Harris & van den Bergh (1974), and Kukarkin (1974); in table 2.1 we present those of Harris & Racine (1979).

Methods involving stellar photometry

Sandage (1969) uses two methods of determining reddenings to globular clusters with well-studied CM diagrams, described in (i) and (ii) below.

(i) *The two-colour diagram.* When clusters have a well-populated HB, hydrogen-line-blanketed models (Mihalas 1966) indicate that 'the theoretical $(U-B, B-V)$ colour–colour lines for $4 \geqslant \lg g \geqslant 3$ agree with the observed class V relation to better than $\Delta(B-V) = 0.02$ mag for $B-V$ bluer than -0.02'. Sandage (1969) uses the luminosity class V intrinsic colour relation given by Eggen (1965) to find reddenings for the four well-studied globular clusters M3, M13, M15 and M92. It is worth noting that for all but M15 the reddenings derived by Sandage (1969) of 0.00, 0.03, 0.12 and 0.02 mag respectively, are significantly lower than reddenings derived by the original 'Q-method' discussed below.

(ii) *Colour of the main sequence turnoff point.* With strict assumptions about the age spread in globular clusters, and after correcting for differential line-blanketing, Sandage (1969) has determined reddenings by comparing the colours of stars at the main sequence turnoff point with the bluest field sub-dwarfs. The colour limit for sub-dwarfs, derived from statistical reddening arguments, is found to be

$$(B-V)_o = 0.36 \pm 0.02 \text{ mag.}$$

This method of dereddening is obviously restricted to the very few systems where a high degree of confidence can be placed in the MS photometry and necessary blanketing corrections.

(iii) *Differential reddening.* For systems having similar metallicities (see below) and available CM diagrams, *differential* reddenings can be obtained by slide-fitting morphological features of the CM diagrams. An independent reddening to one system then provides absolute reddenings for all similar globular clusters. Typical features of the CM diagram that serve as fitting points are the red and blue edges of the RR Lyrae gap (Sturch 1967; McNamara & Langford 1969; Alcaino & Contreras 1971), the RGB (Arp, 1955), and the turnoff of the HB (Arp 1955).

(iv) *Foreground stars and the general run of reddening with distance.* For field stars of known spectral type surrounding a globular cluster field, estimates of distance and reddening can be made in the hope of bracketing the values appropriate to the globular cluster. Of course this method requires that the absorbing material be relatively uniform across the field and along the line of sight. Furthermore, the assumption that the field stars are on the MS yields a simultaneous overestimate of the reddening and underestimate of the distance, should the stars in fact be evolved. This problem can, of course, be resolved if spectral types are supplemented with luminosity classes; however, this is usually not the case. Examples of this method are given by Arp (1958), Feast, Thackeray & Wesselink (1960), and Crawford & Barnes (1975).

Methods involving integrated properties.
With the availability of integrated spectral types (Morgan 1969; Kinman 1959b; Kron & Mayall 1960; van den Bergh 1969) for globular clusters, it is tempting to sidestep the direct determination of reddenings by assuming the existence of some intrinsic colour–spectral type relation. This then allows the comparison of apparent integrated colours with spectral types to provide total foreground reddenings.

The method is by no means foolproof and involves large systematic uncertainties in its calibration and application. When multicolour photometry (such as UBV) is used to plot reddening trajectories of globular clusters in the two-colour diagram, the composite nature of the continuum, combined with variations in metallicity from cluster to cluster, makes both the slope and the zero-points of the trajectories quite uncertain. Early attempts to separate these effects used the 'Q-parameter'

$$Q \equiv (U-B) - X(B-V)$$

where, if $X = E(U-B)/E(B-V)$, Q will be independent of reddening. Van den Bergh (1967a) found $X = 0.72 \pm 0.04$ for clusters with spectral types earlier than F7, and established a correlation of Q both with metallicity parameters (see below) as well as with the intrinsic $(B-V)_o$ colour, viz.:

$$(B-V)_o = Q + 1.00.$$

Through this relation colour excesses

$$E(B-V) = (B-V) - (B-V)_o$$

could be derived (van den Bergh's (1967a), Table IV).

Racine (1973) has argued that X is observationally a function of the integrated spectral type of globular clusters (fig. 2.4), and has introduced a new reddening-free parameter $\mathscr{R}$ defined by:

$$\mathscr{R} = \frac{(U-B)+0.10[1-(B-V)]}{2(B-V)-1}$$

to take this into account. With this transformation, Racine claims that UBV photometry is 'redundant and cannot be used to infer the existence of a second classification parameter' such as metallicity. Kukarkin & Kireeva (1973) dispute this conclusion, suggesting that information on metallicity is in fact contained in the slopes of reddening trajectories, X, derived for the various spectral types of globular clusters due to the strong underlying correlation of spectral type with metallicity. To establish his intrinsic two-colour relation for galactic globular clusters given here in

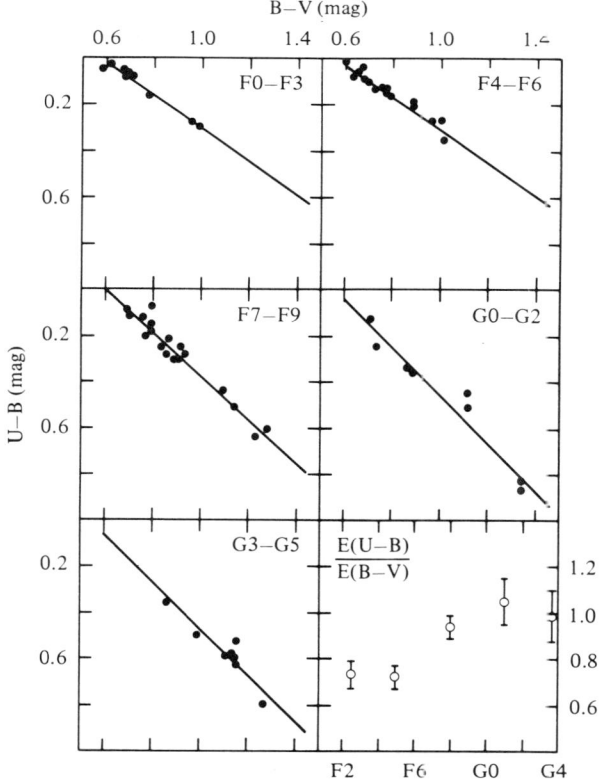

Fig. 2.4. An illustration of the observed variation of $E(U-B)/E(B-V)$ with integrated spectral type of globular clusters (Racine 1973).

table 2.2, Racine (1973) returned to individual estimates of reddenings for globular clusters with well-studied CM diagrams, as discussed above. It is this calibration that forms the basis of the reddenings derived for 94 galactic globular clusters with UBV photometry listed by Harris & van den Bergh (1974).

The cosecant law
Finally, failing all other methods, one may estimate a reddening based on a statistical description of the distribution of the absorbing matter in our galaxy. The validity and calibration of this method have been reviewed by Burstein & McDonald (1975), who derive:

$$E(B-V) = 0.01 + 0.057 \operatorname{cosec} |b|; \quad \sigma = 0.11$$
$$\pm 0.03 \pm 0.006$$

for 36 northern clusters with galactic latitudes $|b| < 12°$. Over ten years earlier Arp (1965) used

$$E(B-V) \approx 0.06 \operatorname{cosec} |b|,$$

while others (for instance, Harris & Racine 1979) prefer a form of the relation which gives no absorption at the poles, viz.

$$E(B-V) \approx 0.06 (\operatorname{cosec} |b| - 1).$$

In any event, the well-known patchiness of the interstellar medium makes these formulae, at best, first approximations to the true reddening.

Table 2.2. *Intrinsic UBV colours for galactic globular clusters (Racine 1973)*

$(B-V)_o$	$(U-B)_o$
0.60	0.010
0.65	0.050
0.70	0.105
0.75	0.175
0.80	0.255
0.85	0.360
0.90	0.50:

2.3.2 True distance moduli

Having decided on an appropriate value for the foreground reddening $E(B-V)$, it is then necessary to convert this to an absorption A_V. Usually this is accomplished through a knowledge of the ratio of total-to-selective absorption $R = A_V/E(B-V)$. For early-type single stars, $R \approx 3.1$; and on observational and theoretical grounds R is expected to increase with advancing spectral type (Schmidt-Kaler 1961; Straižys, Sūdižius & Kurilienė 1976). What the value is for a composite system is even more uncertain, and no corrections have been made to date: a single value of $R \approx 3.1$ is usually adopted for all conversions of $E(B-V)$ to A_V.

With the correction for absorption and reddening applied, it now becomes a question of slide-fitting the CM diagram in the vertical (apparent luminosity) sense so as to correct for distance. The availability and stability of various features in the CM diagram determine the procedure. We examine some of the standard methods below, although for a more thorough treatment see Harris (1976).

RR Lyrae stars and the level of the HB

The level of the HB in the vicinity of the RR Lyrae gap is a natural high-luminosity plateau which might be used as a fitting point for distance determinations. While some theories (Christy 1966) predict a dependence of the luminosity of the HB/RR Lyrae stars on metallicity, this is still controversial (Iben 1971; Stobie 1971; van Albada & Baker 1971), and observational tests are equivocal (see Sandage 1970; Hartwick & Hesser 1974). Harris (1976) adopts the HB luminosity as fundamental in his distance determinations, assuming

$$M_V(\text{HB}) = 0.6 \pm 0.3 \text{ mag.}$$

Extensive photographic photometry of RR Lyrae stars on the old m_{pg} system can be transformed to the photoelectric B system (Arp 1965) so as to provide mean magnitudes of these variables in estimating the level of the HB. Photographic data for the RR Lyrae variables can be found in Sawyer-Hogg (1973), from which Harris (1976) estimates that distance moduli can be found to ± 0.5 mag.

MS fitting

As shown in fig. 2.5, the position of the MS depends strongly upon the metallicity of the stellar population. Furthermore, the slope of the MS is quite steep ($\Delta V/\Delta(B-V) \approx 6$), making the fitting procedure precarious at best and also open to a serious propagation of errors in the case of uncertain reddening. For a full discussion of this technique see Sandage (1970).

Brightest stars

Photometry of the brightest stars has provided one means of estimating the distance to systems where only a handful of members can be resolved. First used by Shapley (1918) and later by Arp (1965), this method has been most recently calibrated by Harris (1976) to relate $\langle B \rangle_{25}$, the average blue magnitude of the 25 brightest stars, to the magnitude of the HB. Harris explicitly corrects for population effects through the integrated magnitude V_T and for metallicity effects (which are known to raise and lower the tip of the RGB) through $(B-V)_0$. He adopts

$$V_{HB} = 0.87\{\langle B \rangle_{25} - 0.14 V_T - 1.57[(B-V)_0 - 0.70] - E(B-V) - 1.15\},$$
$$\sigma \approx \pm 0.5 \text{ mag}.$$

Visual resolution

As a zeroth-order approximation to the brightest-star calibration, van den Bergh (1967b) gives visual estimates of the resolution of a cluster into stars.

Fig. 2.5. The theoretical variation of the CM diagram and especially the position of the MS for globular cluster stars with [Fe/H] = -2.0, -1.5 and -0.5 (Davis Philip 1977).

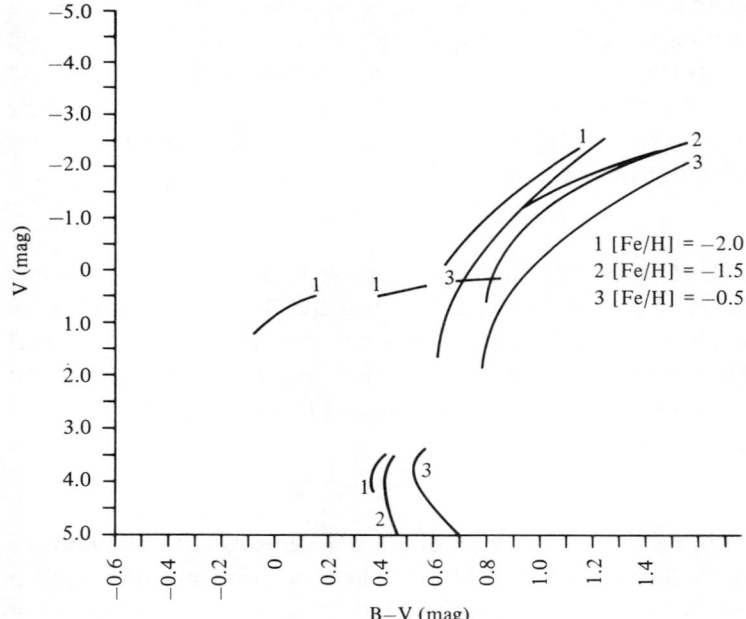

Harris (1976) has calibrated this in terms of the magnitude of the HB. For some clusters this is the only available distance estimate.

Richness index

Kukarkin (1974) showed that a visually determined 'richness index', IR, correlates with absolute integrated magnitude. On the distance scale adopted here, Harris (1976) finds

$$M_{V_T} = -6.94 IR - 4.03 \ (\pm 0.6 \text{ mag}),$$

with no apparent dependence on metallicity.

2.4 Variable stars in globular clusters

An excellent review of the problems of variable star research associated with globular clusters has been published by Rosino (1978); it is highly recommended and will not be repeated here. For source data and references, the 'Third Catalogue of Variable Stars in Globular Clusters', Sawyer-Hogg (1973), has no equal.

Fig. 2.6. Positions of variable stars in the CM diagram of a typical globular cluster (Rosino 1978).

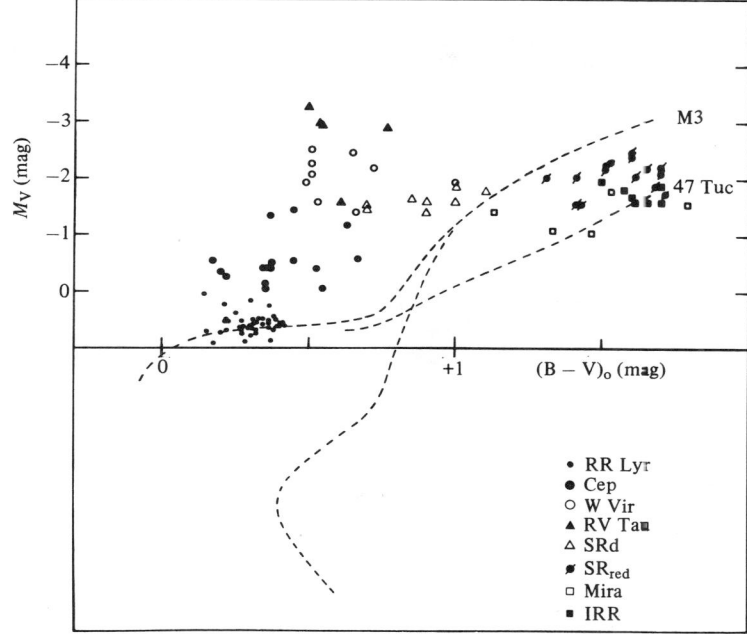

Half a dozen different types of variable stars have been identified in globular clusters. These are identified and plotted in the CM diagram in fig. 2.6. Of the 108 globular clusters so far searched for variables, only 12 show no candidates. In the remaining 96 clusters some 2000 variables have been discovered of which over 90 per cent are RR Lyrae pulsators with periods of less than one day (Rosino 1978). The remaining variables are: cepheids with periods between 1 and 10 days; W Vir (Population II cepheids) stars with periods from 11 to 20 days; RV Tauri stars; yellow and red semi-regular variables; Mira variables ($190^d < P < 212^d$); or irregular variable red giants. Three novae have been found in globular clusters. While eclipsing binaries are rare, they have been found, and may provide some link with the variable X-ray sources which must be added to the list. The eruptive U Gem stars found in the field of M5 (Oosterhoff 1941) and M30 (Rosino 1949) may be field objects. A recent discussion of variable star research with special attention to globular clusters is found in IAU Colloquium No. 21, edited by J. D. Fernie (1973).

2.4.1 RR Lyrae variables

Certainly the most interesting variables found in globular clusters are the RR Lyrae pulsators. Not only are their properties thought to be well

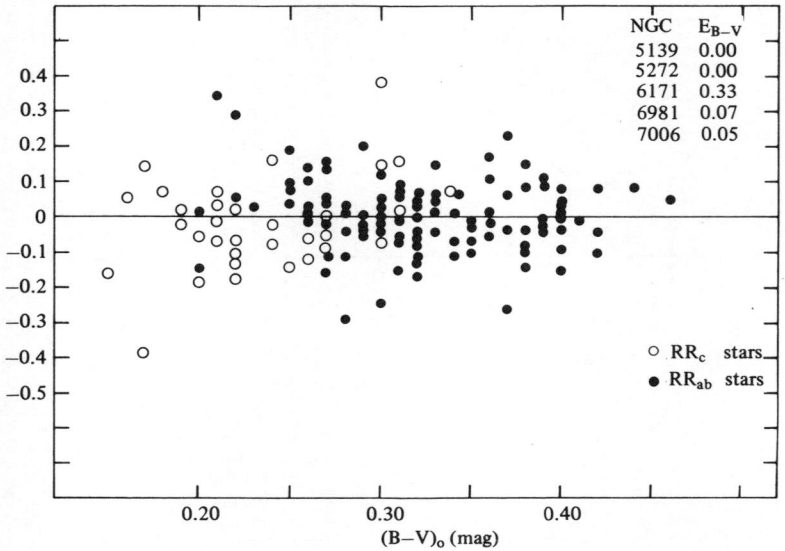

Fig. 2.7. Distribution of ab- and c-type RR Lyrae variables across the instability strip at the level of the HB (Rosino 1978).

understood from a theoretical point of view (Christy 1966), but from a purely phenomenological point of view the behaviour of their group properties is very suggestive.

As classified by their light curves and periods, RR Lyrae variables fall into two main classes:

(i) ab types, with steeply-rising and slowly-declining asymmetric light curves, large amplitudes (up to 1.8 mag), and periods generally in the range 0.4–0.8 days;

(ii) c types, with symmetric light curves, small amplitudes, and periods typically less than 0.4 days.

The c types are generally bluer than the ab types, although some overlap does occur (fig. 2.7).

Oosterhoff Groups

Considering only the ab-type variables, Oosterhoff (1939, 1944) noted that the average periods clustered strongly around 0.55 days or 0.65 days. These two groupings he called I and II respectively; they appear to be weakly correlated with the spectral type/metallicity of the parent cluster in the sense that globular clusters of Oosterhoff Group I are of higher metal abundance, as is also reflected in their later integrated spectral type.

This behaviour has been interpreted by Dickens (1970) in terms of Christy's transition period, P_{tr}, between c- and ab-type variables. A dependence of absolute magnitude on Oosterhoff Group follows. Using $\bar{M}_V = -0.46 - 4.17 \lg P_{tr}$ (Christy 1966), Dickens finds absolute magnitudes distributed as follows:

Class A (very low metal abundance): $\bar{M}_V = 0.51$ mag,
B (low metal abundance): $\bar{M}_V = 0.69$ mag,
C (moderate abundance): $\bar{M}_V = 0.95$ mag,

which Rosino (1978) translates into

Oosterhoff Group I: $\bar{M}_V = 0.60$ mag,
Group II: $\bar{M}_V = 0.94$ mag.

Such a systematic variation of the magnitudes of the RR Lyrae stars (and possibly the whole HB) is fundamental to the globular cluster distance scale *but* sufficiently uncertain to allow most workers to use the mean value of $\bar{M}_V \approx 0.6 \pm 0.35$ mag for practical purposes.

As a matter of interest, the sharp dichotomy into two Oosterhoff period groups does not appear to extend to the dwarf galaxies such as Draco and Ursa Minor (see Zinn, p. 197, this volume) in which the mean periods of the ab types are 0.613 and 0.636 days, respectively (van Agt 1973). For a more complete discussion of the controversial interpretation of the Oosterhoff Groups see Rosino (1978).

Table 2.3. *Metallicity-related parameters*

NGC/IC (1)	Name (2)	Spectral type (3)	(4)	Morgan class (5)	C (6)	S (7)	S (8)	$S_{0.6}$ (9)	ΔV (10)	ΔV (11)	Y,Z (12)
104	47Tuc	G3	G2.6			3.4	3.38	3.77	2.15	2.0	3,3
288			F6					4.88			
362		F8	F8.3		0.009	3.6	3.54	4.32		2.2	4,2
1261		F8	F7.7				4.52			2.6	
	Pal 1										
	AM-01										
0423−21	Eri										
	Pal 2										
1851		F7	F7.0				4.31			2.2	
1904	M79	F6	F5.7								
2298		F7	F6.7				5.55			3.1	
2419		F5	F5.5					5.16			
2808		F8	F7.2				4.63	4.15		3.1	
	Pal 3										
3201			F9				4.93	4.27		2.8	
	Pal 4		F6		0.046		4.56	4.68		2.6	
4147		F2	F3.8		0.092	5.6	5.45		2.45	2.4	1,2
4372			F3				6.10	5.40		3.1	
4590	M68	F2	F2.7					4.99			
4833			F4				5.14	4.70		3.0	
5024	M53	F4	F4.2	II	0.089	5.0	5.07	4.86	3.10	2.8	3,1
5053			F3			4.5	4.73			3.2	4,1
5139	ω Cen	F7	F6.6	II	0.010	6.6	5.26	4.63		2.8	1,1
5272	M3	F7	F6.1	II	0.026	4.9	5.00	4.49	2.64	2.7	3,2
5286		F8	F7.3					4.82			
5466			F5		0.140		5.49	4.90	3.0	3.0	2,
5634		F5	F5.1								
5694		F3	F3.0								
IC 4499											
5824		F5	F4.3								
	Pal 5										
5897			F3		0.017	4.7	5.00	4.53	3.10	2.7	2,2
5904	M5	F5	F6.1	II	0.096	4.3	4.42	4.32	2.58	2.6	3,2
5927		G2	G2.8								
5946			F5								
5986		F8	F6.3								
1608+15	Pal 14										
6093	M80	F7	F5.9					4.28			
6101											
6121	M4		F6.9				3.20	3.98		2.1	
6144			F7								
6139		F8	F8.7								
6171	M107	G0	G0.0		0.141	3.5	4.03	4.54		2.2	2,3
6205	M13	F5	F5.4	III	0.146	5.3	4.86	4.75	2.55	2.6	2,2
6218	M12	F6	F5.7				5.74	4.42		2.8	
6229		F7	F7.3	III							
6235			F5								
6254	M10	F8	F6.4	IV	0.151	5.0	5.38	4.87	2.85	2.6	2,2
6256	TRZ 12										
	Pal 15										
6266	M62	F8	F8.1								
6273	M19	F4	F5.3	IV							
6284		F9	F7.8								
6287			G3								
6293		F3	F4.1								
6304		G2	G4.6								
6316			F9								
6325			G1								
6341	M92	F2	F2.8	I	0.064	6.2	6.79	4.99	2.92	3.1	2,1
6333	M9		F2.8								
6342			G2								
6356		G5	G3.4	VI	0.073	2.6	2.56	3.36	2.15	1.9	4,3
6355			G4								
6352			G2				2.9	3.34		1.6	
	TRZ 2										
6366			G5								
6362			F9				4.61	4.69		2.4	
	TRZ 4										
	HP 1										
1730–33	Liller 1										
6380	Ton 1										
	TRZ 1										

Fundamental data

[Fe/H] (13)	[Fe/H] (14)	[Fe/H] (15)	HB type (16)	IR (17)	IM (18)	$B/(B+R)$ (19)	$(B-V)_{OG}$ (20)	$(B-V)_{OG}$ (21)	$(B-V)_{OG}$ (22)	$(B-V)_{TO}$ (23)
−0.52			7	0.75	0.60	0.07	0.81	0.89	0.95	0.52
−1.37			1	0.30					0.78	
−0.98			5	0.66	0.44	0.05	0.73	0.81	0.82	
−1.25				0.47				0.79		
				0.11						
−1.00				0.33						
−1.46				0.64				0.93		
				0.61						
−1.46				0.40				0.70		
−1.48				0.64					0.70	
−1.38				0.76				0.57	0.64	
				0.10						
−1.36			4	0.39				0.65	0.56	
−1.15				0.17				0.72	0.81	
−1.56	−2.0		3	0.39	0.36	0.86	0.81	0.80	0.79	
−1.59				0.26				0.78	0.68	
−1.95				0.45					0.66	
−1.67			1	0.55				0.71	0.71	
−1.75	−1.6	−1.85	2	0.61	0.30	0.92	0.68	0.71	0.73	
−1.87			2	0.18			0.66	0.64		
−1.55			2	0.86	0.34	0.90	0.67	0.78	0.80	
−1.44	−1.4	−1.57	4	0.64	0.35	0.56	0.74	0.79	0.81	0.40
−1.25				0.71					0.80	
−1.87			2	0.27	0.27	0.77		0.64	0.71	
−1.50				0.56						
−1.92				0.64						
				0.17						
−1.79				0.72						
				0.08						
−1.85			1	0.32	0.32	0.93	0.79	0.76	0.79	
−1.34		−1.01	3	0.62	0.35	0.72	0.71	0.73	0.80	0.43
0.9				0.56						
−1.6				0.53						
−1.39				0.62						
				0.08						
−1.45				0.72					0.72	
				0.29						
−1.11		−1.24	4	0.38				0.73	0.86	
				0.27						
−0.70				0.60						
−0.80		−0.82	6	0.39	0.48	0.07	0.85	0.83	0.98	
−1.53		−1.03	1	0.70	0.35	0.97	0.75	0.83	0.80	0.42
−1.52			1	0.42				0.77	0.77	
−1.36				0.69						
				0.35						
−1.47	−1.7		1	0.49	0.36	0.92	0.78	0.82	0.81	
−0.99				0.67						
−1.23				0.61						
−1.30				0.58						
				0.43						
−1.74				0.63						
0.08				0.57						
−0.89				0.58						
				0.41						
−1.99	−2.2	−2.18	2	0.61	0.20	0.83	0.67	0.69	0.71	
−1.71				0.56						
				0.65						
−0.21			7	0.69	0.62	0.10	0.80	0.85	0.83	
				0.42						
−0.11			7	0.35				0.97	0.92	
0.1				0.16						
−1.08			3	0.42				0.90	0.93	

Table 2.3 (cont.)

NGC/IC (1)	Name (2)	Spectral type (3)	(4)	Morgan class (5)	C (6)	S (7)	S (8)	$S_{0.6}$ (9)	ΔV (10)	ΔV (11)	Y,Z (12)
6388		G3	G2.5								
	Ton 2										
6401			F4								
6397		F5	F4.8		0.092	4.9	5.06	4.88		2.5	2,2
6402	M14	F8	F8.1	IV			3.03			2.0	
	Pal 6										
6426			F6								
	TRZ 5										
6440		G5	G4.6	VII							
6441		G4	G2.6								
	TRZ 6										
6453			F2								
6496											
	TRZ 9										
6517			G2								
6522			F8.7	IV	0.150	3.8	3.57	4.20		2.2	2,3
6535			F5								
6528			G3.4	VII			3.2			1.7	
6539			G3								
6544		F9	F8.8								
6541		F6	F5.4				6.60	4.76		2.8	
6553			G3.2	VII		3.0	3.0	2.79			2,3
6558											
IC 1276	Pal 7										
	TRZ 11										
6569			G0								
6584		F7	G0.2								
6624		G4	G2.5								
6626	M28	F8	F8.5								
6638		G2	G0.7								
6637	M69	G5	G4.8	VII	0.218	3.4	3.16	3.71		1.5	2,3
6642			F9.2	III							
6652		G2	G1.3								
6656	M22	F5	F4.8	II	0.040	3.7	3.72	4.10	2.50	2.6	4,1
	Pal 8										
6681	M70	G0	F7.7								
6712		G4	G0.3	V	0.146	4.5	4.54		2.0	2.2	1,3
6715	M54	F7	F7.9	III							
6717	Pal 9										
6723		G2	F9.8		0.095	3.6	3.79	4.13		2.0	1,3
6749			F8								
6752		F6	F5.6		0.058	6.6	6.58	4.42		2.9	1,1
6760			G0	VI							
	TRZ 7										
6779	M56	F5	F4.6			4.4	4.63	4.90		3.0	4,2
	Pal 10										
1925–30	Arp 2										
6809	M55		F4.7	III				4.92			
	Pal 11										
6838	M71	G5	G2.1	VI		3.4	3.22	3.68	2.1	1.9	2,3
6864	M75	F8	F8.2					4.04			
6934		F7	F5.7				6.81	4.71		2.9	
6981	M72	G0	F7.5	II			4.87	4.57		2.6	
7006		F3	F4.9	II	0.056	4.5	4.44	4.27		2.6	4,2
7078	M15	F3	F3.2	I	0.056	6.8	7.07	5.17	3.10	3.2	1,1
7089	M2	F3	F4.0	II	0.084	6.8	6.78	4.42	2.98	2.9	1,1
7099	M30	F3	F2.8				4.75	4.96		3.0	
	Pal 12										
	Pal 13		F6								
7492			F5				3.20			2.3	

(1) New General Catalogue or Index Catalogue number.
(2) Other Catalogue number or name.
(3) Spectral type from Kinman (1959a) as given by Davis Philip (1977), as derived from integrated blue spectra taken at a dispersion of 86 Å mm^{-1}.
(4) Spectral types by Kukarkin (1974) derived from weighted averages of other published data.
(5) Morgan (1959) metallicity class.
(6) Slope of the SGB at the level of the HB (Spencer Young 1970).
(7) and (8) Slope of the line from the HB/SGB intersection to a point 2.5 mag in V up the GB. From Hartwick (1968) and Kukarkin (1974) as tabulated by Davis Philip (1977).
(9) Slope of the GB at a point 0.6 mag in (B−V) from the intersection of the SGB and HB (Davis Philip 1977).
(10) and (11) V magnitude difference between the HB and GB, measured at $(B-V)_0 = 1.4$ mag. Data from Kukarkin (1974) and Davis Philip (1977), respectively.

Fundamental data

[Fe/H] (13)	[Fe/H] (14)	[Fe/H] (15)	HB type (16)	IR (17)	IM (18)	$B/(B+R)$ (19)	$(B-V)_{OG}$ (20)	$(B-V)_{OG}$ (21)	$(B-V)_{OG}$ (22)	$(B-V)_{TO}$ (23)
−0.10				0.64						
				0.49						
−1.48			1	0.37	0.32	0.95	0.80	0.84	0.74	0.28
−1.14			2	0.65				0.74		
				0.26						
−1.57				0.29						
−0.24				0.54						
−0.02			1	0.79						
				0.70						
				0.21						
−0.5				0.59						
−0.86				0.50	0.51	0.00	0.84	0.85	0.85	
−0.7				0.24						
0.06				0.49				1.0		
0.1				0.31						
−0.94				0.47						
−1.61				0.56				0.81	0.79	
−0.22			7	0.42			1.05	0.98	1.03	
				0.53						
				0.18						
−0.81				0.45						
−1.00				0.50						
−0.24				0.62						
−0.90				0.66						
−0.65				0.68						
−0.18			7	0.59	0.63	0.03	1.01	0.98	1.00	
−0.82				0.58						
−0.68				0.70						
−1.68	−2.0	−1.70	2	0.56	0.32	0.95	0.63	0.64	0.73	
				0.37						
−1.34				0.57						
−0.83		−0.39	5	0.44	0.54	0.15	0.95	0.51	0.91	
−1.17				0.68						
				0.47						
−0.79			5	0.53	0.56	0.26	1.05	0.97	0.89	
				0.28						
−1.52			1	0.56	0.35	0.94	0.80	0.81	0.82	
−0.76				0.42						
−1.77			1	0.43	0.30	0.77	0.53	0.58	0.55	
				0.11						
−1.55				0.38					0.78	
				0.17						
−0.36	−0.3	−0.04	7	0.29	0.63	0.00	0.90	0.87	0.95	0.50
−1.18				0.72					0.77	
−1.46				0.55				0.81	0.76	
−1.38			4	0.46				0.74	0.77	
−1.50	−2.0		5	0.51	0.38	0.25	0.67	0.74	0.71	
−2.02		−2.04	3	0.64	0.22	0.78	0.69	0.73	0.66	0.34
−1.72		−1.43	2	0.71	0.32	0.93	0.72	0.75	0.78	
−1.78		−1.96	1	0.47				0.66	0.65	
				0.13						
−1.25				0.02				0.9		
−1.2			3	0.16				0.74	0.69	

(12) Helium and metallicity parameters from Hartwick (1968) or Mironov (1972).
(13), (14) and (15) lg [Fe/H]$_*$ −lg [Fe/H]$_\odot$ as determined by Kukarkin (1974), Canterna (1976) and Butler (1975), respectively.
(16) HB morphological type (Dickens 1972a).
(17) Index of Richness (Kukarkin 1974).
(18) Index of Metallicity (Kukarkin 1974).
(19) Ratio of the numbers of stars on the HB bluer than the RR Lyrae gap to the total number bluer and redder.
(20), (21) and (22) Intrinsic colour index of the intersection of the HB and SGB; Hartwick (1968), Kukarkin (1974) and Davis Philip (1977).
(23) Intrinsic (B−V) colour of the turnoff of the main sequence (Davis Philip 1977).

2.5 Metallicity: spectral types and other parameters

Early work on the visual classification of globular cluster integrated spectra (Mayall 1946; Morgan 1956, 1959; Kinman 1959b; Kron & Mayall 1960) revealed interesting discrepancies depending on the classification criteria employed. Morgan (1956) noted that the spectral types determined from the iron lines ($\lambda\lambda 4250$–4400) differed systematically from the types derived from inspection of the hydrogen lines, while the G-band of CH provided an intermediate type. Using the G-band strength relative to Hγ as the fundamental, it appears that the metals are weakest in systems that otherwise classify as earliest.

2.5.1 The Morgan metallicity class and the Deutsch metallicity class: halo and disc populations

Morgan's (1956) fundamental types, based on CH/Hγ, range from F2 to G5 (table 2.3). Deviations ΔS from these types when considering the metal lines have been expressed in a number of ways. Morgan (1959) uses Groups I–VIII to indicate progressively increasing metallicity, while Kinman (1959b) uses Deutsch (1955) groups A, B and C to represent progressively diverging types (i.e. decreasing metallicity).

At the most rudimentary level, discussing early-type spectra and late-type spectra, we come into the realm of the galactic distribution and origin of globular clusters. The globular clusters divide into two main populations, as suggested by Morgan (1959) and refined by Woltjer (1975):

H: halo clusters with spectral types F6 and earlier or $(B-V)_o \leqslant 0.66$ mag; clusters with an F7 spectrum *and* $(B-V)_o \leqslant 0.69$ mag plus all clusters of Morgan class I–II.

D: disc clusters with spectral types later than G2 or $(B-V)_o \geqslant 0.73$ mag.

An intermediate halo population 'h' is also considered by Woltjer (1975) for clusters not contained in H or D but with spectra earlier than G1 or $(B-V)_o \leqslant 0.72$ mag or Morgan class IV or Deutsch class B.

Metallicities based on the integrated spectral types have long indicated a radial dependence of type with distance from the galactic centre. While details are still controversial, the overall trend is for the more metal-rich globulars to be more centrally concentrated to the galactic centre than is the case for metal-poor systems. References and discussions can be found in Harris & Racine (1979).

2.5.2 Other measures of metallicity and their absolute calibration to [m/H]

A host of spectrophotometric and photometric indices have been used to determine metallicities of globular clusters. We list some of these below, and draw special attention to Kukarkin's (1974) calibration of each of these indices in terms of a homogeneous determination of true logarithmic metallicities [m/H] relative to the Sun.

(i) Van den Bergh and Henry (1962) define three spectrophotometric parameters, all of which Kukarkin (1974) finds to correlate with metallicity.

Δ is a measure of the continuum discontinuity at $\lambda 4000$ Å (fig. 2.8), and dramatically correlates with metallicity for well-observed clusters. It is subject to large statistical uncertainties for low signal-to-noise observations.

ϕ is a reddening-free parameter defined by combining Whitford's (1958) interstellar reddening curve with two monochromatic colours,

$$\phi \equiv C(41-45) - 0.8\, C(46-51).$$

ψ is also a reddening-free parameter which includes the $\lambda 4000$ discontinuity measured by Δ,

$$\psi \equiv C(39-45) - C(45-51).$$

The values of Δ for 21 globular clusters are contained in table 2.1.

(ii) *The DDO system.* McClure & van den Bergh (1968) found two

Fig. 2.8. A comparison of low-resolution scans of a metal-poor globular cluster (M92) and a metal-rich one (van den Bergh & Henry 1962).

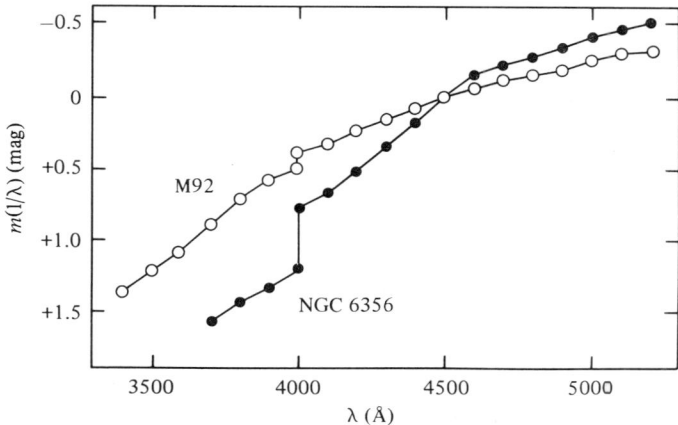

intermediate-band colour indices which, when corrected for reddening, were strongly correlated with one another and with the parameter Q. Kukarkin (1974) indicates that both of these colours, $C^*(38-41)$ and $C^*(41-42)$, are sensitive to metallicity. However, it should be noted that in order to deredden these indices it was necessary to appeal to the reddenings derived from the Q-method (van den Bergh 1967a), which is now known to be in error (Racine 1973).

(iii) *Strömgren photometry*. The narrow-band uvby photometric system of Strömgren (1963) and its associated reddening-free indices

$$[m_1] = (v-b) - 0.82(b-y),$$
$$[c_1] = (u-v) - (v-b) - 0.20(b-y),$$

have been applied to 15 globular clusters, and have been shown to correlate with other metallicity parameters. These data have also been analysed by Kukarkin (1974), and the index $[m_1]$ appears in table 2.1.

(iv) *Line-strength index*. For 43 galactic globular clusters and for 36 globular clusters in M31 van den Bergh (1969) has given a line-strength index, L, based on the strengths of the H and K lines of Ca II, $\lambda 4226$ of Ca I, the G-band and the $\lambda 4325$ Fe I feature, such that

$$L = \tfrac{1}{4}[\mathrm{Sp}(H+K) + \mathrm{Sp}(\lambda 4226) + \mathrm{Sp}(G) + \mathrm{Sp}(\lambda 4325)].$$

(v) *Ten-colour photometry*. Of the five metal-line indices defined by Faber (1973) from her intermediate-band photometry of eleven globular clusters, Kukarkin (1974) finds two of them

$$(Mg)_o = 0.373(45-55)_o - (52-55)_o$$

and $\qquad (38-41)_o,$

independently corrected for reddening, to be most dependent on [m/H]. Faber corrects her indices using individual values of $E(B-V)$ taken from CM diagrams in the literature.

(vi) *UBVI indices*. Kukarkin (1974) gives a direct calibration of UBVI indices in terms of logarithmic [m/H] values; for example,

$$[m/H] = 0.61 + 5.6(U-B) - 3.9(B-V).$$
$$\pm 1.1 \qquad \pm 1.2$$

Other reddening-free indices Q and $\mathscr{R}$ have been discussed on p. 32.

(vii) In the $(U-B)$ versus $(R-I)$ diagram Eggen (1972) defines $\Delta(U-B)$ as the vertical displacement of globular cluster GB stars away from the Hyades mean relation read at $(R-I) = 0.5$ mag. These values, Eggen notes, may be 0.05 mag too large, as they do not include back-warming effects; nevertheless, they do appear to correlate with the Deutsch class, with larger numerical values of $\Delta(U-B)$ indicating lower metallicity.

(viii) *The Washington System: Canterna photometry.* Canterna (1976) has introduced a broad-band photometric system designed to be sensitive to metallicity differences among late-type stars. Canterna & Schommer (1978) have applied this system to individual giants in eight distant globular clusters. Through model-atmospheres, [Fe/H] and T_{eff} have been derived for the cluster stars. It is not clear at this time just how sensitive the Washington System is to extremely low metallicities; however, the calibration of this system in terms of astrophysical quantities is a major step forward.

(ix) ΔS *and* $\Delta S(G)$ *for RR Lyrae variables.* Preston (1959) defined ΔS for RR Lyrae stars as the difference, expressed in tenths of a spectral type, between the hydrogen spectrum and the calcium K-line spectrum observed at minimum light. From coarse curve-of-growth analyses, ΔS was found to correlate with [Fe/H] such that large ΔS indicated low metallicity. Butler (1975) has introduced a modified $\Delta S(G)$ which uses 'a conspicuous absorption feature at the nominal position of the G-band' to replace the K-line measurement. Also using a coarse curve-of-growth analysis, Butler (1975) was able to calibrate ΔS with [Fe/H] such that

$$[\text{Fe/H}] = -0.16 \, \Delta S - 0.23.$$
$$(\pm 0.02) \quad (\pm 0.09)$$

Using this method twelve galactic globular clusters have measurements of [Fe/H].

(x) *Spectral peculiarities of individual stars and 'The weak-G-band effect'.* As detailed spectroscopic observations of individual stars in globular clusters became available, a number of peculiarities in the atmospheric compositions were noted. The CH and CN molecules seem to be the most affected. On the one hand, there appear to be CH stars and CN-strong stars in a number of globular clusters (see Harding 1962; Dickens 1972b; Zinn 1973a; Bessell & Norris 1976, among others), while Zinn (1973b) has found weak-G-band stars in M92 and Norris & Zinn (1977) have reported additional objects of this class in M13, M15 and NGC 6397. While the mixing of processed materials from the nuclear regions to the surface is generally supposed to be responsible for atmospheric enhancement, it appears that mass loss and thermal instabilities on the lower GB are necessary to explain the weakening of features. See Kraft, p. 87, this volume; Iben, p. 125, this volume; and Norris, p. 113, this volume.

2.5.3 CM diagram: morphology as an indicator of metallicity

Morphological characteristics of the CM diagram, especially the form of the RGB and the relative populations of the RHB and BHB, have been correlated with metallicity. Below we outline the various parameters defined most recently in the literature (figs. 2.9 and 2.10). Castellani, Giannone & Renzini (1970) discuss the interrelationship and theoretical interpretations of many of these.

(i) Sandage & Wallerstein (1960) define ΔV as the height of the GB above the level of the HB, measured at $(B-V)_o = 1.4$ mag. These authors find that ΔV correlates with other metallicity indicators; however, its definition is dependent on the choice of reddening; and, moreover, it is sensitive to changes in the colour of the SGB.

(ii) Sandage & Smith (1966) define $(B-V)_{o,g}$ as the intrinsic colour of the intersection of the HB and the SGB. This is another metallicity-sensitive parameter from the CM diagram. Once again, independent colour excesses are required to derive this parameter.

Fig. 2.9. Metallicity-sensitive features of the CM diagram measured after colour excess corrections have been applied (Arp 1962).

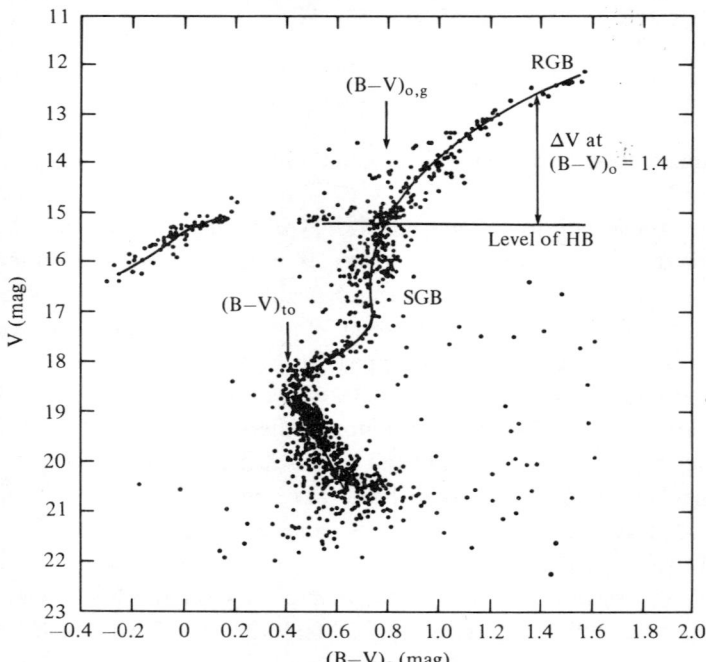

(iii) In an attempt to avoid any reddening dependence in a morphological metallicity parameter, Hartwick (1968) defined S as the slope of the line from the intersection of the HB and the SGB to a point on the GB 2.5 mag above the level of the HB. This procedure followed from theoretical models: Demarque & Geisler (1963) showed that for increased metal abundance (*a*) the SGB becomes redder and (*b*) the slope of the GB decreases. However, an increase in helium abundance has the same effect. It was Hartwick's hope to separate these two parameters and, through a more elaborate classification of the CM morphology in terms of the HB (fig. 2.11), Hartwick has suggested a two-dimensional classification, with four stages in Y (helium abundance) and three in Z (metallicity). Earlier suggestions of these combined effects are found in Sandage & Wildey (1967) and van den Bergh (1961, 1965). Additional values of S are given by Kukarkin (1974).

(iv) In response to the diversity of shapes of the extended SGB, Spencer Young (1970) defines C as the slope of the SGB at the magnitude of the HB. C correlates only weakly with Morgan metallicity class and even more

Fig. 2.10. Slopes S (Hartwick 1968) and C (Spencer Young 1970) are shown, as measured independently of colour excess determinations (Arp 1962).

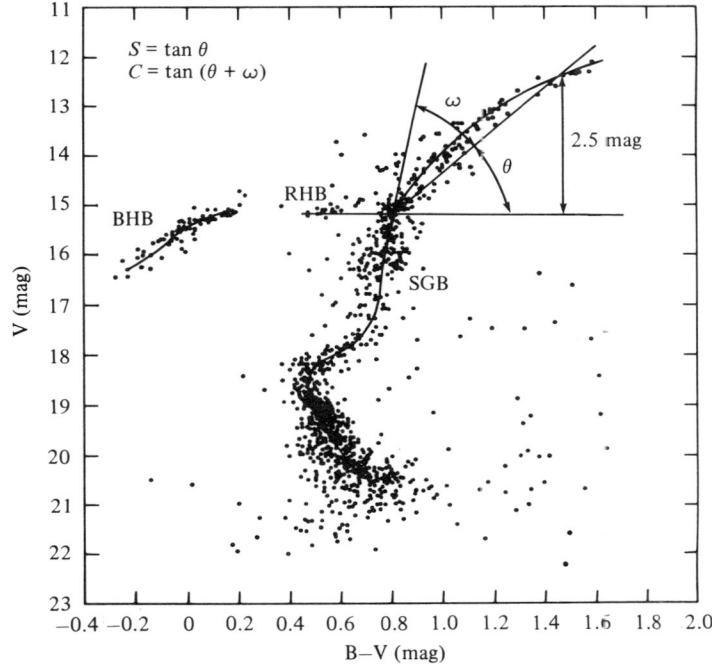

poorly with other metallicity indicators, suggesting that yet another parameter is involved. Whether this additional factor is age, helium abundance or observational error is still not clear.

(v) Another definition of the slope of the GB, introduced by Davis Philip (1977), $S_{0.6}$ accommodates more modest or poorly populated systems. It is defined as the slope of the GB measured from the intersection of the SGB and the extended HB to a point on the GB 0.6 mag to the red in $(B-V)$.

(vi) $(B-V)_{TO}$. As suggested by the work of Simoda & Iben (1970), the effective temperature at MS turnoff becomes hotter and the luminosity becomes higher (fig. 2.5) as the total metal abundance is decreased. Davis Philip (1977) has measured the intrinsic colours at MS turnoff for a number of globular clusters and confirms the correlation, although it is sensitive to both reddening and deblanketing uncertainties.

(vii) *HB morphology*. Castellani *et al.* (1970) and independently Dickens (1972a) have classified globular clusters according to the relative population of the HB. The former authors used theory to suggest the classification shown in fig. 2.12 while Dickens proceeded from a purely observational point of view. Dickens' types 1–7 correspond to Castellani's A–G. Quantitative measures of HB morphology are: (*a*) $B/(B+R)$, the ratio of B, the number of stars on the HB bluer than the RR Lyrae gap, to $(B+R)$,

Fig. 2.11. Variations of HB morphology as a function of metallicity Z and helium abundance Y, as suggested by Hartwick (1968).

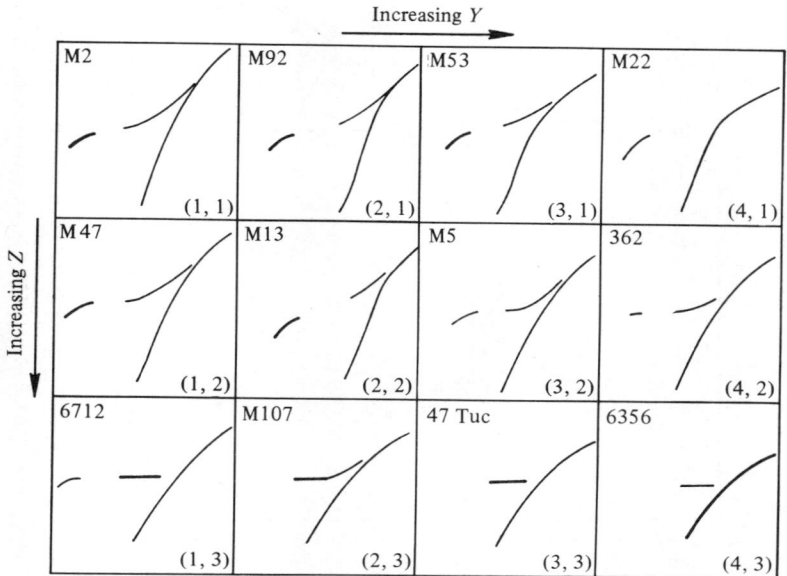

the sum of the number bluer and redder (Davis Philip 1977); and (*b*) N_c/N_{ab}, the ratio of c-type RR Lyrae variables to ab types. However, Dickens (1972a) finds a poor quantitative correlation between these two measures.

Mironov (1972, 1973) and Eigenson (1973) have also investigated the relative population of the HB, and Kukarkin (1974) has amalgamated their results with those of Dickens (1972a) to produce a continuous parameter *K* varying from 0 to 1, with 0 indicating a complete absence of the RHB.

2.6 Structural properties and dynamics

In a series of observational and theoretical papers King (1962, 1965, 1966a, 1966b) and King, Hedemann, Hodge & White (1968) have demonstrated

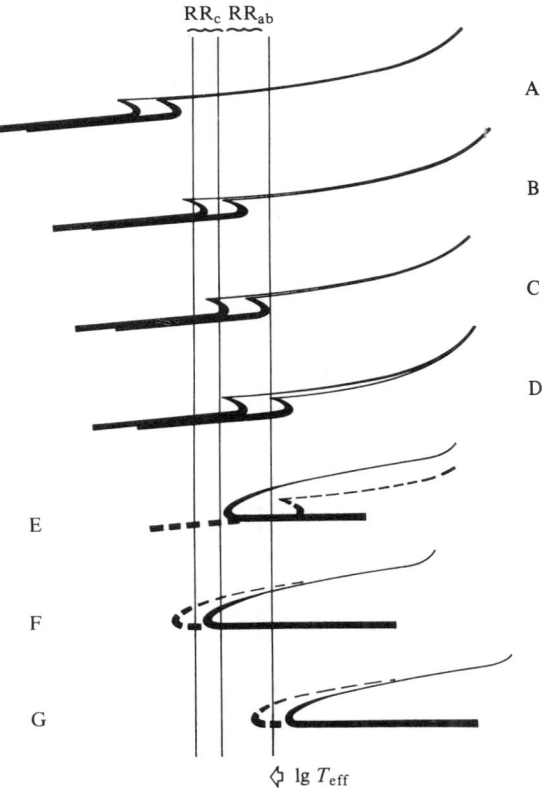

Fig. 2.12. Variations in HB evolution and subsequent population of the RR Lyrae gap, illustrating the classification of Castellani *et al.* (1970).

Table 2.4. *Structural and dynamic properties*

NGC/IC (1)	Name (2)	lg r_c (3)	lg r_t (4)	c (5)	C(S) (6)	Surf. br. (obs.) (7)	ρ (M$\odot$/pc^3) (8)
104	47Tuc	−0.33	1.70	2.03	III	5.55	4.18E+04
288		0.20	1.19	0.99	X		
362		−0.63	1.2		III	6.11	3.74E+04
1261		−0.40	0.90	1.30	II	8.61	1.01E+03
	Pal 1	−0.80	0.70	1.50	XII		
	AM-01						
0423−21	Eri	−0.36	0.54	0.90			1.50E+00
	Pal 2				IX		
1851		−0.87	1.34	2.21	II	5.81	1.01E+05
1904	M79	−0.57	1.03	1.60	V	7.83	4.71E+03
2298		−0.36	0.8		VI		
2419		−0.38	1.03	1.41	II	0.92	4.11E+01
2808		−0.49	1.2		I	6.41	6.13E+04
	Pal 3	−0.34	0.63	0.97	XII		
3201		0.04	1.56	1.52	X	9.49	1.39E+03
	Pal 4	−0.27	0.49	0.76	XII		
4147		−0.65	0.86	1.51	VI:	9.46	7.57E+02
4372					XII		
4590	M68	−0.15	1.47	1.62	X	9.03	1.01E+03
4833		0.06			VIII	9.58	9.18E+02
5024	M53	−0.33	1.34	1.67	V	9.54	7.01E+02
5053		0.39	1.14	0.75	XI	3.33	3.05E+00
5139	ω Cen	0.38	1.74	1.36	VIII	7.84	1.40E+03
5272	M3	−0.31	1.59	1.90	VI	7.42	3.63E+03
5286		−0.51	0.5		V		
5466		0.17	1.32	1.15	XII	1.90	1.26E+01
5634		−0.55	1.1		IV		
5694		−0.90	0.3		VII	8.32	4.50E+03
IC 4499					XI		
5824		−1.21	1.30	2.51	I	6.12	9.33E+04
	Pal 5	0.46	1.0		XII		
5897		0.24	1.06	0.82	XI	1.27	4.47E+01
5904	M5	−0.32	1.46	1.78	V	7.21	5.88E+03
5927		−0.32	0.7		VIII		
5946					IX		
5986		−0.28	1.1		VII	8.53	2.64E+03
1608+15	Pal 1	−0.17	0.6				
6093	M80	−1.05	1.14	2.19	II	5.51	3.31E+05
6101					X		
6121	M4	0.19	1.64	1.45	IX	9.15	3.48E+03
6144		−0.04	0.6		XI	0.91	1.10E+02
6139		−0.71	0.7		II		
6171	M107	−0.15	1.1		X	0.04	6.94E+02
6205	M13	−0.12	1.43	1.55	V	7.83	3.16E+03
6218	M12	−0.05	1.26	1.31	IX	9.34	1.01E+03
6229		−0.66	0.75	1.41	IV	8.49	2.72E+03
6235					X		
6254	M10	−0.11	1.38	1.49	VII	8.72	4.92E+03
6256	TRZ 12						
	Pal 15						
6266	M62	−0.56	0.6		IV	6.59	3.16E+04
6273	M19	−0.15	0.9		VIII	8.22	1.37E+04
6284		−0.99	0.5		IX:	7.13	5.47E+04
6287		−0.34	0.5		VII	9.73	4.17E+03
6293		−0.78	0.6		IV	7.50	5.81E+04
6304		−0.38	1.1		VI	9.02	6.58E+03
6316		−0.59			III		
6325		−0.52	0.5		IV:		
6341	M92	−0.56	1.22	1.78	IV	6.52	2.24E+04
6333	M9	−0.40	0.8		VIII	8.28	1.07E+04
6342		−1.00			IV		
6356		−0.50	0.7		II	8.45	7.18E+03
6355		−0.46	0.5				
6352		0.00			XI:	9.90	1.51E+03
	TRZ 2						
6366		0.34	1.3		XI		
6362		0.20	1.22	1.02	X	0.26	1.51E+02
	TRZ 4						
	HP 1				IX		
1730−33	Liller 1						
6380	Ton 1						
	TRZ 1						

Fundamental data

t_r (9)	σ (km s^{-1}) (10)	v_e (11)	Surf. br. (derived) (12)	ρ (M$\odot$/pc^3) (13)	t_r (14)	σ (km s^{-1}) (15)	v_e (16)
1.22E+08	10.52	44.76	6.02	2.68E+04	1.02E+08	8.42	35.83
			0.83	5.38E+01	2.31E+09	3.01	10.09
9.66E+07	9.28						
4.84E+08	4.73	17.46	8.64	9.13E+02	4.66E+08	4.50	16.64
					1.00E+10		
3.61E+07	9.14	40.37	6.88	3.71E+04	2.43E+07	5.55	24.51
1.36E+08	5.08	19.99	7.86	4.42E+03	1.32E+08	4.92	19.35
7.65E+09	4.25	16.09	1.06	3.41E+01	7.09E+09	3.57	14.66
1.50E+08	12.97		4.79	8.16E+01	5.59E+09	0.98	3.24
2.25E+08	3.99	15.45	9.60	1.21E+03	2.13E+08	3.72	14.41
			4.10	1.37E+00	1.04E+10	1.42	4.19
1.26E+08	2.59	10.00	9.66	6.14E+02	1.16E+08	2.33	9.01
4.29E+08	4.53	17.90	0.38	2.82E+02	2.59E+08	2.39	9.43
6.30E+08	5.04						
1.07E+09	5.57	22.20	9.01	4.81E+02	9.17E+08	4.52	18.39
5.43E+09	1.50	4.41	3.06	3.12E+00	5.48E+09	1.52	4.46
3.54E+09	10.96	41.06	7.93	1.22E+03	3.34E+09	0.22	38.31
4.25E+08	7.07	29.36	8.14	1.82E+03	3.22E+06	5.01	20.61
2.76E+09	1.93	6.84	2.12	9.43E+00	2.46E+09	1.57	5.92
1.11E+08	4.64						
2.50E+07	7.77	36.80	7.89	1.83E+04	1.34E+07	3.44	16.28
3.01E+09	3.11	9.55	1.00	4.79E+01	3.09E+09	3.22	9.88
2.41E+08	6.79	27.61	7.41	4.76E+03	2.21E+08	6.11	24.84
3.64E+08	5.97						
9.38E+06	8.42	37.05	6.01	2.07E+05	7.81E+06	6.66	29.30
1.25E+08	4.42	16.90	9.38	2.66E+03	1.12E+08	3.87	14.79
1.03E+09	2.88						
2.11E+08	3.05						
3.13E+08	6.02	23.45	7.98	2.64E+03	2.91E+08	5.50	21.43
2.73E+08	3.83	14.17	8.79	1.57E+03	3.26E+08	4.78	17.69
1.90E+08	4.75	17.98	8.63	2.27E+03	1.77E+08	4.34	16.45
1.50E+08	5.36	20.64	8.69	4.84E+03	1.50E+08	5.32	20.48
9.73E+07	8.78						
6.46E+07	5.62						
2.57E+07	6.49						
1.17E+08	4.61						
1.47E+07	5.39						
1.15E+08	5.37						
8.79E+07	7.49	30.45	6.82	1.65E+04	7.79E+07	6.44	26.16
1.13E+08	6.34						
3.49E+08	8.35						
2.64E+08	4.36						
1.25E+09	3.45	11.69	1.03	6.67E+01	8.99E+08	2.29	7.76

Table 2.4 (cont.)

NGC/IC (1)	Name (2)	lg r_c (3)	lg r_t (4)	c (5)	C(S) (6)	Surf. br. (obs.) (7)	ρ (M☉/pc³) (8)
6388		−0.65	0.8		III		
6401	Ton 2	−0.22	1.64	1.86	IX	7.92	1.53E+04
6397		−0.54					
6402	M14	−0.10	1.00	1.10	VIII	9.28	2.77E+03
	Pal 6				XI		
	TRZ5						
6426		−0.20			IX	2.50	5.44E+01
6440		−0.88			V		
6441		−0.66	0.6		III		
	TRZ 6						
6453		−0.90			IV		
6496					XII		
	TRZ 9						
6517		−1.0	0.5		IV		
6522		−0.92	0.7		V	6.89	1.24E+05
6535					XI:		
6528		−0.73	0.4		V	8.17	3.21E+04
6539		−0.31	1.2		X		
6544		−0.50					
6541		−0.50	1.50	2.00	III	6.98	2.61E+04
6553		−0.16	0.6		XI	9.46	3.27E+03
6558							
IC 1276	Pal 7	−0.10	1.3		XII		
	TRZ 11						
6569		−0.42	0.5		VIII		
6584		−0.40	0.5		VIII		
6624		−0.80	0.5		VI		
6626	M28	−0.42	0.6		IV	7.50	5.21E+04
6638		−0.53	0.5		VI	8.89	4.81E+03
6637	M69	−0.48	0.6		V	7.80	5.97E+03
6642							
6652		−0.97	0.5		VI:		
6656	M22	0.28	1.52	1.24	VII	8.43	4.65E+03
	Pal 8				X		
6681	M70	−1.00	0.7		V		
6712		−0.09	0.9		IX	9.87	1.40E+03
6715	M54	−0.81	1.2		III	6.32	4.50E+04
6717	Pal 9				VIII		
6723		−0.10	1.10	1.20	VII	8.82	6.28E+02
6749							
6752		−0.30	1.54	1.84	VI	6.93	1.23E+04
6760		−0.27	0.8		IX:	0.09	1.13E+04
	TRZ 7						
6779	M56	−0.17	0.7		X	9.59	6.55E+02
	Pal 10				XII		
1925−30	Arp 2						
6809	M55	0.24	1.27	1.03	XI	9.78	2.93E+02
	Pal 11	0.12	0.87	0.75	XI		
6838	M71	−0.09	0.7			9.99	1.12E+03
6864	M75	−0.93	0.5		I	6.81	2.00E+04
6934		−0.40	0.95	1.35	VIII	8.62	1.27E+03
6981	M72	−0.25	0.94	1.19	IX	0.19	1.99E+02
7006		−0.58	0.80	1.38	I	9.72	1.79E+02
7078	M15	−0.64	1.32	1.96	IV	6.20	3.94E+04
7089	M2	−0.40	1.21	1.61	II	6.97	6.99E+03
7099	M30	−0.91	1.20	2.11	V	6.23	8.08E+04
	Pal 12	−0.34	0.7		XII		
	Pal 13	−0.40	0.1		XII		
7492		−0.10	0.88	0.98	XII		

All data from Peterson & King (1975) as given by Davis Philip (1977) except as otherwise noted.
(1) New General Catalogue or Index Catalogue number.
(2) Other catalogue number or name.
(3) Logarithm of the core radius measured in minutes of arc.
(4) Logarithm of the limiting/tidal radius, measured in minutes of arc.
(5) Concentration parameter = lg r_t/r_c.
(6) Shapley's Concentration Class (Arp 1965).
(7) Central surface brightness in mag/arcmin².
(8) Density in M☉/pc³ derived from (7).

t_r (9)	σ (km s^{-1}) (10)	v_e (11)	Surf. br. (derived) (12)	ρ (M$\odot$/pc^3) (13)	t_r (14)	σ (km s^{-1}) (15)	v_e (16)
1.78E+07	3.59	14.82	8.21	1.15E+04	1.59E+07	3.11	12.83
7.69E+08	7.98	27.87	9.01	3.24E+03	8.21E+08	8.64	30.16
3.40E+08	1.48						
1.44E+07	6.98						
3.58E+07	6.09						
6.68E+07	7.14	30.20	7.55	1.52E+04	5.39E+07	5.44	23.02
4.28E+08	6.83						
4.45E+07	7.83						
1.38E+08	5.15						
1.96E+08	6.33						
5.47E+08	8.46	30.77	8.69	3.41E+03	4.82E+08	7.24	26.33
3.44E+08	4.67						
1.03E+08	10.13						
7.47E+08	4.70	16.93	8.87	5.58E+02	7.12E+08	4.43	15.95
9.66E+07	6.27	25.77	7.14	9.86E+03	8.86E+07	5.63	23.12
9.35E+07	6.02						
5.36E+08	4.22						
1.11E+09	4.16	14.15	9.36	3.48E+02	1.23E+09	4.76	16.20
8.14E+07	2.52						
1.82E+08	9.42						
5.45E+08	5.35	20.01	9.51	5.25E+02	3.83E+08	3.45	12.88
7.19E+08	3.09	11.10	0.33	1.62E+02	6.62E+08	2.79	10.01
1.22E+09	3.63	13.67	0.26	1.03E+02	9.80E+08	2.76	10.37
1.21E+08	10.25	43.06	6.59	2.68E+04	1.03E+08	5.46	35.55
4.61E+08	9.15	36.08	7.33	4.81E+03	3.95E+08	7.59	29.93
1.46E+07	6.04	26.10	6.83	4.62E+04	1.18E+07	4.57	19.75
			2.95	8.57E+00	9.34E+08	1.12	3.74

(9) Relaxation time (in years) derived from (7).
(10) Velocity dispersion (in km s^{-1}) derived from (7).
(11) Escape velocity (in km s^{-1}) derived from (7).
(12) Central surface brightness in mag/arcmin^2 derived from the total magnitude and the concentration parameter.
(13) Density in M$\odot$/pc^3 derived from (12).
(14) Relaxation time (in years) derived from (12).
(15) Velocity dispersion (km s^{-1}) derived from (12).
(16) Escape velocity (km s^{-1}) derived from (12).

that a physically reasonable fit to the density/intensity profiles $f(r)$ of globular clusters can be made using only three parameters: a core radius r_c; a limiting radius r_t; and a richness scaling factor K such that

$$f(r) = K\{[1+(r/r_c)^2]^{-\frac{1}{2}} - [1+(r_t/r_c)^2]^{-\frac{1}{2}}\}^2.$$

This work has recently culminated in the compilation of measured and derived structural parameters for over one hundred galactic globular clusters (Peterson & King 1975). These data are reproduced in table 2.4, while definitions and explanations of related parameters are given in the following section. See King, p. 249, this volume, for further discussion.

2.6.1 Structural parameters

(i) *Core radius*. This parameter specifies the size of the region of approximately constant density near the centre of a globular cluster. The core radius, r_c, is related in a model-dependent way to r_{hb}, the radius at which the surface brightness drops to one half of its central value (Peterson & King 1975). For all but the most diffuse clusters (where lg $r_t/r_c < 1$) $r_c \approx r_{hb}$.

(ii) *Limiting or tidal radius*. Acted upon by the differential tidal force of the Galaxy, globular clusters possess a physical limiting radius, r_t, beyond which member stars are no longer bound to the parent cluster. Beyond this radius, stars are free to join the field population. The limiting radius can only be determined from star counts that go well beyond the obvious extent of the cluster. Major uncertainties are found in establishing the correct background level for field stars and in noting that while the tidal limit may be well-defined in theory it may not be accurately delineated at any instant when stars that are unbound are still in the process of leaving the cluster.

(iii) *Richness factor*. A final scaling in terms of the central surface brightness/density completes the King-model description of globular clusters. While this measure of the nuclear density is directly related to the dynamics of the system, it is not easily accessible to observation. However, if only three independent parameters determine the structure of a globular cluster then the total magnitude, which is much more easily measured, can be used as an alternative scaling parameter. It should be noted, however, that use of the total magnitude compared with an independently determined central surface brightness does not in fact always result in the same richness factor, K. This is due to a combination of observational errors and in some cases due to failure of the simple model.

(iv) *Concentration parameter*. Independently of diameter or richness, it is possible to classify the cluster models by a concentration parameter c. This is defined by King (1966a) as the ratio of the two characteristic radii

$$c \equiv r_t/r_c,$$

while Peterson & King (1975) prefer to use its logarithmic definition

$$c \equiv \lg|r_t/r_c|.$$

(v) *Concentration class*. Shapley (1930) provides a determination of concentration class, C (S), based on the visual appearance of the globular clusters on photographic plates. Concentration Class I clusters are the most compact while Class XII clusters are the most diffuse. Peterson & King (1975) surprisingly find only a very weak correlation between their concentration parameter and that of Shapley.

2.6.2 Dynamical properties and mass-to-light ratios

The form of a cluster is related to its dynamics, and the dynamics themselves are amenable to theoretical and, now more than ever, direct observational investigations. Without any claim to completeness the following discussion outlines some of the more important definitions of time scales and velocities used in globular cluster studies.

(i) *Relaxation time*. Various definitions apply, but in terms of the energy transfer due to gravitational encounters one relaxation time passes for a star after the sum of the exchanges of energy is equal to the original kinetic energy of the star. Peterson & King (1975) explicitly define it as follows

$$t_r = [2\pi Gm^2 \nu j^3 \ln(\tfrac{1}{2}n)]^{-1},$$

where the number-density is $\nu = \rho/m$ and the modulus of precision $j^2 = 9/(8\pi Gr_c^2\nu_0)$ is related to the (observable) velocity dispersion $\sigma^2 = 1/2j^2$. With minor adjustments, t_r is of the same form as the 'reference time' defined by Spitzer & Härm (1958). For a Maxwellian velocity distribution it is of interest to note that 0.74 per cent of the stars will be lost from the system in one relaxation time.

(ii) *Central escape velocity*. In order to become unbound from the cluster, a star at the centre must exceed the central escape velocity defined (Peterson & King 1975) by

$$v_{e,o} = [W + (9/4\pi)(r_c/r_t)\mu]/j^2,$$

where W is a model-dependent dimensionless potential, scaled to the true potential V by

$$W = -2j^2(V + GM/r_t),$$

and where μ is a dimensionless mass such that

$$M = \rho_0 r_c^3 \mu.$$

(iii) *Mass-to-light ratios*. The central velocity dispersions $\langle v_r^2 \rangle^{\frac{1}{2}}$ have been measured in ten globular clusters (Illingworth 1975) using Fourier power-spectrum-analysis techniques. Masses and mass-to-light ratios, M/L, have been calculated from models discussed in Illingworth & Freeman (1974), Illingworth (1976) and Illingworth & Illingworth (1976). These are given in table 2.5. The average is $\langle M/L_V \rangle = 1.7$, with a range from 0.9 to 2.8, while deduced masses range from 1.6×10^5 $M_\odot$ to 1.1×10^6 $M_\odot$.

2.6.3 Kinematics, distribution, and total population

The most recent and extensive discussions of the system of galactic globular clusters are by Woltjer (1975), Harris (1976), Sharov (1976), Oort (1977) and de Vaucouleurs (1977). The importance of studying globular cluster populations as a whole is emphasized by Harris & Racine (1979) and Hanes (p. 213, this volume) and will not be repeated here, other than to point out that in studies of the early dynamical and chemical evolution of our own galaxy and of other galaxies, globular clusters continue to play a fundamental role. Fig. 2.14, from Harris (1976), shows in a striking way the very different spatial distributions of early and late-type globular clusters in our own galaxy, while fig. 2.15, from Hanes (1977), shows the strong dependence of total globular cluster population on the luminosity

Table 2.5. *Central velocity dispersions, masses and M/L for 10 globular clusters (Illingworth 1975)*

NGC	$\langle v_r^2 \rangle^{\frac{1}{2}}$ (km s^{-1})	Mass ($M_\odot \times 10^6$)	$(M/L_V)_\odot$
104	10.5±0.4	0.54±0.07	1.4±0.2
362	7.5±0.9	0.19±0.06	0.9±0.3
1851	7.9±0.7	0.16±0.05	0.9±0.3
2808	14.2±1.3	0.92±0.24	1.4±0.4
6093	12.5±2.5	0.39±0.18	2.8±1.3
6266	13.7±1.1	0.55±0.14	1.3±0.4
6388	18.4±0.8	1.12±0.26	2.0±0.5
6441	17.6±0.8	0.74±0.18	2.6±0.6
6715	14.2±1.0	1.03±0.23	1.8±0.4
6864	10.3±1.5	0.40±0.13	1.9±0.6

and type of the parent galaxy. These are fundamental aspects of galaxy evolution and globular cluster formation.

From a study of the three-dimensional distribution of galactic globular clusters Harris (1976) estimates that the total galactic population may be as high as 200, up nearly a factor of two over the estimates by Arp (1965) a decade earlier. In this context note that few new distant members of the galactic population have been discovered even with the advent of the southern sky surveys. One new system, AM-01, may be a very distant member or perhaps the first representative of an intergalactic globular cluster class (Madore & Arp 1979).

Fig. 2.13. A comparison of the galactic distribution of open clusters (upper panel) and globular clusters (lower panel) projected against the galactic plane (Payne-Gaposchkin 1979).

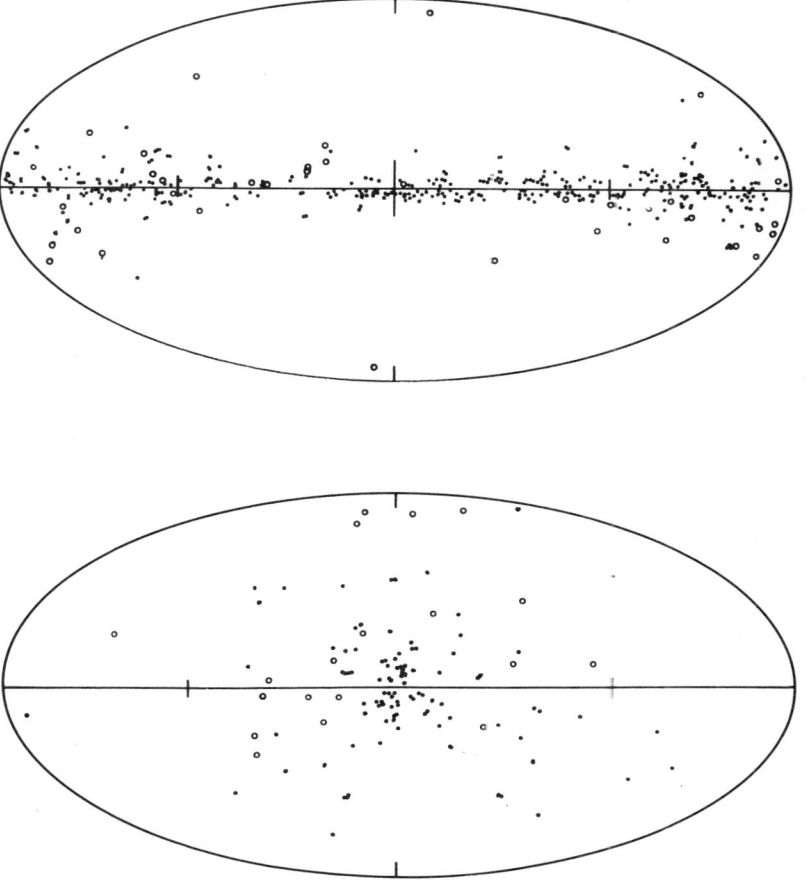

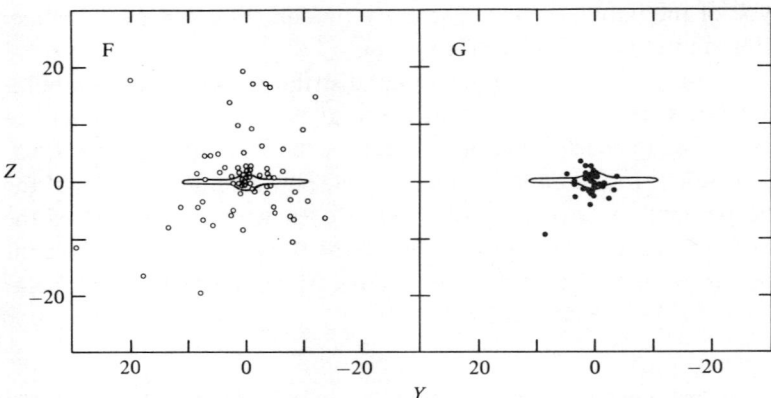

Fig. 2.14. Demonstration of the preference of G-type globular clusters (right panel) to be found more concentrated to the galactic centre than F-type clusters (left panel) (Harris 1976).

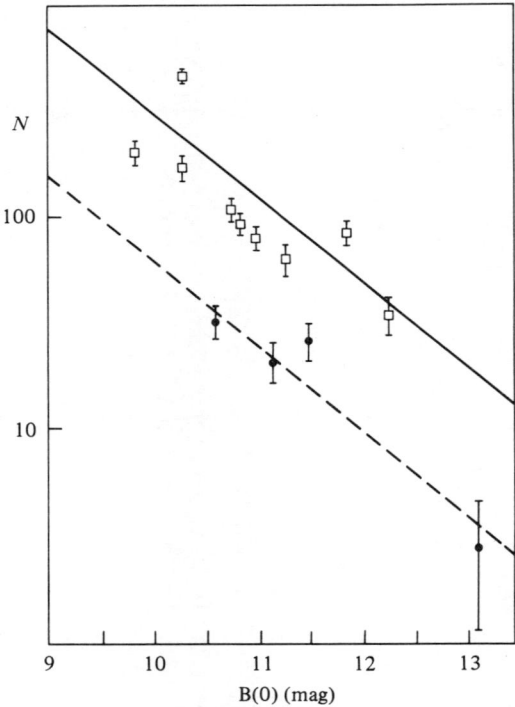

Fig. 2.15. Variation of globular cluster total population with intrinsic magnitude of the parent galaxy (Hanes 1977). Ellipticals are fitted by the upper line, spirals by the same line, shifted to lower total population.

While the sample is reasonably large, its distribution is amenable to several descriptions: Harris (1976) suggests that the radial density distribution of globular clusters falls off with increasing galactocentric distance as a power law (of index $n = 3$ to 4), while de Vaucouleurs (1977) suggests that the usual Population II $r^{\frac{1}{4}}$-law applies.

Mayall (1946), Kinman (1959a) and Hartwick & Sargent (1978) provide the basic data on globular cluster radial velocities. These have been discussed by the same authors and by Woltjer (1975) and Harris (1976), who conclude that the globular cluster system is in rotation, despite the lack of any apparent flattening with respect to the plane of the galaxy. It should be noted, however, that many of the radial velocities of globular clusters have never been confirmed, and that revisions in the globular cluster distance scale and total observed population could easily change these conclusions.

References

Alcaino, G. (1973). *Atlas of Galactic Globular Clusters with Colour–Magnitude Diagrams.* Universidad Católica de Chile, Santiago.
Alcaino, G. & Contreras, C. (1971). *Astron. Astrophys.* **11**, 14.
Alter, G., Balász, B. & Ruprecht, J. (1970). *Catalogue of Star Clusters and Associations*, 2nd edn. New York: Adler.
Arp, H. C. (1955). *Astron. J.* **60**, 317.
Arp, H. C. (1958). *Astron. J.* **63**, 118.
Arp, H. C. (1962). *Astrophys. J.* **135**, 311.
Arp, H. C. (1965). In *Galactic Structure*, ed. A. Blaauw & M. Schmidt, p. 401. University of Chicago Press.
Baade, W. (1944). *Astrophys. J.* **100**, 137.
Bessell, M. S. & Norris, J. (1976). *Astrophys. J.* **208**, 369.
Burstein, D. & McDonald, L. H. (1975). *Astron. J.* **80**, 17.
Butler, D. (1975). *Astrophys. J.* **200**, 68.
Canterna, R. (1976). *Astron. J.* **81**, 228.
Canterna, R. & Schommer, R. A. (1978). *Astrophys. J. Lett.* **219**, L119.
Castellani, C., Giannone, P. & Renzini, A. (1970). *Astrophys. Space Sci.* **9**, 418.
Christy, R. F. (1966). *Astrophys. J.* **144**, 108.
Crawford, D. L. & Barnes, J. V. (1975). *Publ. astron. Soc. Pacific*, **87**, 65.
Davis Philip, A. G. (1977). *Vistas Astron.* **21**, 407.
Demarque, P. & Hirshfeld, A. W. (1975). *Astrophys. J.* **202**, 346.
Demarque, P. & Geisler, J. E. (1963). *Astrophys. J.* **137**, 1102.
Deutsch, A. J. (1955). *Principes Fondamentaux de Classification Stellaire*, p. 25. Paris.
de Vaucouleurs, G. (1977). *Astron. J.* **82**, 456.
Dickens, R. J. (1970). *Astrophys. J. Suppl.* **22**, 249.
Dickens, R. J. (1972a). *Mon. Not. Roy. Astron. Soc.* **157**, 281.
Dickens, R. J. (1972b). *Mon. Not. Roy. astron. Soc.* **159**, 7P.
Eggen, O. J. (1965). *Ann. Rev. Astron. Astrophys.* **3**, 235.
Eggen, O. J. (1972). *Astrophys. J.* **172**, 639.
Eigenson, A. M. (1973). *Astrophysics*, **9**, 58 (= *Astrofizika*, **9**, 107).

Faber, S. M. (1973). *Astrophys. J.* **179**, 731.
Feast, W. M., Thackeray, A. D. & Wesselink, A. J. (1960). *Mon. Not. Roy. astron. Soc.* **121**, 337.
Fernie, J. D., ed. (1973). *Variable Stars in Globular Clusters and in Related Systems*, IAU Colloquium no. 21. Dordrecht: D. Reidel.
Hanes, D. A. (1977). *Mem. Roy. astron. Soc.* **84**, 45.
Harding, G. A. (1962). *Observatory*, **82**, 205.
Harris, W. E. (1976). *Astron. J.* **81**, 1095.
Harris, W. E. & Racine, R. (1979). *Ann. Rev. Astron. Astrophys.* **17**, 241.
Harris, W. E. & van den Bergh, S. (1974). *Astron. J.* **79**, 31.
Hartwick, F. D. A. (1968). *Astrophys. J.* **154**, 475.
Hartwick, F. D. A. (1970). *Astrophys. J.* **161**, 845.
Hartwick, F. D. A. & Hesser, J. E. (1974). *Astrophys. J. Lett.* **194**, L129.
Hartwick, F. D. A. & Sargent, W. L. W. (1978). *Astrophys. J.* **221**, 512.
Iben, I., Jr (1971). *Publ. astron. Soc. Pacific*, **83**, 697.
Illingworth, G. (1975). In *Dynamics of Stellar Systems*, IAU Symposium no. 69, ed. A. Hayli, p. 151. Dordrecht: D. Reidel.
Illingworth, G. (1976). *Astrophys. J.* **204**, 73.
Illingworth, G. & Freeman, K. C. (1974). *Astrophys. J. Lett.* **188**, L83.
Illingworth, G. & Illingworth, W. (1976). *Astrophys. J. Suppl.* **30**, 227.
Johnson, S. L. & McNamara, D. H. (1969). *Publ. astron. Soc. Pacific*, **81**, 415.
King, I. R. (1962). *Astron. J.* **67**, 471.
King, I. R. (1965). *Astron. J.* **70**, 376.
King, I. R. (1966a). *Astron. J.* **71**, 64.
King, I. R. (1966b). *Astron. J.* **71**, 276.
King, I. R., Hedemann, E., Hodge, S. M. & White, R. E. (1968). *Astron J.* **73**, 456.
Kinman, T. D. (1959a). *Mon. Not. Roy. astron. Soc.* **119**, 157.
Kinman, T. D. (1959b). *Mon. Not. Roy. astron. Soc.* **119**, 538.
Kron, G. E. & Mayall, N. U. (1960). *Astron. J.* **65**, 581.
Kukarkin, B. V. (1974). *The Globular Star Clusters*. Moscow: Sternbergh State Astron. Inst., Nauka. English version, NASA Technical Translation TT F-16, 157 (1975).
Kukarkin, B. V. & Kireeva, N. N. (1973). *Astron. Circ.* no. 797.
Kukarkin, B. V. & Kireeva, N. N. (1974). *Soviet Astron.* **18**, 346. (= *Astron. Zh.* **51**, 588).
Kukarkin, B. V. & Kireeva, N. N. (1976). *Soviet Astron.* **20**, 44 (= *Astron. Zh.* **53**, 83).
McClure, R. D. & van den Bergh, S. (1968). *Astron. J.* **73**, 313.
McNamara, D. H. & Langford, W. R. (1969). *Publ. astron. Soc. Pacific*, **81**, 141.
Madore, B. F. & Arp, H. C. (1979). *Astrophys. J. Lett.* **227**, L103.
Mayall, N. U. (1946). *Astrophys. J.* **104**, 290.
Mihalas, D. (1966). *Astrophys. J. Suppl.* **13**, 1.
Mironov, A. V. (1972). *Soviet Astron.* **16**, 105 (= *Astron. Zh.* **49**, 134).
Mironov, A. V. (1973). *Soviet Astron.* **17**, 16 (= *Astron. Zh.* **50**, 27).
Morgan, W. W. (1956). *Publ. astron. Soc. Pacific*, **68**, 509.
Morgan, W. W. (1959). *Astron. J.* **64**, 432.
Norris, J. & Zinn, R. (1977). *Astrophys. J.* **215**, 74.
Oort, J. H. (1977). *Astrophys. J. Lett.* **218**, L97.
Oosterhoff, P. Th. (1939). *Observatory*, **62**, 104.
Oosterhoff, P. Th. (1941). *Ann. Sternwarte Leiden*, **17**.
Oosterhof, P. Th. (1944). *Bull. astron. Inst. Netherlands*, **10**, 55.
Payne-Gaposhkin, C. (1979). *Stars and Clusters*, p. 159. Harvard University Press.
Peterson, C. J. & King, I. R. (1975). *Astron. J.* **80**, 427.
Preston, G. (1959). *Astrophys. J.* **130**, 507.
Racine, R. (1973). *Astron. J.* **78**, 180.

Rosino, L. (1949). *Pubblicazioni dell'Osservatorio Astronomico Universitario di Bologna*, **5**, no. 9.
Rosino, L. (1978). *Vistas Astron.* **22**, 39.
Sandage, A. R. (1957). *Astrophys. J.* **125**, 422.
Sandage, A. R. (1969). *Astrophys. J.* **157**, 515.
Sandage, A. R. (1970). *Astrophys. J.* **162**, 841.
Sandage, A. R. & Smith, L. L. (1966). *Astrophys. J.* **144**, 886.
Sandage, A. R. & Wallerstein, G. (1960). *Astrophys. J.* **131**, 598.
Sandage, A. R. & Wildey, R. (1967). *Astrophys. J.* **150**, 469.
Sawyer-Hogg, H. (1959). *Hbd. Physik*, **53**, 129.
Sawyer-Hogg, H. (1973). *Publ. David Dunlap Obs.* **3**, no. 6.
Schmidt-Kaler, Th. (1961). *Astron. Nachr.* **286**, 113.
Shapley, H. (1918). *Astrophys. J.* **48**, 89.
Shapley, H. (1930). *Star Clusters.* New York: McGraw-Hill.
Sharov, A. S. (1976). *Soviet Astron.* **20**, 397 (= *Astron. Zh.* **53**, 702).
Simoda, M. & Iben, I., Jr (1970). *Astrophys. J. Suppl.* **22**, 81.
Simoda, M. & Kimura, H. (1968). *Astrophys. J.* **151**, 133.
Simoda, M. & Tanikawa, K. (1972). *Publ. astron. Soc. Japan*, **24**, 1.
Spencer Young, P. (1970). *Publ. astron. Soc. Pacific*, **82**, 619.
Spitzer, L. & Härm, R. (1958). *Astrophys. J.* **127**, 544.
Stetson, P. (1979). *Astron. J.* **84**, 1149.
Straižys, V., Sūdžius, J. & Kurilienė, G. (1976). *Astron. Astrophys.* **50**, 413.
Stobie, R. S. (1971). *Astrophys. J.* **168**, 381.
Strom, S. E., Strom, K. M., Rood, R. T. & Iben, I., Jr (1970). *Astron. Astrophys.* **8**, 243.
Strömgren, B. (1963). In *Basic Astronomical Data*, ed. K. Aa. Strand, p. 123. University of Chicago Press.
Sturch, C. (1967). *Astrophys. J.* **148**, 477.
van Agt, S. (1973). In *Variable Stars in Globular Clusters and in Related Systems*, IAU Colloquium no. 21, ed. J. Fernie, p. 35. Dordrecht: D. Reidel.
van Albada, T. S. & Baker, N. H. (1971). *Bull. Amer. astron. Soc.* **3**, 241.
van den Bergh, S. (1961). *Publ. astron. Soc. Pacific*, **73**, 135.
van den Bergh, S. (1965). *J. Roy. astron. Soc. Canada*, **59**, 151.
van den Bergh, S. (1967a). *Astron. J.* **72**, 70.
van den Bergh, S. (1967b). *J. Roy. astron. Soc. Canada*, **61**, 179.
van den Bergh, S. (1969). *Astrophys. J. Suppl.* **19**, 145.
van den Bergh, S. (1975). *Astrophys. J.* **201**, 585.
van den Bergh, S. & Henry, R. C. (1962). *Publ. David Dunlap Obs.* **2**, 281.
White, R. E. (1970). *Astrophys. J. Suppl.* **19**, 343.
Whitford, A. E. (1958). *Astron. J.* **63**, 201.
Woltjer, L. (1975). *Astron. Astrophys.* **42**, 109.
Zinn, R. (1973a). *Astron. Astrophys.* **25**, 409.
Zinn, R. (1973b). *Astrophys. J.* **182**, 183.

3
Stellar evolution and globular clusters

V. CASTELLANI

3.1 Introductory remarks

The following is designed to provide an overview of stellar evolution theory in the context of globular star clusters. While there is no doubt that we can thank observations for improving our general understanding of stellar evolution and for throwing light on many theoretical uncertainties, there is now a sufficient body of elegant work to convince everyone that the converse is also true and we are beginning to use theory to understand our observations.

In discussing globular clusters we are dealing with the best sites that we have for testing the theory of stellar evolution. In globular clusters we find an exceedingly large sample of stars in advanced evolutionary phases, such as those related to the occurrence of hydrogen-shell burning or to the subsequent evolution from central to shell helium burning. This is simply a direct consequence of the large *total number* of stars (10^5–10^6) populating most galactic globular clusters. Consider the following order-of-magnitude discussion. We know that the lifetimes for late evolutionary phases (the luminous red giant and horizontal branch (HB) phases) are roughly on the order of 3×10^8 yr. In order to get a feeling for the range of masses covered by those late evolutionary stages we can compare this with the mass–lifetime relation for main sequence stars. Adopting the relation for main sequence lifetimes derived by Rood (1972) or Sweigart & Gross (1978) for Population II stars

$$d \lg M \sim 0.28 \, d \lg \tau_{MS},$$

we find that the range of masses for the evolving stars is

$$d \lg M \sim 0.003.$$

This means that if the main sequence in globular clusters is uniformly populated between $0.5 \, M_\odot$ and $0.8 \, M_\odot$ then

$$dM \sim 2.5 \times 10^{-3} \, M_\odot,$$

and accordingly only 1 star out of 120 is expected to be in an advanced evolutionary stage. In a typical globular cluster then we can expect a thousand highly evolved stars. Of course many of these will be hidden in the densely populated cluster nucleus.

As a further illustration it is of interest to estimate the number of *bright* red giants in a cluster. From the quoted evolutionary computations we find that a red giant spends about 5×10^6 yr within one magnitude of its maximum luminosity (i.e. before the onset of helium flash). Thus we would expect to find one very luminous red giant for every 5×10^3 stars, or a few dozen luminous giants in a typical globular cluster of 10^5 stars. It is reassuring that this is indeed the typical number of luminous giants found in nature: a short exposure plate of M92 reproduced in fig. 3.1 reveals the giants in this typical globular cluster, while for contrast the colour–magnitude diagram (fig. 3.2) of the poorly populated globular cluster Palomar 13 (Ciatti, Rosino & Sussi 1965) shows some systems are so poor that the most luminous stars are a few RR Lyrae variables.

Fig. 3.1. A short exposure V-plate (5 min at Cima Eckar telescope) of the galactic globular cluster M92. Plate limit is roughly 13 mag, i.e. $\approx$ 1 mag below the red giant tip.

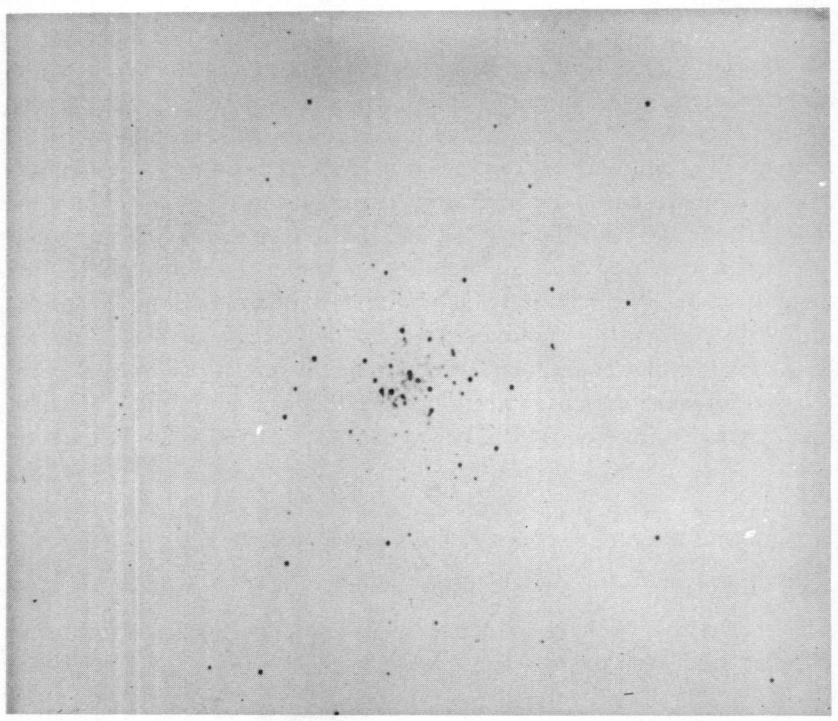

Stellar evolution and globular clusters 67

Fig. 3.2 (a) RR Lyrae variables (1–3) are among the most luminous stars in the remote globular cluster Palomar 13. (b) The HR diagram of stars in the field of Palomar 13 shows the lack of luminous red giants in the cluster. Open dots show the location of RR Lyrae variables, crosses represent probable field stars (from Ciatti, Rosino & Sussi 1965).

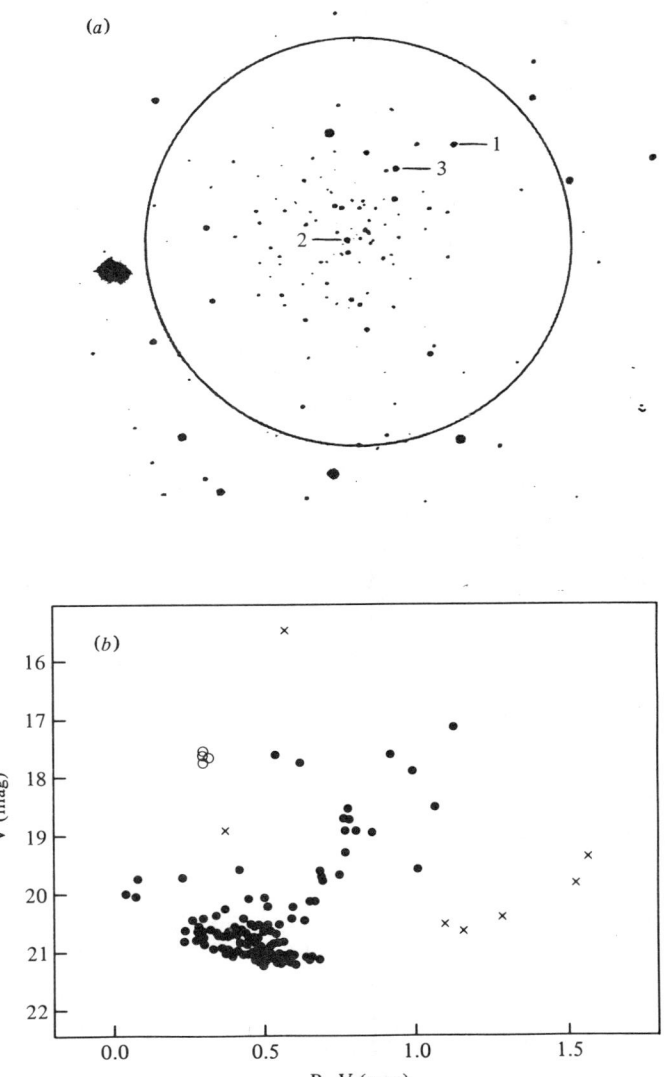

3.2 An introduction to the evolution of Population II stars

The evolution of low-mass stars away from the main sequence and through the subsequent phases of hydrogen-shell burning, central helium burning, and ultimately double-shell (H+He) burning is well understood, at least in its main expected features (see Iben 1974, and p. 132, this volume; Renzini 1977). Table 3.1 summarizes the main evolutionary phases together with a rough indication of the appropriate evolutionary time scales, while fig. 3.3 illustrates these preliminary phases of evolution in terms of a colour–magnitude diagram. Further, after the second ascent of the giant branch during the asymptotic giant branch (AGB) evolution, globular cluster stars are unable to ignite carbon in their degenerate cores. As a result they evolve toward very high effective temperatures, reaching highly degenerate configurations at the beginning of their careers as hot white dwarf stars. Unfortunately this final stage is at a sufficiently low intrinsic luminosity that observations, to date, have failed to detect this white dwarf population.

Understanding globular cluster stars is a recent achievement indeed. Following the pioneering work by Nishida (1960), the real start to the subject came when Faulkner (1966), following a suggestion by Hoyle & Schwarzschild (1955), showed that double-energy-source models qualitatively fit the observed HB. On the basis of much theoretical work it soon became clear that it would be possible to constrain the fundamental parameters governing the evolution of stars in globular clusters; that is, it would be possible to specify the age and original composition of the cluster. Indeed the existence of a well-defined 'cluster age' and 'original composition' is immediately suggested by the mono-parametric behaviour of the HR diagrams of cluster stars as far as the large majority of clusters is concerned.

Table 3.1. *Lifetimes and model structures for the evolutionary phases observed in globular clusters*

Evolutionary phase	Model structure	Lifetime
Main sequence (MS)	Central H-burning	10^{10} yr
First giant branch (FGB)	H-shell burning	10^8 yr
Horizontal branch (HB)	Central He-burning + H-shell burning	10^8 yr
Asymptotic giant branch (AGB)	Double-shell burning	10^7 yr

As an example of the chain of theoretical arguments that lead to constraints on the age and evolution of a globular cluster let us consider the problem of the luminosity level of the HB. If we define the 'HB luminosity level' as the luminosity of HB stars measured at the effective temperature of RR Lyrae variables then numerical computations tell us that the luminosity of zero-age-horizontal-branch (ZAHB) stars within the instability strip is a function of two parameters: (i) the mass of the helium core (M_c) at the beginning of HB evolution and (ii) the helium content of the hydrogen-rich envelope of the star. At the same time, evolutionary computations show that M_c is itself a function of the *original* helium content (with only a slight dependence on the cluster age and original abundance of heavy elements). As a result, the quoted luminosity level turns out to be primarily a function of the original helium content alone.

Furthermore, computations indicate that first-giant-branch (FGB) *lifetimes* are only slightly dependent on age and chemical composition. And the same holds for HB lifetimes. It turns out that the number ratio of FGB stars (above the HB luminosity level) to HB stars is, once again, primarily

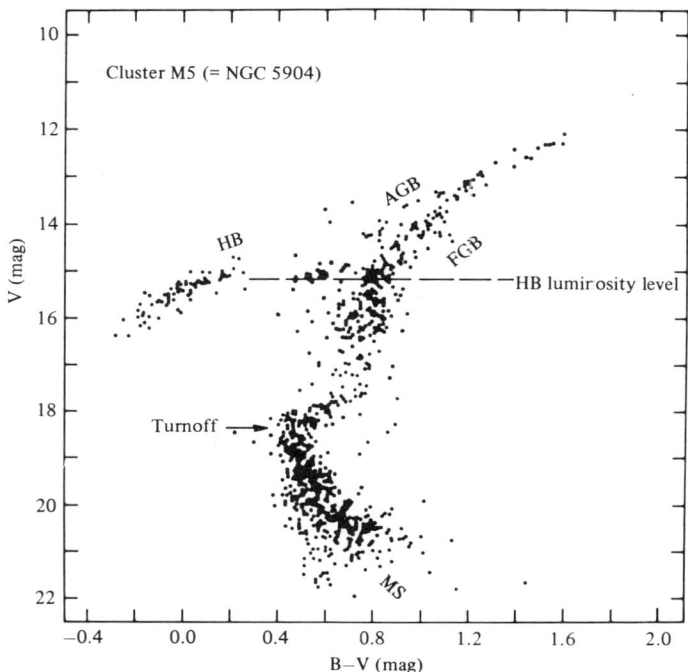

Fig. 3.3. The HR diagram location of the various evolutionary phases in the typical globular cluster M5 (symbols as in table 3.1).

a function of the original helium abundance as was first suggested by Iben (1968). This all follows from the simple fact that by increasing the helium abundance, the HB luminosity level increases, which in turn decreases the portion of the FGB we are looking at.

3.3 Limits to the present theory

Some of the outstanding problems in the theory of Population II stars are listed in the final column of table 3.2. In the following we will have a closer look at these problems so as to see just how far we can presently push theoretical interpretations.

3.3.1 The main sequence

Perhaps the best way to assess the reliability of theoretical models for main sequence stars is first to intercompare the results of recent computations. On the whole no dramatic discrepancies are apparent; differences are within $\Delta m \sim 0.25$ mag in luminosity and $\Delta(B-V) \sim 0.08$ mag in colour. Nevertheless one can see, as shown in fig. 3.4, that even these differences can mask variations in the original helium content, Y, as large as $\Delta Y = 0.10$. On the other hand variations in the zero-age-main-sequence (ZAMS) locus with Y are very similarly defined by all computations. For the ZAMS luminosity, measured at fixed effective temperature,

Table 3.2. *A sketch of some evolutionary constraints on the observed evolutionary parameters for globular clusters stars*

Parameter	Main evolutionary constraints	Major problems
1 Main-sequence location	$f(X_i)$	External convection, observations
2 Turnoff luminosity	$f(X_i,t)$	X_C, X_O, observations
3 Turnoff temperature	$f(X_i,t)$	Convection, X_C, X_O, observations
4 Red giant branch location	$f(Z-Z_{CNO})$	Bolometric correction, colour–temperature relation, convection
5 HB luminosity level	$f(Y)$	
6 HB colour	$f(X_i,t)$	Mass loss, rotation, mixing...
7 N_{HB}/N_{FRG}	$f(Y)$	Induced semi-convection, observations

$$\Delta \lg L \sim -1.3\,\Delta Y.$$

That is, if the amount of original helium varies by $\Delta Y \sim 0.10$ between two clusters with the same metallicity, we expect that the ZAMS of the helium-enriched cluster will be shifted by ~ 0.3 mag toward lower luminosities. It should be pointed out that in this context we are still at the stage where it is more reliable to look for relative differences among globular clusters, rather than looking for absolute determinations of their evolutionary parameters. To aid in this, general relations for the dependence of ZAMS loci on the original chemical composition have been published by Caputo (1977).

In trying to understand how reported differences arise, we are faced with the problem that differences may be due to the general treatment of the physics throughout the stellar model or due to specific assumptions, such as those concerning the efficiency of superadiabatic convection in the external layers of the star. As far as the first problem is concerned it is generally a difficult task to understand in detail the origin of discrepant

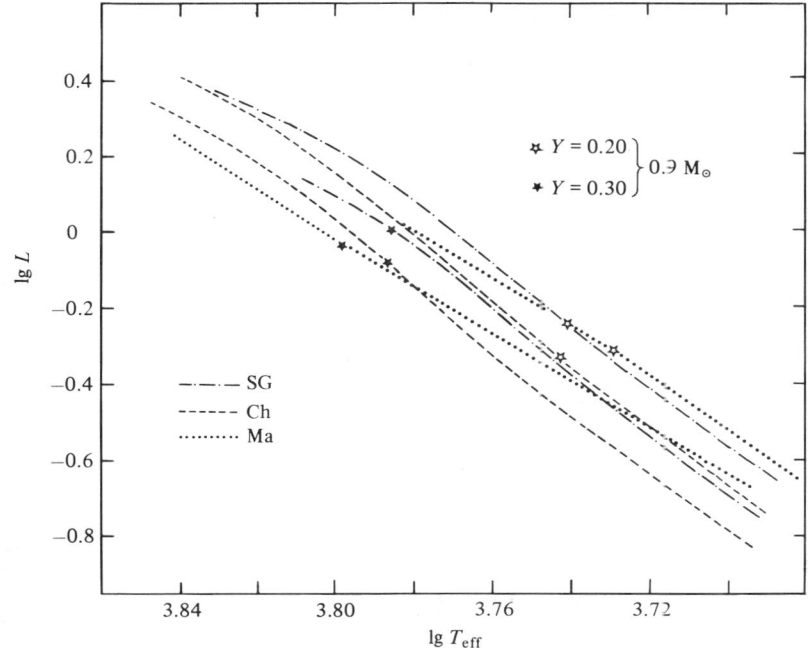

Fig. 3.4. Theoretical main sequences for $Z = 4 \times 10^{-3}$ and various Y as computed by Sweigart & Gross (1978) (SG) or C. Chiosi (private communication) (Ch) compared with very recent computations given in Frascati by I. Mazzitelli (private communication) (Ma).

results without a careful examination of the computing programmes. However, there is no difficulty in finding very general indications. For example, by comparing the ZAMS of Sweigart & Gross (1976, 1978) with that of C. Chiosi (private communication) as in fig. 3.4, we find that systematic discrepancies start at effective temperatures lower than $\lg T_{\text{eff}} \sim 3.82$. We then suspect that we are dealing with differences in the treatment of something characteristic of low-mass stars: either superadiabatic convection or energy generation via the p–p chain. On the other hand we know that variations in the treatment of external convection tend to preserve a model's luminosity, affecting only the stellar radius; the contrary appears in fig. 3.4 and on this basis alone a best guess is that something is different in the energy generation. Fortunately this conclusion can be further supported by the fact that in both computations the same mixing length to pressure scale height ratio, l/H_p, of unity was chosen.

As far as external convection is concerned, the best we can do to choose a value for the free parameter l ($=$ mixing length) in the treatment of superadiabatic convection is to require that the models fit the present

Fig. 3.5. The evolution of the 'solar' model fitting the observed characteristics of the real Sun. $Z_\odot = 0.02$ and $t_\odot = 4 \times 10^9$ yr have been assumed: $Y = 0.22$, $l = 2 H_p$ have been derived from the fitting procedure. Small stars indicate the expected variations in the effective temperature when the assumed mixing length is varied as labelled. (I. Mazzitelli, private communication).

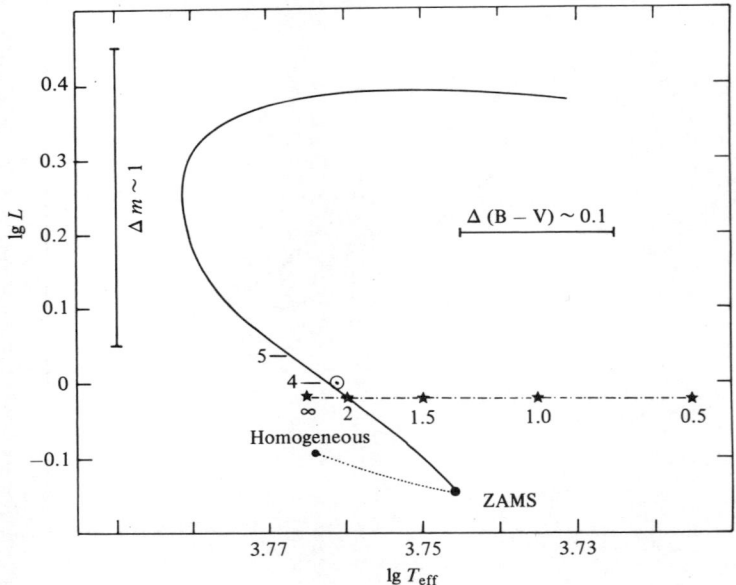

radius and luminosity of the Sun. Unfortunately we have no idea whether l can be assumed as constant with varying evolutionary phase and/or the original chemical composition. However, if we assume $Z_\odot = 0.02$ and $t_\odot \sim 4 \times 10^9$ yr, we find a fit is obtained for the values

$$Y_\odot = 0.23 \text{ and } l = 2H_p.$$

in full agreement with the values given by P. M. Hejlesen (private communication). It must be emphasized that the luminosity is roughly governed by helium content whereas the radius is fixed by the mixing length (fig. 3.5), such that these values are both strongly constrained in fitting to the Sun. In other words, as far as present computations are concerned the Sun must have a 'theoretical helium abundance' in the range $Y \sim 0.21$–0.24. Such a result will force us into a curious situation, as we will find later, where at least in some galactic globular clusters there is evidence for an ever larger 'theoretical Y' than in the Sun.

The mixing length necessary to fit the Sun is twice that generally adopted in evolutionary computations. This can be interpreted as a larger efficiency of convection in the Sun. We will see in the following some of the consequences of the choice of the value of mixing length as far as the main sequence is concerned. Fig. 3.5 shows the range of temperatures that can be covered in varying the mixing length between 0.5 (low convective efficiency) and ∞ (complete convection).

The relative abundance of CNO elements is expected to be of minor importance in determining the ZAMS location for Population II stars. This is because when $M < 0.8$ $M_\odot$ the nuclear energy is largely supplied by the p–p cycle. In some ZAMS structures central carbon is depleted toward an equilibrium value, but in no case does the CNO efficiency exceed a few per cent of the total nuclear energy released. More generally, theoretical evaluations suggest that the ZAMS location does not depend on the detailed evaluation of the evolution of some pseudo-primary elements, such as ^{13}C or ^{7}Be. Correct distributions of elements throughout a main-sequence structure can easily be determined by the so-called 'relaxation methods' (Moretti 1978).

3.3.2 The later phases of hydrogen-burning evolution

Evolution off the main sequence through the subgiant phase and along the red giant branch results as a consequence of hydrogen depletion at the centre of a main sequence star and the formation of a hydrogen-burning shell surrounding the hydrogen-depleted core.

As a general warning, it should be remembered that during these later evolutionary phases the star becomes more and more dependent upon external convection and large differences can occur as a consequence of the choice of mixing-length values. Once again we find that luminosities are little affected whereas effective temperatures depend strongly on the assumed mixing length. Evidence for such an effect is shown in fig. 3.6 where the first phases of evolution off the main sequence are shown for different choices of the mixing length. As a result we can roughly estimate that while the luminosity of the turnoff point along the evolutionary track is known to within 0.25 mag, we can estimate the star's radius only if a value of a mixing length is firmly established.

As is well known, evolutionary tracks can be manipulated to obtain

Fig. 3.6. The evolution off the main sequence on the basis of available recent computations, on the assumption that $Z = 4 \times 10^{-3}$ and $Y = 0.30$. Evolutionary paths for 1 $M_\odot$ models are from P. M. Hejlesen (private communication) (He) and C. Chiosi (private communication) (Ch) and largely reflect the difference in the assumed mixing length, as quoted in the text ($l_{He} = 2 H_p$; $l_{Ch} = 1.24 H_p$). The main sequence location as given by Mazzitelli (private communication) (Ma) is in full agreement with Hejelsen's previous computations. The evolution of 0.9 $M_\odot$ models (Chiosi, private communication; Sweigart & Gross 1978 (SG)) discloses the differences occurring between models computed under the same assumptions but with different evolutionary codes.

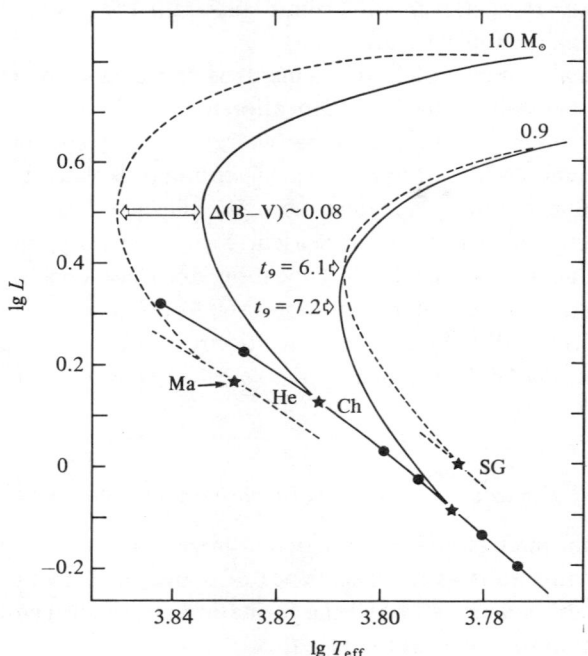

theoretical 'isochrones' which predict the expected distribution of stars with different masses but of common age (a cluster-like situation). Without entering into the fine details, it is worth remembering that comparisons with observations (e.g. the observed luminosity functions for red giant stars) show general agreement, so that we can hopefully believe theoretical predictions on the mass size of the evolving helium core, a parameter of extraordinary importance in determining the subsequent evolution through helium burning.

Following hydrogen depletion, central temperatures increase and CNO burning becomes more and more important. For a typical model ($M = 0.8$ $M_\odot$, $Y = 0.25$ and $Z = 10^{-3}$) we find that at the turnoff of the track both ^{12}C and ^{16}O have been depleted to their equilibrium values (a few per cent of ^{14}N; fig. 3.7) and that at the centre, at least 99 per cent of the nuclear energy is generated by the CNO cycle. As a consequence, many morphological features of these phases of evolution are expected to depend strongly on the original amount of C, N and O (and of course, on the correct treatment of the CNO burning!).

Inspection of the available literature shows that turnoff points computed *assuming* CNO equilibrium rates accord with each other in being about 0.25 mag less luminous than those actually following the evolution of the quoted elements. As far as the influence of the original chemical abundance of CNO is concerned we are, unfortunately, still lacking exhaustive computations in this respect. The majority of evolutionary tracks have been computed under the hypothesis of a 'universal' distribution of heavy

Fig. 3.7. The distribution through the stellar model of the abundance by mass of CNO elements in a typical model (see text) for a globular cluster star at the turnoff point. Abundances are normalized to their maximum values.

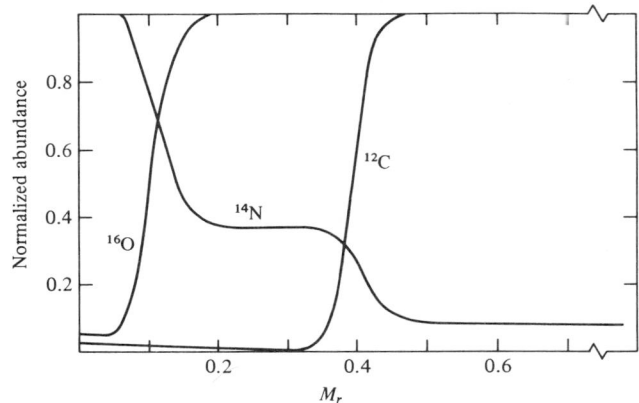

elements. That is, the *relative* abundances of CNO with respect to Fe have been assumed to be everywhere the same as in the Sun. Simoda & Iben (1970) showed that increasing the CNO abundance results in a decreased turnoff luminosity. More recently, Renzini (1977) discussed the expected influence of CNO elements on the location of subgiant and red giant stars.

As for the location of the red giant branch in the HR diagram, it was realized very early on that the location does not depend on the efficiency of nuclear energy generation mechanisms: it does depend on the mass of the evolving star and on the chemical composition of the convective stellar envelope (and, of course, once again on the assumed mixing length). Quantitative estimates of such a dependence can be found in the quoted works and can be derived by interpolation of the results of C. Chiosi (private communication) or Sweigart & Gross (1976, 1978). This approach has not yet been extensively used in interpreting observed differences among globular clusters. Perhaps we will learn much about early galactic evolution simply by interpreting the observed properties of red giant branches and in particular by focusing attention on extragalactic clusters with peculiar $(B-V)_{o,g}$ values.

At present the theoretical framework developed under the 'solar CNO hypothesis' and adopting canonical assumptions on the efficiency of convection has been extensively used in discussing the age and composition problems of globular clusters. Many of the results obtained in this way look very promising. In addition we know, at least in principle, which theoretical constraints depend on the two quoted assumptions. However, as far as the range of uncertainty found by comparing the results of different authors is concerned, I believe these differences will only be reduced by relying on accurate computations with up-to-date input physics. Relative differences between theoretical cluster loci will, however, remain more reliable than 'absolute' characteristics since the differential characteristics are probably less susceptible to the remaining ambiguities in the physical framework.

3.3.3 Helium-burning phases

It appears that the region of the HB phase in globular clusters is most sensitive to varying the evolutionary parameters. At this stage variations are amplified and manifest themselves as observable differences in the HB morphology. For example, while it is suspected that consistent mass loss is occurring during the FGB evolutionary phase, no one expects any clear effect as far as the location of the red giants in the HR diagram is concerned. On the other hand, the colour of an HB star is dramatically dependent on the amount of mass lost by the red giant progenitor. It was in this way that the first suggestions of mass loss were inferred from the HB morphology. Likewise there is reason to believe that CNO anomalies are more clearly recorded in the colours of stars on the HB than elsewhere (Castellani & Tornambè 1977).

Basically, we know that after the onset of helium flash a red giant quickly settles down onto its ZAHB location, where (roughly) the luminosity is dependent on the mass of the central helium core. Here the effective temperature depends on the mass of the hydrogen-rich envelope (and thus in turn is dependent upon the amount of mass loss). The larger the helium core, the more luminous the star is; the larger the envelope, the redder the star is. Fortunately, there is general agreement among various computations as far as the dependence of M_c on cluster age and chemical composition is concerned. On this basis alone much progress can be made mainly in understanding the origin of the observed differences among globular clusters.

Further evolution during the HB phase is strongly dependent on whether or not the so-called 'induced semi-convection' (Castellani, Giannone & Renzini 1971) occurs. Aside from the purely theoretical support (Renzini 1977), recent computations on the expected number of AGB stars (Caputo, Castellani & Wood 1978b) appear to support a large efficiency of semi-convection. The occurrence of semi-convection mainly affects HB lifetimes where the helium-burning phase lasts longer if semi-convection introduces a fresh supply of helium into the centrally convective region.

3.4 Results

In discussing the status of evolutionary theories we have to remember that for every assumed cluster age and original chemical composition we believe ourselves capable of predicting the distribution of stars throughout the HR diagram with only a few adjustments of the (free?) parameters, mixing length and mass loss. All observational evidence agrees with this belief. But just how accurately can observations of HR diagrams themselves prescribe the age and chemical composition as well? In my opinion, when attempting interpretations of this kind for a large sample of globular clusters, the largest difficulty at present comes from a lack of complete and reliable observational data. I believe that theory can give a clear indication at least of the age and the original helium content for *exhaustively observed* clusters, such as NGC 5272 ($=$ M3).

Supposing that we know only the HR diagram of the cluster stars, then from the number ratio of HB to FGB stars we can derive a good estimate of the original helium content. Furthermore, from the observed difference in luminosity between the HB and the turnoff, we can derive the cluster age. Also, possible anomalies in the CNO abundances can be taken into account. The unfortunate point is that completely reliable data for the starting parameters (e.g. N_{HB}/N_{FGB} ratio, etc.) as yet seem to be unavailable for all but a few clusters. Most observations have been concerned with the form of, rather than the population distribution within, the HR diagram. As a consequence we are still at the stage of speculating. Here I am tempted to claim that if half of the effort that has been put into the subject of quasars and other exotic extragalactic objects had been concentrated on galactic globular clusters we would know much more about the evolution of our galaxy (and of the universe) than we do now.

Nevertheless, present results indicate, in a very general way, that a somewhat 'normal' helium abundance ($Y = 0.20$–0.30) is present in globular clusters and that ages of the order of 10^{10} yr are appropriate (e.g. Iben 1974; Caputo & Castellani 1975a, b; Castellani 1978). As far as more precise determinations of the original helium are concerned, the situation is intriguing. If we take the observational data given by Lee (1977a, b) for the ratio $R = N_{HB}/N_{FGB}$ in the clusters 47 Tuc, NGC 3201 and M4, and adopt a recent calibration for R as a function of Y (Castellani 1977) we derive helium abundances between $Y = 0.20$ and $Y = 0.25$, the larger value being derived for 47 Tuc. At the same time, by relying on the canonical analysis of the properties of RR Lyrae variables, we find a clear indication of 'low' helium abundance in Oosterhoff II type clusters and higher values ($Y \sim 0.26$–0.30) in well-known Oosterhoff I type clusters (Caputo, Castellani & Tornambè 1978a).

The quoted investigations now suggest that both the helium content and the age do differ among galactic globular clusters so that for the large majority of these objects the range of values

$$Z = 10^{-3}\text{--}10^{-4},$$
$$Y = 0.26 \pm 0.04,$$
$$t = 12 \pm 3 \times 10^9 \text{ yr},$$

appears to be appropriate, where the range is due to real differences among the various objects.

If this is true, comparison with the 'theoretical helium abundance' for the Sun poses the question of why old, metal-poor globular cluster stars appear to be more helium-rich than the metal-rich Population I Sun. My guess is that previous values will in fact be confirmed; so in the end must we conclude that the chemical evolution of globular clusters is not directly correlated with the chemical evolution of the galactic gas as a whole? Some evidence indicates a direct correlation between Y and Z, in the sense that lower metallicity clusters can also be less helium-rich: the old dilemma posed by Stobie (1971) looks like being overcome and possibly even being reversed.

Possible variations in the adopted evolutionary scheme will probably improve our knowledge of some details of the framework, such as the mechanism of mass loss, the effects of rotation and the origin of the observed spread of stars along the HB. However, I do not feel that the results discussed above will be drastically varied by their inclusion.

3.5 RR Lyrae pulsators: a progress report

The occurrence of RR Lyrae pulsators during the HB evolutionary phase provides us with an exciting opportunity to obtain information on the true values of the HB evolutionary parameters. In this way we can obtain an independent check on the values obtained from the long and intricate procedures of stellar evolution codes by using independently developed pulsation theory.

In the first instance, pulsation theories tell us that the period of pulsation depends on the mass, luminosity, effective temperature and helium abundance in the outer layers of the star. All four of these quantities are intimately related to the evolutionary calculations. Such a connection is of particular interest because variable stars are easily discovered in a cluster and their periods are well-defined quantities, unaffected by the usual observational uncertainties related to reddening, distance modulus, etc. Unfortunately, the majority of galactic globular clusters have few or no RR Lyrae variables, so that these stars can give information on only a select class of RR Lyrae-rich clusters.

By combining stellar evolution with pulsation theories we find, for example, that under very wide assumptions there are severe constraints on the possible masses for evolving RR Lyrae variables (in a cluster where c-type pulsators are found). In fact for $0.20 < Y < 0.30$ and

$$Z = 10^{-3}, \quad M_{RR} = 0.61\text{--}0.64 \, M_\odot,$$

or for $\quad Z = 10^{-4}, \quad M_{RR} = 0.70\text{--}0.76 \, M_\odot,$

as discussed by Caputo et al. (1978a).

The technique of combining evolutionary theory and pulsation theory was essentially started by the work of Iben & Huchra (1972). As a general rule we expect the properties of RR Lyrae variables to be extremely sensitive to helium abundance, so that their observed properties should be excellent indicators of helium variations. Information on RR Lyrae variables has been extensively used in recent studies of the evolutionary status of some particularly interesting clusters (Caputo & Castellani 1975a, c; Castellani 1975). In the following we will discuss some preliminary results based on computations of synthetic HBs following the precepts of Rood (1973).

The basic procedure is quite simple: for every adopted age and chemical composition for a cluster, theory gives a distribution of stars along the HB. This is true, of course, only if some added information (or assumption) is used concerning the occurrence of mass loss. In this way Rood was able to reproduce the HB morphology observed in galactic globular clusters but unfortunately, as stated in his paper, 'the effort to combine evolutionary theory with pulsation theory was largely unsuccessful'. Of course this was a rather disturbing result which could raise serious doubts about our understanding of evolution or pulsation or both.

In Frascati we have recently started a new investigation of the subject, encouraged by the improved understanding of both HB evolution and RR Lyrae pulsation. As far as the evolutionary input is concerned we used the

Fig. 3.8. The theoretical variation with time (in 10^8 yr) of the effective temperature of evolving HB models of various masses and assuming $Y = 0.30$, $M_c = 0.475$ $M_\odot$. Full lines (and dots) refer to the computations by Sweigart & Gross (1976). Open squares and dashed-dotted lines refer to computations by Caputo et al. (1978b). The agreement between the two different computations is excellent at least as far as the major phase of HB lifetime is concerned.

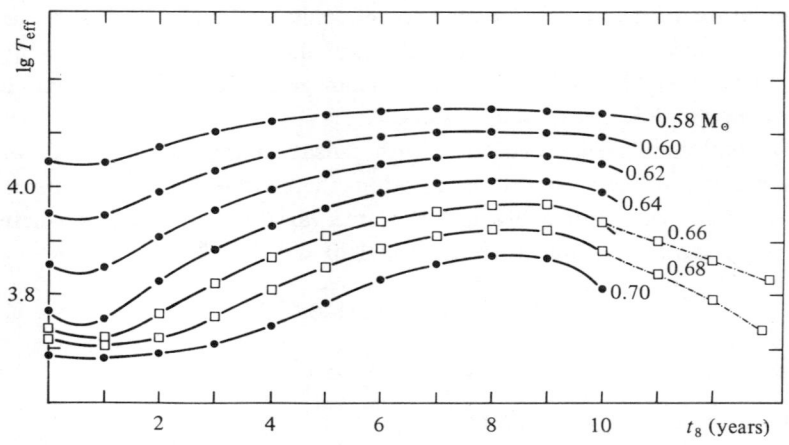

Stellar evolution and globular clusters

models of Caputo *et al.* (1978b) to follow the HB evolution up to the AGB for the case $Z = 10^{-3}$. Comparisons with previous results by Sweigart & Gross (1976) show general agreement (fig. 3.8) at least as far as the major phase of helium burning is concerned. Thereafter, for every choice of sets of cluster parameters we were able to derive the expected distribution along the HB, again provided that information on the amount of mass loss was added.

Since the cluster age is of minor importance in determining the RR Lyrae properties (Caloi, Castellani & Tornambè 1978) we fixed $t = 10^{10}$ yr and studied the expected properties of these variables by using the mean mass and spread in mass along the HB as free parameters in order to look for pulsation properties (if any) which are relatively independent of the assumptions.

Before discussing the results it is worth recalling the general relations in the theoretical scheme adopted here. Fig. 3.9 presents a plot of the various regions of pulsational instability expected in the HR diagram. With reference to this figure we find:

(i) A strip of instability. All stars within the strip are expected to pulsate

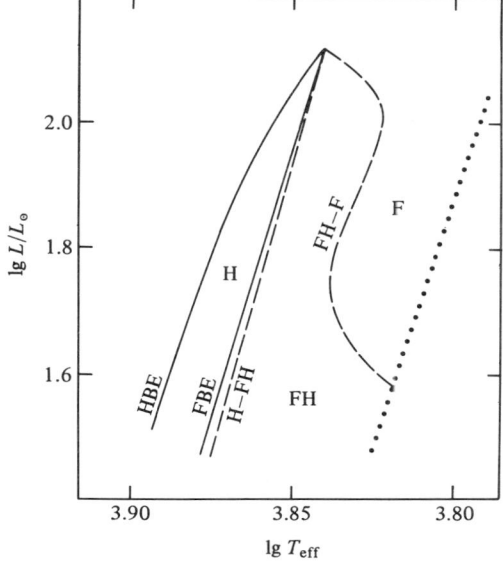

Fig. 3.9. The different regions for pulsational instability as expected on the basis of Stellingwerf's (1975) work. HBE and FBE represent the blue edge for instability in the first harmonic and in the fundamental mode, respectively. H–FH and FH–F represent the two possible transition lines between fundamental and first harmonic pulsators (see text). Dotted line represents the red boundary of the instability strip.

either in the fundamental or in the first harmonic mode. (There are some problems with the width of the strip, especially with the location of the red boundary.)

(ii) Regions F and H where only the fundamental or first harmonic pulsators, respectively, are allowed.

(iii) A region FH where a star can retain its previous pulsation mode, according to the suggestion of van Albada & Baker (1973) as confirmed by Stellingwerf (1975).

Finally we adopted Stellingwerf's suggestion on the conservation of the relative topology of the three quoted regions in varying the chemical composition and masses of the pulsators. As a result, in the case of

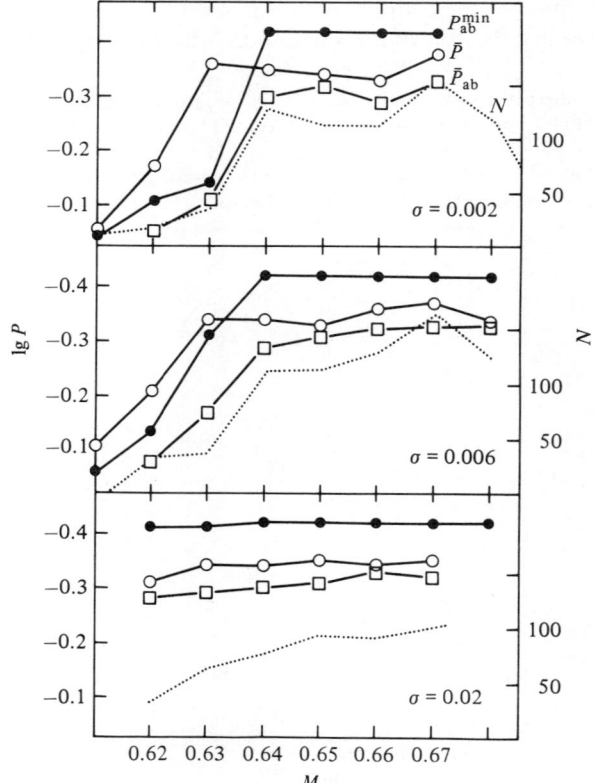

Fig. 3.10. The behaviour of selected pulsational properties as a function of the mean mass of the HB stars and for different assumptions on the mass dispersion (see text), in the case $Y = 0.25$. Dotted line represents the expected number N of RR Lyrae pulsators assuming 400 stars are observed in the HB phase.

$Z = 10^{-3}$ and $(Z_{CNO}/Z) = (Z_{CNO}/Z)_\odot$, for every assumed original helium content and for every choice of the mean mass and spread of masses along the HB, we can construct synthetic HBs and study the behaviour of the pulsators therein. In particular we will discuss below results concerning:

(i) N, the total number of variable stars,

(ii) $\bar{P}$, the mean period of all of the pulsators after first harmonic pulsators are reduced to the fundamental periods.

(iii) $\bar{P}_{ab}$, the mean period of fundamental pulsators, and

(iv) P_{ab}^{min}, the minimum period of fundamental pulsators.

Parameters (i) and (iii) depend (slightly) on the width of the instability strip while (iii) and (iv) depend on the location of the first harmonic boundary.

Synthetic HBs have been computed assuming 400 stars in this phase. The width of the instability strip was $\Delta \lg T_{eff} = 0.06$, and M_{RR} and σ, the dispersion of masses about the mean mass, were left as free parameters. Fig. 3.10 shows the behaviour of the modelled parameters as a function of the mean mass if $Y = 0.25$ and under different assumptions on the mass dispersion. It is now possible to derive many interesting things. We find general support for a recent suggestion about the connection between pre-HB evolution and the Oosterhoff dichotomy (Caputo et al. 1978a). Here I will only draw attention to the fact that for an RR Lyrae-rich HB P_{ab}^{min}, $\bar{P}$ and $\bar{P}_{ab}$ reach asymptotic values which are fairly well defined and do not depend on either the mean evolving mass or the mass spread. Further computations show convincingly that these asymptotic values are also insensitive to the form of the mass distribution and to the width of the strip. By way of contrast it is worth emphasizing that such values depend sensitively on the assumed helium abundance, as shown in table 3.3 where the results for $Y = 0.25$ and $Y = 0.30$ are reported. For RR Lyrae-rich globular clusters we find, for every value of Z, predicted values of the three quoted parameters as a function of the initial chemical

Table 3.3. *Theoretically expected values for selected pulsational properties in RR Lyrae-rich globular clusters when $Z = 10^{-3}$ and for different assumptions on the original helium content*

Parameter	$Y = 0.25$	$Y = 0.30$
$\lg P_{ab}^{min}$	-0.42	-0.35
$\lg \bar{P}_{ab}$	-0.30 to -0.34	-0.21 to -0.27
$\lg \bar{P}$	-0.33 to -0.38	-0.23 to -0.29

composition Y. *For moderately metal-poor clusters the quoted parameters can be assumed then to be helium abundance indicators.*

For comparison with observation we selected from the literature all globular clusters of intermediate metallicity containing more than thirty RR Lyrae variables. Data for these clusters are given in table 3.4. It is interesting to see the agreement of all three observed parameters among the six clusters. Comparison with the theoretical values in table 3.3 shows that each of these three parameters agrees with a value of $Y \approx 0.30$. A rather satisfying result.

Of course, beyond the quoted agreement, one has to look carefully at the general agreement with other observational parameters, such as period–frequency histograms, number of c-type pulsators, etc. Preliminary analysis seems to indicate that something is very clearly wrong with the adopted evolutionary tracks. Lifetimes in the last phases of HB evolution ought to be reduced in order to get complete agreement with observations. This may not be surprising considering how difficult the last phases of semi-convection are to compute. In addition, the stability criterion adopted (Schwarzschild or Ledoux) may also play a relevant role, as may the 'spike' in the computations of Sweigart & Demarque (1973) found at the end of central helium burning. Nevertheless we must note that the values reported in table 3.3 are rather insensitive to the treatment of these last phases. Furthermore, among the pulsational parameters discussed earlier, P_{ab}^{min} cannot be expected to depend on the treatment of these last evolutionary phases. In fact P_{ab}^{min} depends on the location of the fundamental–first harmonic transition line. This in turn can be regarded as well known, falling very near to the blue boundary for instability in the fundamental

Table 3.4. *Observed values of the selected pulsational properties in variable-rich globular clusters of intermediate metallicity. N represents the number of RR Lyraes found in the various clusters*

Cluster (NGC)	N	$\lg \bar{P}_{ab}$	$\lg \bar{P}$	$\lg P_{ab}^{min}$
5272 = M3	179	−0.26	−0.27	−0.34
5904 = M5	90	−0.26	−0.29	−0.35
6121 = M4	39	−0.27	−0.30	−0.35
6402 = M14	34	−0.25	−0.26	−0.33
3201	83	−0.25	−0.27	−0.33
6934	30	−0.28	−0.26	−0.34

mode. In all, then, the conclusions concerning the helium abundance look unavoidable.

3.6 Concluding remarks

While possible variations in the adopted evolutionary scheme for treating globular cluster stars will be introduced and will refine our understanding of the various mechanisms involved, the general framework appears sound. The overall agreement between evolutionary and pulsation theories is the most impressive piece of confirming evidence for such an opinion.

References

Caloi, V., Castellani, V. & Tornambè, A. (1978). *Astron. Astrophys. Suppl.* **33**, 169.
Caputo, F. (1977). *Astrophys. Space Sci.* **49**, 113.
Caputo, F. & Castellani, V. (1975a). *Astrophys. Space Sci.* **38**, 39.
Caputo, F. & Castellani, V. (1975b). *Mem. Soc. Astron. It.* **46**, 455.
Caputo, F. & Castellani, V. (1975c). *Mem. Soc. Astron. It.* **46**, 303.
Caputo, F., Castellani, V. & D'Antona, F. A. (1974). *Astrophys. Space Sci.* **28**, 303.
Caputo, F., Castellani, V. & Tornambè, A. (1978a). *Astron. Astrophys.* **67**, 107.
Caputo, F., Castellani, V. & Wood, P. R. (1978b). *Mon. Not. Roy. astron. Soc.* **184**, 377.
Castellani, V. (1975). *Mon. Not. Roy. astron. Soc.* **172**, 59P.
Castellani, V. (1976). *Astron. Astrophys.* **48**, 461.
Castellani, V. (1977). In *Chemical and Dynamical Evolution of our Galaxy*, IAU Colloquium no. 45, ed. E. Basinska-Grezesik & M. Mayor, p. 133. Geneva Observatory Publ.
Castellani, V. (1978). *Chemical Inhomogeneities in the Galaxy*, Proceedings of the Frascati workshop (in print).
Castellani, V., Giannone, P. & Renzini, A. (1971). *Astrophys. Space Sci.* **10**, 355.
Castellani, V. & Tornambè, A. (1977). *Astron. Astrophys.* **61**, 427.
Ciatti, F., Rosino, L. & Sussi, M. G. (1965). *The Position of Variable Stars in the Hertzsprung-Russell diagram*, 3rd Colloquium on Variable Stars, Bamberg. Remeis-Sternw. no. 40, 3045.
Faulkner, J. (1966). *Astrophys. J.* **144**, 978.
Hoyle, F. & Schwarzschild, M. (1955). *Astrophys. J. Suppl.* **2**, 1 (no. 13).
Iben, I., Jr (1968). *Nature*, **220**, 143.
Iben, I., Jr (1974). *Ann. Rev. Astron. Astrophys.* **12**, 215.
Iben, I., Jr & Huchra, J. (1971). *Astron. Astrophys.* **14**, 293.
Lee, S. W. (1977a). *Astron. Astrophys. Suppl.* **27**, 367.
Lee, S. W. (1977b). *Astron. Astrophys. Suppl.* **28**, 409.
Moretti, M. (1978). *Lab. for Space Astrophys. Report*, 17/78.
Nishida, M. (1960). *Prog. theor. Phys.* **23**, 896.
Renzini, A. (1977). In *Advanced Stages in Stellar Evolution*, 7th Course of the Swiss Society of Astronomy and Astrophysics, Saas-Fee, ed. P. Bouvier & A. Maeder. Geneva Observatory Publ.
Rood, R. T. (1972). *Astrophys. J.* **177**, 681.
Rood, R. T. (1973). *Astrophys. J.* **184**, 815.
Simoda, M. & Iben, I., Jr (1970). *Astrophys. J. Suppl.* **22**, 81.
Stellingwerf, R. F. (1975). *Astrophys. J.* **195**, 441.
Stobie, R. S. (1971). *Astrophys. J.* **168**, 381.

Sweigart, A. V. & Demarque, P. (1973). In *Variable Stars in Globular Clusters and in Related Stellar Systems*, IAU Colloquium no. 21, ed. J. D. Fernie, p. 221. Dordrecht: D. Reidel.
Sweigart, A. V. & Gross, P. G. (1976). *Astrophys. J. Suppl.* **32**, 367.
Sweigart, A. V. & Gross, P. G. (1978). *Astrophys. J. Suppl.* **36**, 405.
van Albada, T. S. & Baker, N. (1973). *Astrophys. J.* **185**, 477.

4
Abundance anomalies in globular clusters

ROBERT P. KRAFT[†]

4.1 Introduction

In this short chapter I would like to make some general comments on the topic of abundance anomalies – a nebulous topic at best, since what may strike one person as anomalous may seem perfectly natural to another. It would require a person of the stature of Bengt Strömgren to present a lucid and comprehensive review of everyone's work, so I will set myself the more modest aim of giving a brief review of some interesting work done by others and, towards the end, discuss the work we have been doing recently at the Lick Observatory. I will deliberately not try to be encyclopaedic; rather, I will simply point out in the most general terms some findings of special interest or importance. For a more thorough review with more detailed discussions, see Kraft (1979).

At least on the spectroscopic side of this subject there has been a real explosion of activity just lately. There are, I think, three basic questions to be considered. I will discuss each of them in turn in the following sections.

4.2 Composition differences between stars

The first question one might ask is: 'In a given cluster, are the stars chemically identical?' The answer to that question is no (Zinn 1973; Dickens & Bell 1976; Carbon *et al.* 1980; Bell, Dickens & Gustafsson 1978; Hesser, Hartwick & McClure 1977; Hesser 1978; Frogel, Persson & Cohen 1979; Freeman & Rodgers 1975; Mallia 1977; Norris, p. 113, this volume; and many other authors cited in Kraft 1979) and it is thus natural to enquire why not. There are two basic ideas: (i) stellar evolution and mixing

---

[†] This article has been thoroughly revised and greatly improved by the Editors, who composed it entirely from a recorded transcript of my talk. I am deeply indebted to them for their kindness in undertaking this difficult task. R.P.K.

of processed material to the stellar surfaces may account for the observed differences between stars: and/or – for the suggestions are not mutually exclusive – (ii) there may be primordial abundance variations of some kind, perhaps predating the cluster formation, or perhaps induced by self-pollution through early generations of stars, or finally perhaps via external pollution of some kind (see Iben, p. 127, this volume).

Let us consider the evidence for the second of these first. There are several points. Freeman & Rodgers (1975) and Butler, Bell, Dickens & Epps (1978) have shown that analysis of the wide variations observed in Preston's (1959) Δs metallicity-sensitive parameter for RR Lyrae stars in ω Cen implies a range of about 1 dex for the heavy-metal abundance in that cluster. Mallia (1977) showed within ω Cen that RGO 65 has a significantly lower [Fe/H] than does RGO 40. Rodgers (1978) has used the Zinn–Searle technique of studying ultraviolet blanketing (Searle & Zinn 1978) to deduce variations in [Fe/H] which accord with the results of Butler et al. (1978). Hesser et al. (1977), using DDO photometry, inferred [Fe/H] Variations within ω Cen. Norris (p. 133, this volume) has found that the CN variations reported by Hesser et al. (1977) correlate with variations in Ca II abundances, which must be primordial: no reasonable evolution/mixing process can lead to enhancements in atmospheric Ca for these low-mass cluster stars. Judith Cohen (1978) has detected significant scatter in Na and Ca abundances in giants in M3 and M13. And finally, Freeman and his associates (see Freeman, p. 105, this volume) have found what seem to be real radial abundance gradients in ω Cen. It seems to me then that at least in ω Cen there is no doubt that there are primordial abundance variations involving not just the elements in the CNO group but also the primary elements such as iron.

On the other hand, there is also some evidence in favour of the stellar evolution and mixing explanation. CH stars in ω Cen have been studied by Harding (1962), Dickens (1972) and Dickens & Bell (1976). These stars have abundance ratios of: [C/Fe] $\approx$ +0.5, [N/Fe] $\approx$ +1.3, [^{12}C/^{13}C] $\approx$ 10, [Fe/H] = [O/H] = −1.3. The ^{12}C/^{13}C ratio is close to its equilibrium value, indicating that material has been put through CN processing and then dredged up. As expected, nitrogen abundances are significantly up; and the iron abundance is low because, after all, this star is a member of ω Cen. Of equally great interest is the C/Fe ratio, up from solar, indicating perhaps that some triple-α processed material has been brought up from the core.

The cluster 47 Tuc has produced some interesting findings. Freeman (p. 105, this volume) and various other workers have discussed pro and con the evidence for abundance gradients within 47 Tuc; that question is still open. Jim Hesser's (1978) commendable study, some results of which are reproduced here as fig. 4.1, showed that at virtually every point in the colour–magnitude diagram – even very low on the subgiant branch (SGB) – there are quite significant variations in the strengths of the CN features. These variations are found even in the regions of very little stellar evolution so are unlikely to be caused by mixing. Norris (1978) found within 47 Tuc that the distribution of CN strengths is bimodal. The easy interpretation of this in terms of two generations of stars is perhaps *too* easy; we would like to see more observations.

Let us turn back to ω Cen to consider an interesting property: its

Fig. 4.1. A sample of spectra for stars in 47 Tuc. The colour–magnitude array represents the observed luminosity function. The wavelength scale is approximate, and the resolution is ~ 15–20 Å. As can be seen, even stars low on the SGB evidence vastly different CN band strengths. (Reproduced from Hesser (1978), wherein details of the observations are given.)

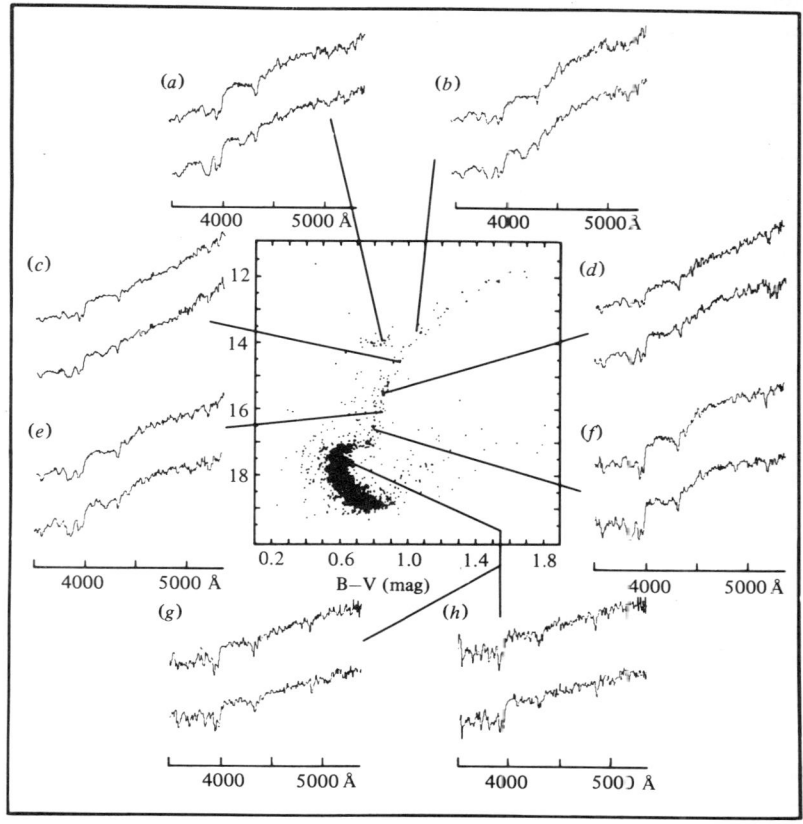

well-known wide giant branch (Cannon & Stobie 1973). Renzini (1977) has summarized the astrophysical explanation, shown graphically in fig. 4.2. If the abundances of the iron peak group are varied, little effect is seen at the cluster turnoff but the giant branch location changes, moving to the red as the abundances go up. On the other hand, the CNO abundances control both the opacity and the energy generation of the hotter stars, so changes in Z_{CNO} affect the turnoff stars and the horizontal branch (HB). The wide giant branch in ω Cen thus bespeaks the range of iron-peak or primary element abundances in that cluster.

This kind of observation can be considered in reverse, too: in other clusters, are the giant branches wider than the observational errors alone can account for? In general, the answer is no (see e.g. Sandage & Katem 1977) for those clusters with sufficiently accurate photometry to permit the test, so Z_{Fe} cannot vary widely (though of course Z_{CNO} may). There are problems, though: BV photometry is not the best approach, the B−V colour being rather insensitive in this regard. Moreover, many of the indices used – not just B−V – are not sensitive to changes in Z_{Fe} when this value is already quite low (see Kraft, Trefzger & Suntzeff 1978). For example, M92 and M15 can safely be said to contain very few stars with [Fe/H] ⩾ −2, but it is more difficult to decide whether these clusters contain stars that are significantly more metal-poor.

Fig. 4.2. A qualitative representation of the effect of composition changes upon the position of the main sequence and giant branch in a globular cluster lg L versus lg T_{eff} diagram. Note that changes in Z_{CNO} predominantly affect the turnoff (and also the HB, not shown) while changes in Z_{Fe} affect the giant branch most strongly. (From Renzini 1977.)

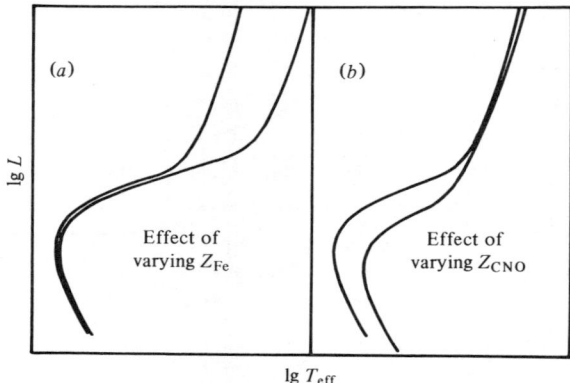

4.3 Elemental abundance ratios

Let me now turn to the second basic question. Consider mean elemental abundances (the unevolved ones) averaged over the stars in any one cluster. Then we can ask, 'For various elements A, is the mean value of [A/H] independent of A?' Think, for example, of A taking on the identity of Fe, representative of the iron-peak group; of O, representative of the CNO group; of Ca, representative of the α-process group; and of Ba, representative of the s-process elements. We are asking whether a cluster is just as deficient in each element separately as it is in the iron-peak elements.

A closely related question can also be put: 'For two clusters with identical [Fe/H], do you invariably find that, for example, [O/H] is the same for each cluster?' This is related to the so-called second-parameter problem, considered further in the next section.

Anticipating the observations I will describe, let me briefly answer these questions by saying that the balance of evidence for the first question suggests that the answer is probably no,† though little is known for certain; while for the second question the evidence is that at least the CNO group is decoupled from the iron group.

What then is the available evidence?

(i) Consider calcium as representative of the α-process elements. Judith Cohen has shown (Cohen 1978) in her high-dispersion studies that [Ca/Fe] = +0.4 for five stars in M13, while [Ca/Fe] > 0 for three stars in M3, albeit with a lot of scatter. Her observation that stars in M3 at given [Fe/H] have wide-ranging values of [Ca/H] suggests that the material out of which M3 was formed had non-homogeneous mixing in the early stages. Moreover, she finds that in M92 [Ca/Fe] $\sim$ 0.4–0.5 as well. Additional evidence concerning calcium abundances comes from the study carried out by Manduca & Bell (1978), who compared the work of Freeman & Rodgers (1975) on [Fe/H] values in RR Lyrae stars in ω Cen with the work of Butler et al. (1978) on [Ca/H] values in these stars. Manduca & Bell's suggestion is that $\langle$[Ca/Fe]$\rangle \gtrsim 0.4$. That is, in the old most-metal-poor stars calcium is somewhat overabundant. The only observational difficulty I can see here is that of the treatment of the interstellar Ca lines, which may make the findings still rather uncertain.

(ii) Let us turn now to the CNO group. Judith Cohen (1978) has found that in M3 and M13 [O/Fe] $\leq$ 0.2, while in M92 the non-appearance of

† Very recent work by Wallerstein & Pilachowski suggest that in the case of oxygen in M3 and M13, the answer is definitely no!

the [O I] $\lambda 6364$ line allows one to set a rough upper limit of [O/Fe] ≤ 0.5. So it does look as if the oxygen abundances are essentially normal in these clusters, while calcium, as we have seen, really does seem to be overabundant. Additional data are available from Hawley & Miller's (1978) study of the planetary nebula in M15. For the cluster as a whole, [Fe/H] = -2.0, but, in the planetary nebula, Hawley & Miller find a very high nitrogen abundance ([N/Fe] = 1.5) combined with a neon abundance that may be compatible with normal ([Ne/Fe] = 0.5). The oxygen abundance looks a bit high ([O/Fe] = 0.8), in disagreement with the findings of Cohen (1978) in other clusters. Unfortunately [C/Fe] is not measured, so the interpretation in terms of primordial variations or processing in stellar interiors remains difficult. I would very much like to see this ratio measured, but that may have to wait for the Space Telescope.

(iii) Let us now look at the s-process elements. Mallia (1977) has studied 23 giants in NGC 6397, 6656 (M22) and 6752, clusters in which [Fe/H] = -1.7 to -2.0. He finds that $\langle$[Ba/Fe]$\rangle \approx 0.0$. Similar results were found by Judith Cohen (1978) in M3 and M13 for both barium and yttrium, but the lanthanum series (La II, Ce II, Pr II, Nd II) seemed *over*abundant; however, she found [Ba/Fe] = -0.4 in M92. These results overall seem rather strange since at least for the very old stars in the field there is some evidence that the s-process elements are low relative to iron – the so-called 'classical aging effect' (Peterson 1976). Yet the stars studied by Mallia (1977) and Cohen (1978) show no particular effect in that direction. I will comment further on this in section 4.6.

As preliminary as a lot of these data may be, they seem to indicate that clusters are *not* equally deficient in all elements separately, for what reasons we can so far only speculate.

4.4 The second-parameter problem

Let us restate the question: 'For two clusters with identical values of [Fe/H], do you invariably find that, for example, [O/H] is the same for each cluster?' An instructive way of approaching this is first to examine some globular cluster colour–magnitude diagrams. Fig. 4.3 shows the colour–magnitude diagrams of M3 and M13. They differ in that in M3 the whole instability strip is full of RR Lyrae stars whereas there are almost none at all in M13, most of the HB stars being left of the strip. How can these clusters, which have very nearly the same iron abundance, have such different HB morphology?

Classical theory (Faulkner 1966) tells us that, everything else being

Abundance anomalies

Fig. 4.3. The colour–magnitude diagrams for (a) M3 and (b) M13 (from Alcaino 1973). The two clusters are of closely similar [Fe/H] but differ greatly in HB morphology: the instability strip in M3 is full of RR Lyrae stars (conventionally not plotted, but here represented by a light stippling) while there are almost none at all in M13 – the gap is nearly vacant.

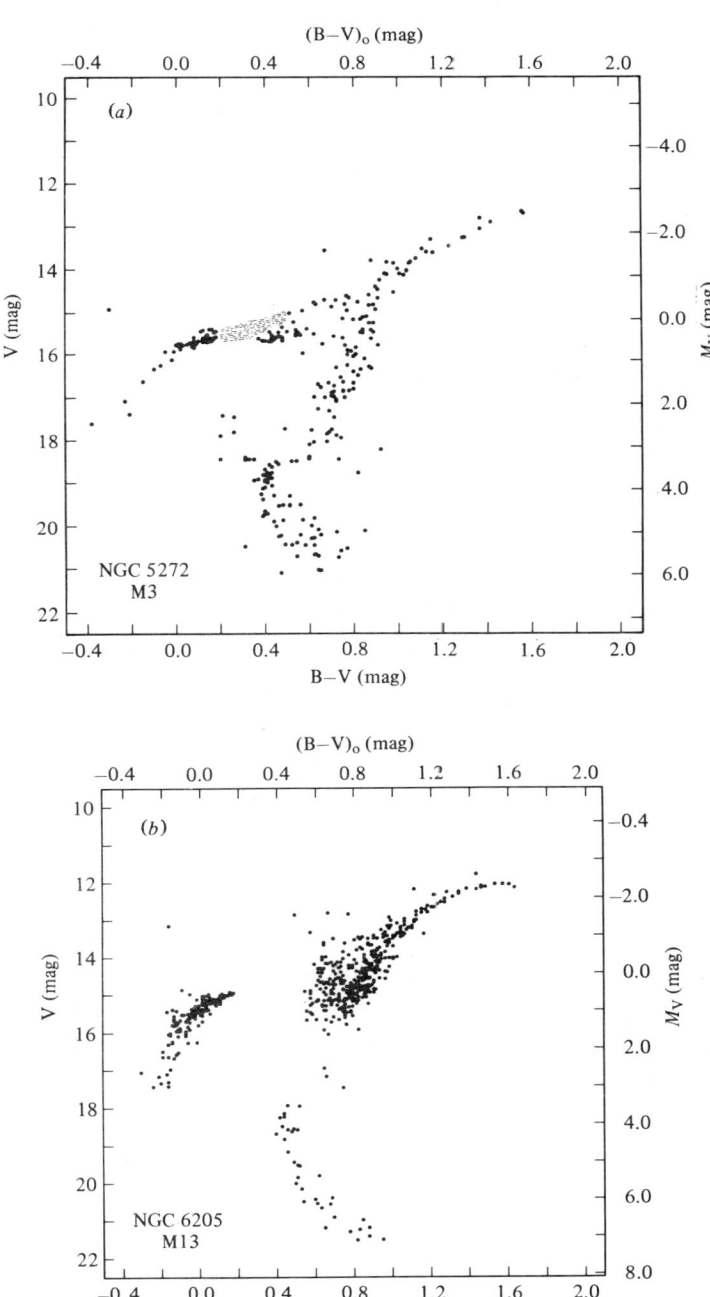

equal, the HB is governed by two parameters: (i) the ratio of total stellar mass to core mass – if the total mass of the star is large at the time of the helium flash, then the star will wind up at the red end of the HB; and

(ii) the metal abundance – if this is high, then stars again wind up at the red end of the HB.

So if the metallicities are roughly the same the different HBs may be due to a difference in age (Renzini 1977). Examination of Castellani & Tornambè's (1977) work reveals that the requisite age differences would manifest themselves by changing the turnoff location by only about 0.2 mag, a difference hard to detect at $V \approx 19$–20. Perhaps it would be more fruitful to start by ascribing the various HB morphologies to a different cause – the decoupling of Z_{CNO} from Z_{Fe} – and investigate this first.

Indeed there is evidence supporting just this point of view. Recently Cohen, Frogel & Persson (1978) pointed out that while M3 is slightly more metal-poor than M13 in [Fe/H] (-1.8 versus -1.6), when studied via a near-infrared CO band at 2.4 μm they are oppositely ranked. That is, M3, which possesses the redder HB, is relatively more abundant in CO than is M13. Casting further back, I believe that the first real mention of this problem is in the work of Hartwick & McClure (1972) in the cluster NGC 7006, which has an unusually red HB for its [Fe/H]. On the basis of DDO photometry, Hartwick & McClure suggested that NGC 7006 had unusually high nitrogen abundance. Still more evidence is found when one examines the Palomar clusters, as McClure (1978) has done. He finds suggestions of CN overabundances, and we know already that these distant Palomar clusters have rather red HBs.

Naturally though it is too naive simply to accept that CNO is decoupled from the iron group: there are troubles. First, Schommer (1978) has pointed out that in these distant clusters the ratio of red to blue stars is *too* high, and the colour at the junction of the HB and the red giant branch, denoted $(B-V)_{o,g}$ is *too* red by 0.1 mag to accord with simple CNO variations. Moreover, Harris (1978), in a study of Palomar 12, finds a red giant branch similar to that in 47 Tuc but a turnoff like that in M3: but changing the CNO should not move the giant branch at all! Additional changes – such as assuming enhanced CNO *plus* reduced helium abundance in Palomar 12 – may solve this problem but at the cost of requiring extreme ranges of ages of globular clusters in the Galaxy and so forth. It may be true that the attractive notion of explaining away the second-parameter problem simply by invoking decoupled Z_{CNO} and Z_{Fe} is too optimistic.

4.5 Recent investigations at the Lick Observatory

Before I consider the third fundamental question let me digress briefly to describe some of the work we have been doing at the Lick Observatory recently, in particular in relation to the two questions we have asked so far. The Lick group comprises Duane Carbon, now at Kitt Peak; Ed Langer, of Colorado College; Dennis Butler, now at Yale; Charles Trefzger of the Basel Institute; Lick students Ed Kemper and Nick Suntzeff; and myself. Our interests were kindled by Bob Zinn's discovery – the 'Zinn effect' – of some years ago (Zinn 1973). If you look at SGB stars, say in M92, and compare their spectra with the spectra of asymptotic giant branch (AGB) stars, then any differences must be due to differences in abundances since they have approximately the same surface temperature and very small differences in surface gravity; of course, the stars on the AGB are in a later stage of stellar evolution than those on the SGB. Now the Zinn effect is this: the G-band of CH is systematically weaker for the AGB stars than for the SGB stars. Similar effects have been found in other clusters as well.

At about the same time as Bob Zinn discovered this interesting effect, Chris Sneden published his thesis results (Sneden 1974). He had studied 7 field giant stars for which the mean value of [Fe/H] was closely the same as the classical value for M92, so the stars can be directly compared. Sneden derived carbon and nitrogen abundances from the lines of the CH bands at 4300 Å, the CN bands at 3883 Å, and the NH bands at 3360 Å. In two of the stars, he found [C/H] = [N/H] = [Fe/H], while for the other five the carbon abundance was down and the nitrogen abundance was up, just as if the products of CN processing in the region around or ahead of the hydrogen-burning shell had been convected into the atmosphere. But, interestingly, if the Sneden stars are plotted in the HR diagram for M92 they separate in exactly the way one would expect under this picture, the two stars with no evidence of CN processing lying (roughly) on the SGB and the other five taking up positions near the observed AGB. So at once you might think that this is the explanation of the Zinn effect: at some stage between the SGB and the AGB, carbon is converted to nitrogen and mixed into the atmosphere – an evolutionary effect. There is in fact yet more evidence pointing this way, namely the results of Lambert & Sneden (1977). In one of Sneden's stars, HD 122563, the well-known extremely metal-poor ([Fe/H] = -2.7) field giant, they showed that the ratio of carbon isotopes is $^{12}C/^{13}C \sim 5$, which seems a clear indication of CN processing having occurred.

Now we decided at Lick that a natural way to test this was to study the M92 stars a bit further and see whether there was an anticorrelation between the carbon and nitrogen abundances as the simple picture I have outlined would require. Our approach was to use the Wampler scanner to obtain low-dispersion spectra (~ 125 Å mm^{-1}, $\Delta\lambda \sim 10$ Å) of giant stars in M92 and in particular to deduce nitrogen and carbon abundances from the NH and CH bands respectively. We hoped to use the Sneden stars as a sieve: compare the strengths of features on some arbitrary scale and rely on Sneden's calibrations to set the zero-points. Unfortunately this failed for two reasons. First, the feature strengths are very sensitive to temperature and Sneden's set of seven stars gave us too coarse a grid. Secondly, most of what we did in the nitrogen domain was an extrapolation: the nitrogen features in the M92 stars are on the average much stronger than in the Sneden field giants at the same [Fe/H]. This is of course an important finding and one I will return to later.

A problem we encountered early was the question of cluster membership. We explored the use of photometry in discriminating between field stars and cluster members, but for various astrophysical reasons there is a tendency to reject AGB stars along with the field if one uses the usual photometric indices. A useful piece of information is the observed proper motion of any suspected member, and here Kyle Cudworth's (1976) work is extremely valuable. In very oversimplified terms, any star with measurable proper motion cannot be a cluster member. This of course is not to say that any star which stands still *is* a member, but the contamination is expected to be quite low – perhaps one star in our study of M92 is in fact a halo star. The long and the short of it is that we acquired 169 scans of 71 stars, 45 of which turned out to be cluster members.

The question now was how to analyse these data, given the insufficiency of the grid provided by Sneden's work on field stars. Our approach has been actually to model the spectra (and here I must say that this is entirely Duane Carbon's work) using spectrum synthesis techniques. Without a detailed description of the techniques, which will be described elsewhere (Carbon *et al.* 1980), let me briefly summarize the most important findings as follows:

(i) On the SGB of M92 there are wild variations in the strengths of the nitrogen features. One can find variations of 1 dex in the nitrogen abundance even for stars in this lowest stage of evolution. The *mean* nitrogen abundance is extraordinarily high. (Incidentally, I should point out that when we give Sneden's stars our treatment and compare our results with his we deduce closely similar temperatures and values of

[Fe/H], slightly higher carbon abundances, and significantly lower nitrogen abundances – by a factor of 2 or so. Thus the really large nitrogen abundances we see in the M92 stars certainly seem to be real, not an artifact of the scheme.)

(ii) A comparison of SGB and AGB stars shows the Zinn effect again, but there are exceptions to it, as Bob Zinn has himself noted (Norris & Zinn 1977). There are some SGB stars that have extremely weak G-bands indeed, so whatever physical effect it is that causes this weakening is already at work low on the SGB. Just as for nitrogen, one can find variations in the carbon abundances in stars that sit side by side in the HR diagram, even low on the SGB; but these variations are smaller than is the case for nitrogen.

(iii) On average, the carbon abundance goes down towards the red giant tip (RGT) and remains down along the AGB. However, this trend is not accompanied by a corresponding increase in N abundances with advancing evolutionary stage. There is no C–N anticorrelation, star by star; and there is no mean C–N anticorrelation as a function of evolutionary stage.

(iv) The Zinn–Norris effect (Norris & Zinn 1977), whereby the G-band strength rises near the RGT, is an artifact of the temperature, not of the abundances. Here we agree with the analysis of two giant-branch-tip stars by Bell *et al.* (1978). A star on the AGB with a weak G-band may show a considerable G-band at the RGT for the same carbon abundance.

These then are the bare bones of our findings, detailed elsewhere (Carbon *et al.* 1980). The simple picture put forward earlier of straightforward CN processing and mixing is obviously not going to work here. What, if anything, will? A few possibilities come to mind.

(i) Perhaps a slightly different evolution-and-mixing process could explain these confusing observations. If at the helium-core flash some mechanism could mix hydrogen into the core and bring carbon out (which is then converted to nitrogen and mixed into the atmosphere), then the star might jump back down the SGB to climb the giant branch again, this time with enhanced nitrogen. Such processes were considered by Bob Rood some years ago (Rood 1970), but, as suggested by my colleague Ed Langer, there seem to be time scale problems. If you mix in too much hydrogen, then the star migrates rather far to the blue from the observed SGB; but if you mix in only a little, the time scales are too short – a serious failing, for we already find a considerable fraction of SGB stars with high nitrogen.

(ii) Sweigart & Mengel (1979) have suggested a mixing driven by internal differential rotation. In a large region ahead of the hydrogen-burning shell there can be conversion of carbon to nitrogen, but rather close to the shell there is the possibility of oxygen being converted to nitrogen. Mixing could thus lead to a significant enhancement in the atmospheric nitrogen abundance and would also explain the observed lack of a C–N anticorrelation. This is an attractive idea, the pros and cons of which I have discussed in more detail in Kraft (1979).

(iii) Perhaps there was simply more nitrogen to begin with than we suspect in the primordial material. I need hardly say that this is not a very attractive notion.

(iv) Slightly more plausibly, it may be that if we looked even lower on the SGB than we have done to date we would find the carbon abundance very high indeed. We may simply be 'tuning in' on the SGB at a level where 60–70 per cent of the carbon has already gone to nitrogen, so if 20–30 per cent more is processed in later stages we cannot tell to within the accuracy of the measurements. This explanation, which is the approach Ed Langer and I have taken, would require that the C and O abundances on average be up in the primordial material out of which the cluster formed.

Clearly any complete understanding of the run of abundances in M92 is some way off through we do seem to be making considerable progress.

Fig. 4.4. The [Ba/Fe] ratio plotted as a function of [Fe/H] for metal-poor halo stars (from Spite & Spite 1978). The original data are plotted as filled circles. The mean relation is shown as a thick dashed line. The thin lines represent the solar ratio, as indicated, and a reference line of slope 45°. Also shown are some results for globular cluster giants: Mallia's (1977) results averaged over four similar clusters (an open circle); Cohen's (1978) determinations for [Ba/Fe] in M3 (a cross) and in M13 (a plus sign); and Cohen's (unpublished data) value for [Ba/Fe] in M92 (an open square).

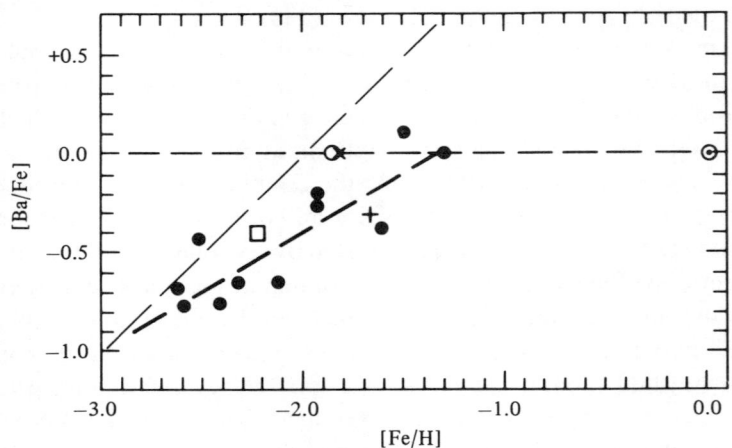

4.6 Field stars versus globular cluster members

In this last section I should like to pose the third fundamental question: 'Are the field halo giants at a given [Fe/H] spectroscopically identical to the cluster giants?' We do not know the answer, but I will briefly mention some of the relevant evidence.

Consider first the s-process elements. At one time I believed these to be differently distributed in cluster stars and field stars, but I am not now so sure. Fig. 4.4 presents some results from the work of Spite & Spite (1978) for barium-to-iron ratios in metal-poor halo stars; the 'classical aging effect' (Peterson, 1976) shows up in the correlation between [Ba/Fe] and [Fe/H] for [Fe/H] $\leqslant -1.25$. I have superimposed on the figure Mallia's (1977) results. He found [Ba/Fe] ≈ 0 in four clusters whose average [Fe/H] was ≈ -1.85. Judith Cohen's results for M3 and M13 (Cohen 1978) and for M92 (Cohen, unpublished data) are shown as well. The departures from the mean relation may be small enough for there to be no problem here.

As for the α-process elements, differences in (for example) calcium abundances should make themselves known because the cluster calibration and the field-star calibration for the RR Lyrae stars should separate. That is, the correlation between [Fe/H] and Δs, a spectral classification parameter sensitive to calcium abundance (Preston 1959; Butler 1975), should be different in field stars and cluster variables. There is, I think, no evidence for such a separation.

Finally, there is the question of the abundance of nitrogen. As I discussed in the previous section, in M92 the average nitrogen abundance is considerably enhanced over that seen in Sneden's (1974) sample of field halo giants of comparable [Fe/H]. This I think is a fairly well-established observational fact and seems to be the only strong evidence that the field star chemical enrichment history is different from that of cluster stars.

4.7 Conclusions

Perhaps the safest conclusion we can draw is that we need more observations!† For the clusters, it would be very important to get more proper motions and radial velocities to be sure of cluster membership, to increase the numbers of stars and clusters studied, and to attempt to reach the main sequence. In the field, let us find more halo giants of low metal-abundance and pursue the studies there as well.

I have not intended this to be a complete or closely argued review.

† I realize that this statement is ubiquitous and plattitudinous – even so, it is true.

Rather, I have tried to indicate what sorts of questions are being asked and the general directions the data collected so far seem to be leading us. It is clearly a fruitful and challenging area of research.

References

Alcaino, G. (1973). *Atlas of Galactic Globular Clusters with Colour–Magnitude Diagrams*. Universidad Católica de Chile, Santiago.
Bell, R. A., Dickens, R. J. & Gustafsson, B. (1978). Presented at symposium on Important Advances in 20th Century Astronomy, Copenhagen.
Butler, D. (1975). *Astrophys. J.* **200**, 68.
Butler, D., Bell, R. A., Dickens, R. J. & Epps, E. (1978). In *The HR Diagram*, IAU Symposium no. 80, ed. A. G. Davis Philip & D. S. Hayes, p. 183. Dordrecht: D. Reidel.
Cannon, R. D. & Stobie, R. S. (1973). *Mon. Not. Roy. astron. Soc.* **162**, 227.
Carbon, D. F., Langer, G. E., Butler, D. Kraft, R. P., Trefzger, Ch. F., Suntzeff, N., Kemper, E. & Nocar, J. (1980) *Astrophys. J.* (in press).
Castellani, V. & Tornambè, A. (1977). *Astron. Astrophys.* **61**, 427.
Cohen, J. G. (1978). *Astrophys. J.* **223**, 487.
Cohen, J. G., Frogel, J. A. & Persson, S. E. (1978). *Astrophys. J.* **222**, 165.
Cudworth, K. M. (1976). *Astron. J.* **81**, 975.
Dickens, R. J. (1972). *Mon. Not. Roy. astron. Soc.* **159**, 7P.
Dickens, R. J. & Bell, R. A. (1976). *Astrophys. J.* **207**, 506.
Faulkner, J. (1966). *Astrophys. J.* **144**, 978.
Freeman, K. C. & Rodgers, A. W. (1975). *Astrophys. J. Lett.* **201**, L71.
Frogel, J. A., Persson, S. E. & Cohen, J. G. (1979). *Astrophys. J.* **227**, 499.
Harding, G. A. (1962). *Observatory*, **82**, 205.
Harris, W. E. (1978). Results reported at NATO Advanced Study Institute on Globular Clusters, Cambridge.
Hartwick, F. D. A. & McClure, R. D. (1972). *Astrophys. J. Lett.* **176**, L57.
Hawley, S. A. & Miller, J. S. (1978). *Astrophys. J.* **220**, 609.
Hesser, J. E. (1978). *Astrophys. J. Lett.* **223**, L117.
Hesser, J. E., Hartwick, F. D. A. & McClure, R. D. (1977). *Astrophys. J. Suppl.* **33**, 471.
Kraft, R. P. (1979). *Ann. Rev. Astron. Astrophys.* **17**, 309.
Kraft, R. P., Trefzger, C. & Suntzeff, N. (1978). In *The Large-Scale Characteristics of the Galaxy*, IAU Symposium no. 84, ed. W. B. Burton, p. 463. Dordrecht: D. Reidel.
Lambert, D. L. & Sneden, C. (1977). *Astrophys. J.* **215**, 597.
McClure, R. D. (1978). Results reported at NATO Advanced Study Institute on Globular Clusters, Cambridge.
Mallia, E. A. (1977). *Astron. Astrophys.* **60**, 195.
Manduca, A. & Bell, R. A. (1978). *Astrophys. J.* **225**, 908.
Norris, J. (1978). In *The HR Diagram*, IAU Symposium no. 80, ed. A. G. Davis Philip and D. S. Hayes, p. 195. Dordrecht: D. Reidel.
Norris, J. & Zinn, R. (1977). *Astrophys. J.* **215**, 74.
Peterson, R. C. (1976). *Astrophys. J.* **206**, 800.
Preston, G. W. (1959). *Astrophys. J.* **130**, 507.
Renzini, A. (1977). In *Advanced Stages in Stellar Evolution*, 7th Course of the Swiss Society of Astronomy and Astrophysics, Saas-Fee, ed. P. Bouvier & A. Maeder. Geneva Observatory Publ.
Rodgers, A. W. (1978). Results reported at NATO Advanced Study Institute on Globular Clusters, Cambridge.

Rood, R. T. (1970). *Astrophys. J.* **162**, 939.
Sandage, A. R. & Katem, B. (1977). *Astrophys. J.* **215**, 62.
Schommer, R. (1978). Results reported at NATO Advanced Study Institute on Globular Clusters, Cambridge.
Searle, L. & Zinn, R. (1978). *Astrophys. J.* **225**, 357.
Sneden, C. (1974). *Astrophys. J.* **189**, 493.
Spite, M. & Spite, F. (1978). *Astron. Astrophys.* **67**, 23.
Sweigart, A. V. & Mengel, J. G. (1979). *Astrophys. J.* **229**, 624.
Zinn, R. (1973). *Astrophys. J.* **182**, 183.

5
Populations in globular clusters

K. C. FREEMAN

5.1 Introduction

Populations in the Galaxy are classes of objects (e.g. young disc, old disc, halo) with similar age, chemical abundance, kinematics and space distribution. These parameters, which change systematically from population to population, are correlated, and from their correlations comes most of what we now understand about the formation and evolution of the Galaxy.

Globular clusters are simpler: there is no disc component, and all the stars have similar ages. However, for a few clusters, we know now that the bright end of the luminosity function, the integrated colours, and apparently the heavy element abundance, do change with radius (i.e. distance from the centre of the cluster), just as in some elliptical galaxies. This relatively new concept of populations in globular clusters is the subject of this review.

5.2 History

The first real indication of radial changes in the stellar content of a globular cluster was Martin's (1937) demonstration that the radial distributions of red giants and RR Lyrae stars in ω Cen were clearly different: the red giants are more centrally concentrated than the RR Lyrae stars. Oort & van Herk (1959) interpreted this as the effect of dynamical relaxation (the masses of the RR Lyrae stars being less than the red giant masses). But this interpretation now seems unlikely, because the relaxation time in this cluster (3×10^9 yr at the cluster centre) is significantly longer than the time for a star to evolve up the red giant branch and through the RR Lyrae stage. How else can one explain this radial change in the ratios of RR Lyrae to red giant numbers? A radial gradient in chemical abundance seemed a likely alternative, because the morphology of the horizontal branch (HB) is sensitive to abundance. Although we will see later that this explanation is probably too simple, it was Martin's original discovery that led to much of the recent work on populations in globular clusters.

Next came Gascoigne & Burr's (1956) observation that the integrated $P-V$ colour of 47 Tuc changes with radius. Again this was attributed at the time to dynamical effects, but G. Da Costa's recent (unpublished) work shows that the observed colour changes are much too large to be explained by dynamical effects alone.

Then in the 1960s and early 1970s came evidence for the chemical inhomogeneity of some globular clusters. The wide giant branch of ω Cen was discovered (RGO 1966) (and ignored, but see Iben 1972), and the G-band and CN anomalies were found (Zinn 1973; see Kraft, p. 87, this volume) and attributed to evolutionary effects within the stars themselves. Next, Cannon & Stobie (1973) showed beyond doubt that the GB of ω Cen is intrinsically wide, which strongly suggested a spread in the metal abundance. Freeman & Rodgers (1975) subsequently showed that the [Ca/H] values for 25 RR Lyrae stars in ω Cen have an intrinsic spread of more than 1 dex: although the sample is small, it showed some evidence for a mean radial decrease of [Ca/H] in this cluster. Calcium variations are particularly interesting, because calcium cannot be produced in these low-mass stars, so here was the first fairly clear direct evidence that a globular cluster has chemical inhomogeneities dating back to the time of its formation.

More recently there has been much work on (i) chemical inhomogeneities in globular clusters, and (ii) radial changes of both the integrated colour and the bright end of the luminosity function in clusters. Are these structural and chemical aspects related? We would really like some direct evidence about the radial dependence of abundance in clusters, and the picture is just starting to appear.

This review will concentrate mainly on the structural aspects of populations in clusters. Other chapters discuss chemical inhomogeneities in more detail. But, just as in galactic populations, the chemical and structural properties of these cluster populations appear to be intertwined. Much of this review will be about ω Cen and 47 Tuc; they are massive and nearby, so the stellar statistics are good and they have been studied in some detail.

5.3 47 Tuc and ω Cen

5.3.1 Colour and luminosity function gradients

For 47 Tuc, Gascoigne & Burr's colour gradient in $P-V$ was confirmed by Chun's (1976) UBV measures. These show how the inner parts of the cluster ($r < 2$ arcmin) are significantly redder than the outer parts, by about 0.1 in $B-V$ and about 0.2 in $U-B$. What radial change in the luminosity function produces this radial colour change? Observationally, this question is fairly straightforward, because most of the cluster light comes from the brightest few magnitudes of the luminosity function, and 47 Tuc is close and rich. Chun compared the radial distribution of several kinds of stars (giants in the top 1.5 mag of the giant branch, fainter giants, HB stars) with the radial distribution of V light. It turns out that the colour gradient is produced by an *excess of the brightest giants* per unit luminosity in the inner parts of the cluster. This excess had already been noticed by Lloyd-Evans (1974).

For ω Cen, the colour data is not so good because the surface brightness of the cluster is relatively low. However, we know already from Martin (1937) that the bright giants and RR Lyrae stars are differently distributed. My counts for ω Cen show again the excess of bright giants per unit luminosity in the inner parts of the cluster; the fainter giants and the HB stars follow fairly closely the radial distribution of the integrated V light.

I should repeat that this central excess of bright giants is unlikely to result from mass segregation by two-body relaxation, because (i) G. Da Costa's (unpublished) work shows that the effects expected for a cluster in thermal equilibrium are much smaller than those observed; (ii) only the brightest giants are in excess in the inner parts of the cluster, and the difference in mass for the brighter and fainter giants is small; (iii) for ω Cen in particular, the relaxation time is much longer than the relevant stellar evolution times.

So, to summarize this section: the bright end of the luminosity function changes with radius in ω Cen and 47 Tuc, and this produces the colour gradient observed in 47 Tuc. These two clusters are the most thoroughly studied; we will see later that a few other clusters also show colour gradients.

5.3.2 Abundance gradients

Now I will review the direct evidence for radial abundance gradients in ω Cen and 47 Tuc: first ω Cen, which is the best studied.

(i) For the RR Lyrae stars, Freeman & Rodgers (1975) found that [Ca/H] varied by more than 1 dex from star to star. Butler, Dickens & Epps (1978) confirm this spread from a larger sample of RR Lyrae stars. If we divide their sample radially and compare histograms of [Ca/H] for the inner and outer half of the sample, we find that the inner RR Lyraes have a higher mean [Ca/H] value: the histograms are significantly different at the 95 per cent confidence level. In *both* halves of the sample, however, we see the whole range of [Ca/H] values, from -0.5 to -2.2 on Butler's scale. The histogram of [Ca/H] for the RR Lyraes in the outer part of ω Cen is qualitatively rather like the distribution of [Fe/H] for the globular clusters in the outer part of the Galaxy (Searle 1978): there is a peak near [Ca/H] $= -1.7$ (compared with [Fe/H] $= -1.4$ for the globular clusters), and a long tail in the distribution towards higher [Ca/H].

(ii) Now consider the giants in ω Cen: these define the famous wide giant branch in the colour–magnitude plane. Norris (p. 113, this volume) has reported that the Ca H and K lines and the CN features vary greatly in strength from giant to giant, for a sample in the restricted magnitude interval $12.2 < V < 12.5$. In particular, the Ca and CN features are well correlated, and the mean Ca and CN strengths increase *across* the giant branch, from blue to red. Now we know that the colour of the giant branch at the magnitude of the HB, $(B-V)_{o,g}$, correlates well with [Fe/H]. The calibration derived by Norris & Bessell (1975) is [Fe/H] $= 5.3$ $(B-V)_{o,g} +$ constant. So we can use the *colour* of a giant at some magnitude as an estimate of its abundance, and we can check whether the observed spread in $B-V$ for the giants in ω Cen is consistent with the observed spread in [Ca/H] for the RR Lyraes.

In the colour–magnitude plane, the giant branch of ω Cen has a well-defined blue envelope, well represented by the giant branch of M92. The [Fe/H] value for M92 is about -2.2, from Butler (1975). So we can use the displacement $\Delta(B-V)$, in $B-V$, of an ω Cen giant from this M92 giant branch, to estimate its [Fe/H], via [Fe/H] $= 5.3 \Delta(B-V) - 2.2$. For giants with reliable photoelectric photometry, the histogram of $\Delta(B-V)$ was transformed to a histogram of [Fe/H] by using this calibration. These giants are all in the outer parts of the cluster, so we should compare this histogram of [Fe/H] with that of [Ca/H] for the RR Lyraes in the outer regions. The two histograms are very similar, in spread and in shape. This means that the width of the giant branch in the colour–magnitude (CM) diagram *and* the distribution of stars across it is consistent with the directly observed abundance spread from the RR Lyrae stars.

We can now go back to the abundance gradient questions. RGO (1966) gives photographic photometry for many stars in ω Cen. We can use this photometry to see whether the mean giant colours (and so the mean abundance) change with radius. There are about 1000 giants with $13 < V < 14$: this interval avoids most of the asymptotic giant branch (AGB) stars. If we divide this sample into red (large $\Delta(B-V)$) and blue (small $\Delta(B-V)$) parts, the ratio of red to blue stars clearly decreases with radius. So this is more evidence that the mean abundance is also decreasing with radius.

(iii) For 47 Tuc, there are few RR Lyraes and the giant branch is narrow in the CM diagram. But again there is some direct evidence for a radial abundance gradient. Norris (1977) has reported C(41–42) data for about 60 giants: one sample lies in the inner parts of the cluster (1 arcmin $< r <$ 3 arcmin), the other in the outer parts ($r >$ 10 arcmin). The histograms of CN in these two regions are different at the 92 per cent confidence level: the inner region stars show stronger CN. Although these *radial* CN changes are established only for the brighter giants, we recall from Norris (1977) and Hesser (1978) that CN variations from star to star persist all the way down the subgiant branch. There is also evidence for inhomogeneity in the heavier elements (A. Rodgers, unpublished), but there is no information yet about its radial behaviour.

In summary, for ω Cen and 47 Tuc, the stars in the inner parts are more metal-rich in the mean. However, the stars in the outer parts of these clusters have a wide range in abundance, rather like the abundance spread among the globular clusters in the outer parts of the Galaxy and M31.

5.3.3 Abundance gradients and luminosity function gradients

In these two clusters we see radial changes in the bright end of the luminosity function *and* in the chemical abundance. What is the connection? We need to relate the central excess of bright giants per unit luminosity to the central increase in abundance. Here are some speculations.

First giant branch effects are probably irrelevant. (i) Although abundance changes do affect the evolution times up the first giant branch, this effect is weak and is most unlikely to produce the central excess of the brightest giants. (ii) An increase in abundance does not result in brighter (first) giants. Although the metal-richer giants are bolometrically brighter at helium flash, they are fainter in V, as observed.

We should consider the possibility that the excess giants near the centre are AGB stars. Then for stars at the upper part of the giant branch, the ratio of AGB to first giants would increase towards the centre. How could this come about? It may be associated with radial changes in the mean mass of the HB stars. For example, take a first giant with $M = 0.72\,M_\odot$,

$Z = 10^{-3}$, $Y = 0.3$ and age 16×10^9 yr. At the helium flash, its core mass is about $0.47\,M_\odot$. It then descends to the HB with a total mass of about $0.6\,M_\odot$, the rest having been lost in the 'mass loss'. After core helium burning ceases, the star evolves towards the first giant branch and then up it, until the envelope mass is reduced to about $0.02\,M_\odot$. The star then evolves rapidly to the blue. So if the envelope is originally (i.e. on the ZAHB) massive enough, then the star can evolve in its second ascent to the tip of the first giant branch and even brighter. For example, Gingold's (1976) models for $Z = 10^{-3}$, $Y = 0.3$, $M_c = 0.47\,M_\odot$ show that a star with total mass greater than $0.55\,M_\odot$ on the HB will evolve up the AGB to these luminosities. This all means that the ratio of AGB to red giant stars at the upper part of the giant branch is sensitive to changes in the mean envelope mass on the ZAHB.

Is it sensitive enough? Our star counts show the central excess of giants in about the top 1.5 mag of the giant branch. Over this top 1.5 mag, the typical evolution time on the first giant branch is about 7×10^6 yr. From Gingold (1976) an AGB star with enough envelope mass to evolve into this region crosses it in about 3×10^6 yr. So the total number of stars (red giants and AGB) in this part of the CM diagram can be significantly affected by the distribution of envelope masses on the ZAHB.

We are trying to understand why the central increase in abundance leads to a central excess of bright giants. If the argument so far is correct, then we now need to ask whether the abundance gradient can produce a corresponding radial gradient in the mean envelope mass on the ZAHB, in the sense that both the envelope mass and the abundance are higher at the cluster centre. Although the processes that determine the envelope mass are not yet properly understood, here are three observations that give some support to this argument.

(i) At the helium flash, the envelope mass is already higher for stars of higher abundance at a given age. For example, for $Z = 10^{-2}$, the total mass at helium flash for an age of 16×10^9 yr is $0.79\,M_\odot$, the core mass is $0.47\,M_\odot$, so the envelope mass is $0.32\,M_\odot$. The corresponding numbers for $Z = 10^{-4}$ are 0.71, 0.49 and $0.22\,M_\odot$ (*Yale Transactions* 1977, No. 33).

(ii) There is some evidence that the mass loss which occurs before a star reaches the ZAHB may depend on Z in the right sense. From Hα observations of bright cluster giants by E. Mallia & B. Pagel (preprint) and Carla Cacciari and myself, Hα emission is seen most often in the metal-weakest systems. For example, all stars so far observed in the metal-weak cluster M22 show emission; none do in 47 Tuc. In ω Cen, where there is a wide range of abundance, two giants on the blue (metal-weak) side of the giant branch show Hα emission, while one star on the red side does not. If Hα emission is really evidence for mass loss (see Cohen 1976), then the observations suggest that the most metal-weak giants are losing mass at the greatest rate. So from this, and considering (i), we could expect the envelope mass on the ZAHB to decrease with abundance in the sense required.

(iii) This radial change in the envelope mass on the ZAHB should produce a radial change in the population of the HB in the CM diagram. Chun's (1976) photometry of 47 Tuc stars shows just this effect: the HB stars in the inner 2 arcmin of the cluster are about 0.1 mag redder in the mean (in $B-V$) than those in the outer parts of the cluster.

5.4 Other clusters

For clusters other than 47 Tuc and ω Cen, there is only data on colour gradients. Chun (1976) found gradients in $U-B$, $B-V$ for 8 out of 24 clusters. For one cluster, NGC 2808, the colour changes extend over the whole core region. In the others (apart from 47 Tuc and ω Cen), the colour changes are more confined to the cluster centre, and G. Da Costa (private communication) argues that they may not be significant.

It is interesting that the three clusters with fairly clear colour gradients (47 Tuc, ω Cen and NGC 2808) are all massive (about 10^6 $M_\odot$) and have long central relaxation times. Abundance gradients set up at the time of cluster formation could then survive to the present time, against the homogenizing effect of dynamical relaxation.

Several other fairly massive clusters with shorter relaxation times show chemical inhomogeneities (e.g. Cohen 1978). These clusters may originally have had abundance gradients.

For the dwarf elliptical (dE) galaxies, there is good evidence for chemical inhomogeneity (see Zinn, p. 191, this volume), but so far there are no data about radial abundance, colour or luminosity function gradients.

5.5 Conclusion

For ω Cen and the dE galaxies, the relatively large masses and relatively high ellipticities (J. E. Norris & M. S. Bessell, preprint) support Searle's (1977) picture of globular cluster formation from several merging cells. But, on the other hand, most clusters (including 47 Tuc and NGC 2808 which show radial inhomogeneities) are nearly spherical. The origin of the radial abundance gradients in particular is not clear yet. The most obvious picture is that chemical enrichment occurred *within the cluster itself*, during its formation, just as we believe it occurred in galaxies. However, there are the obvious problems of containing this enriched material, ejected presumably from supernovae. On the other hand, if a significant fraction by mass of the globular cluster is to be enriched, then the enrichment events probably took place while the cluster was still mainly gaseous. (See also Iben, p. 127, this volume.) Then the problems of containment are not so great, *particularly near the cluster centre*.

It is interesting to compare globular clusters and normal elliptical galaxies. Although there are obvious differences, some of the properties of ellipticals are reproduced in miniature by the clusters. For example, the more concentrated clusters like 47 Tuc are structurally very similar to the standard elliptical galaxy. And now we see galaxy-like colour gradients in a few massive clusters with long relaxation times.

For the galaxies there is now a lot of data on colour gradients and gradients in features of the integrated light spectrum. However, the interpretation is difficult, because one has *only* the integrated light to work with. For example, it would not be easy to disentangle the direct contribution of abundance gradients (i.e. on the luminosity function for the brightest giants, as in the globular clusters). And it is certainly not easy to measure reliably the actual abundances of elements at different radii in these galaxies. On the other hand, for globular clusters the problem is observationally much easier. We can determine directly the radial changes in the luminosity function that produce the colour gradients, and we can now measure directly chemical abundances in individual stars, at various distances from the cluster centres. So, for studying the detailed chemical evolution of stellar systems, it seems that these globular clusters are probably more profitable places to work than the elliptical galaxies.

References

Butler, D. (1975). *Astrophys. J.* **200**, 68.
Butler, D., Dickens, R. J. & Epps, E. (1978). *Astrophys. J.* **225**, 148.
Cannon, R. D. & Stobie, R. S. (1973). *Mon Not Roy. astron. Soc.* **162**, 207.
Chun, M. S. (1976). Australian National University thesis.
Cohen, J. G. (1976). *Astrophys. J. Lett.* **203**, L127.
Cohen, J. G. (1978). *Astrophys. J.* **223**, 487.
Freeman, K. C. & Rodgers, A. W. (1975). *Astrophys. J. Lett.* **201**, L71.
Gascoigne, S. C. B. & Burr, E. J. (1956). *Mon. Not. Roy. astron. Soc.* **116**, 570.
Gingold, R. A. (1976). *Astrophys. J.* **204**, 116.
Hesser, J. E. (1978). *Astrophys. J. Lett.* **223**, L117.
Iben, I., Jr. (1972). *The Evolution of Population* II *Stars*, Dudley Observatory Report no. 14, ed. A. G. Davis Philip, p. 1.

Lloyd-Evans, T. (1974). *Mon. Not. Roy. astron. Soc.* **167**, 393.
Martin, W. Ch. (1937). Leiden University thesis.
Norris, J. E. & Bessell, M. S. (1975). *Astrophys. J. Lett.* **201**, L75.
Norris, J. E. (1977). In *The HR Diagram*, IAU Symposium no. 80, ed A. G. Davis Philip & D. S. Hayes, p. 195. Dordrecht: D. Reidel.
Oort, J. H. & van Herk, G. (1959). *Bull. astron. Inst. Netherlands*, **14**, 299.
RGO (1966). *Roy. Obs. Ann.* no. 2.
Searle, L. (1977). *The Evolution of Galaxies and Stellar Populations*, ed. B. M. Tinsley & R. B. Larson, p. 219. New Haven: Yale University Observatory.
Searle, L. (1978). Results reported at NATO Advanced Study Institute on Globular Clusters, Cambridge.
Yale Transactions (1977). No. 33.
Zinn, R. (1973). *Astrophys. J.* **182**, 183.

6
The correlation of cyanogen, calcium, and the heavy elements on the giant branch of Omega Centauri

JOHN NORRIS

6.1 Introduction

In recent years it has become clear that there is a large range in abundance in ω Cen, and that the most likely explanation of the observed spread lies in some combination of primordial abundance variations and mixing on the giant branch (Freeman & Rodgers 1975; Norris & Bessell 1975, 1977; Dickens & Bell 1976; Bessell & Norris 1976; Mallia 1976; Lloyd Evans 1977a, b; Hesser, Hartwick & McClure 1977; Kraft, p. 87, this volume). It is commonly accepted that calcium variations are almost certainly primordial; variations in cyanogen, strontium and barium, on the other hand, are generally taken to indicate mixing. This chapter describes in preliminary form some efforts to place constraints on the relative importance of these two processes. The results have been obtained in collaboration with M. S. Bessell and K. C. Freeman and will be more fully reported elsewhere. We find that for the majority of stars on the giant branch of ω Cen the absorption in the region of Ca II H and K has a strong positive correlation with the violet CN absorption. Furthermore, when CN and Ca are enhanced, there is a general strengthening of lines of the heavy elements, together, in some cases at least, with an overenhancement of the s-process elements. *For the bulk of material it thus appears that variations in the primordial indicators are inextricably connected with those in the mixing indicators.*

6.2 Observational material

K. C. Freeman has obtained spectra of 100 members of ω Cen using the RGO spectrograph – Image Photon Counting System attached to the Anglo-Australian Telescope as part of a programme to obtain accurate radial velocities in this cluster. These data also contain valuable information on the abundance distribution in ω Cen and form the basis for the present investigation. The spectra cover the wavelength range $\lambda\lambda 3600$–4600, with resolution ~ 1.0 Å FWHM. The stars were chosen to lie on the giant branch in the ranges $12.2 \leqslant V \leqslant 12.9$ and $1.00 \lesssim B-V \lesssim 1.4$. It should be noted for future reference that the sample is incomplete with respect to colour; an unbiased sample would necessarily include some 10 stars redder than $(B-V) \sim 1.4$.

A second set of higher resolution spectra (~ 0.8 Å FWHM) has been obtained by M. S. Bessell and myself using the same equipment. This sample comprises several stars with $V \sim 12$–13 and $1.0 \lesssim B-V \lesssim 1.70$.

6.3 Calcium and cyanogen distributions

To obtain information on cyanogen and calcium, two parameters, CN and Ca, as defined in fig. 6.1, have been measured for each star in the lower resolution sample. CN is a colour which compares the intensity in the violet CN band with that in the adjoining continuum; CN increases as the cyanogen absorption increases. Ca is a measure of the absorption in the wavelength region $\lambda\lambda 3916$–3986; as seen from the figure it is essentially a measure of the Ca II H and K absorption. Ca also increases as H and K increase. The advantage of this parameter is that it includes a completely objective estimate of the 'continuum' in the region of the features being

Fig. 6.1. Composite spectrum of several stars in ω Cen, showing the definition of the parameters CN and Ca.

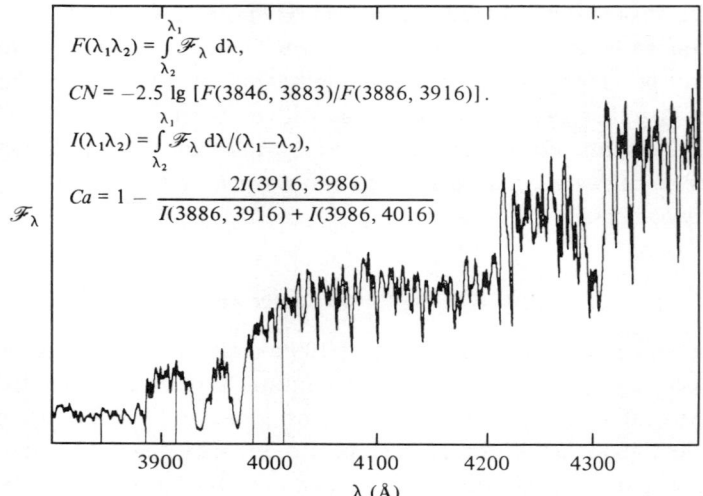

measured. The standard deviations for *CN* and *Ca* in the present sample are 0.03 mag and 0.01 mag respectively.

In fig. 6.2 *CN* and *Ca* are plotted against V magnitude for the 100 stars in our sample. There is a large range in both parameters. In order to obtain the distribution of cyanogen and calcium two further parameters, δCN and δCa, were measured as defined in fig. 6.2. The solid lines in the figure were drawn by hand in an attempt to take account of the slight increase of *CN* and *Ca* with increasing brightness on the giant branch.

Fig. 6.3 shows the generalized histograms (cf. Searle 1977) of δCN and [Ca/H]. The latter distribution was obtained on the hypothesis that the wide giant branch of ω Cen may be regarded as a composite of several giant branches of different metal abundance, and that one may therefore use other globular clusters to calibrate δCa in terms of [Ca/H]. The present preliminary calibration is [Ca/H] = 6.52 δCa − 1.50. The halfwidths of the

Fig. 6.2. *CN* and *Ca* as a function of V magnitude for 100 stars on the giant branch of ω Cen.

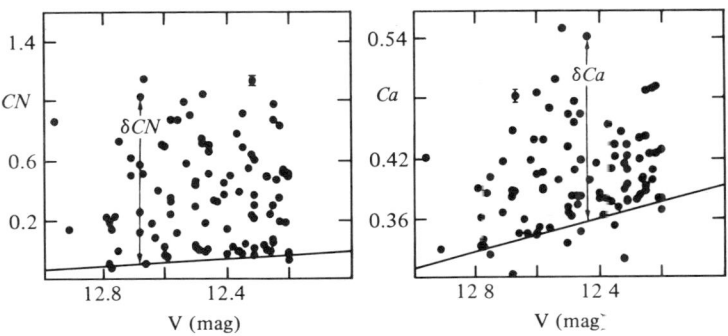

Fig. 6.3. The distribution function ϕ of δCN and [Ca/H] for 100 stars in ω Cen. Note that the sample is not complete; some 10 relatively red stars are needed for completeness.

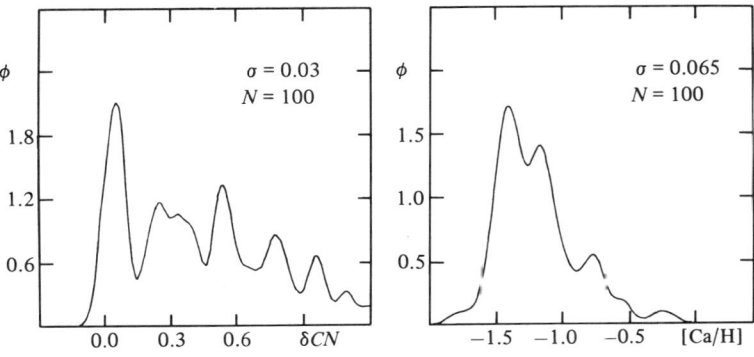

Gaussian kernels used in forming the two distributions were set equal to the estimated observational errors, 0.03 and 0.065, respectively. As is clear from fig. 6.3 there is a wide range in cyanogen and calcium. In the case of the calcium distribution one sees that the bulk of stars in the cluster have $-1.6 \leqslant [Ca/H] \leqslant -1.0$, with a long tail stretching up towards solar abundance. Three further points should be made. First, as noted in section 6.2, the sample is incomplete with respect to large $B-V$; this probably means that some ~ 10 stars are missing from the high abundance tail. Second, the distribution is not unlike that of the globular clusters with distances greater than ~ 8 kpc from the galactic centre discussed by Searle (1978), suggesting that a simple model of cluster enrichment, together with prompt initial enrichment and mass loss, might prove a reasonable model for ω Cen. Third, the distribution is not smooth. At present I do not wish to pursue this problem, but regard it a starting point for further investigation.

Fig. 6.4 shows that there is a strong positive correlation between CN and Ca. We are thus faced with the problem, already encountered by Norris & Bessell (1977), that changes in the primordial abundance variation indicator, Ca, are inextricably connected with changes in the mixing indicator CN. A prime example of the problem is afforded by the star ROA 253 (marked in fig. 6.4) which is a typical 'CN strong' star. According to

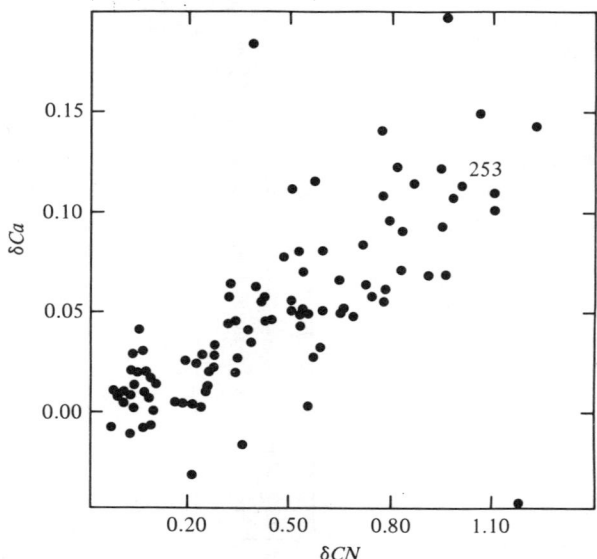

Fig. 6.4. The correlation between δCN and δCa.

Bessell & Norris (1976) ROA 253 has cyanogen as strong as observed in Population I supergiants, but possesses a large ultraviolet excess implying an abundance [Fe/H] ~ −1.0. (The present results indicate [Ca/H] ~ −0.7). The spectrum synthesis analysis of this star by Dickens & Bell (1976), who assume a heavy element abundance [A/H] = −1.3, 'indicates overabundances of N certainly greater than 10 and probably not more than 100 times... a factor of 40 appears to be the best representation of the data'. It will be noted that there is some question as to the appropriate value of [A/H] for ROA 253 and that this will affect the derived nitrogen overenhancement. A more detailed analysis of this star, as very high resolution data become available, would be invaluable.

6.4 The heavy elements

To obtain a first estimate of the behaviour of the heavy elements with increasing cyanogen strength three composite spectra were formed by adding together the spectra of stars in three ranges of cyanogen strength: $-0.2 < \delta CN \leqslant 0.2$, $0.2 < \delta CN \leqslant 0.6$, $0.6 < \delta CN \leqslant 1.2$. The resulting spectra are shown in fig. 6.5, where one sees a general enhancement of

Fig. 6.5. Composite spectra for three groups of stars chosen according to CN strength. Note the general enhancement of spectral features as CN increases.

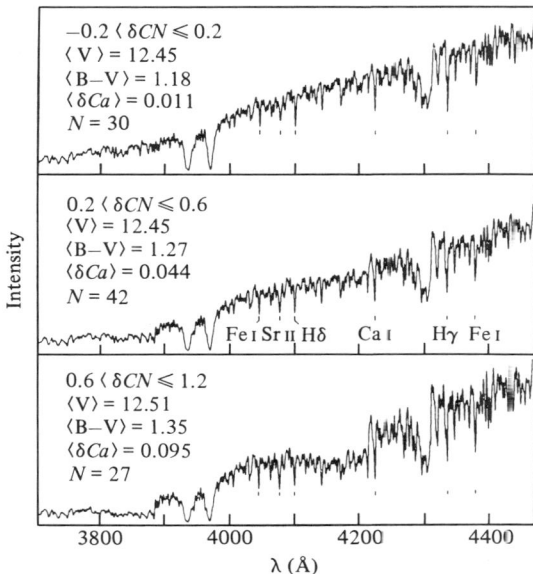

spectral features as the cyanogen increases. (Note, too, that $\langle B-V \rangle$ and $\langle \delta Ca \rangle$ increase with increasing δCN.) In particular it is clear that the lines of Fe I, Ca I, and Sr II all increase, with some indication that Sr II increases more rapidly than Fe I. More work needs to be done, however, on individual stars before the significance of the apparent behaviour of Sr II can be understood.

Fig. 6.6 contains higher resolution spectra of five stars which span the giant branch. In ROA 213 and 159 cyanogen is weak; ROA 139 and 162 are good examples of 'CN strong' stars; and ROA 513 is a very red star first noted by Lloyd Evans (1977a). Note especially the region around $\lambda 4500$, which is free of molecular features, and in which one sees an increasing enhancement of the metallic lines as one moves redward across the giant branch. For three of these stars fig. 6.7 shows a comparison of the observations with synthetic spectra computed on the assumption that

Fig. 6.6. Higher resolution spectra for several stars across the giant branch of ω Cen. Note the increasing line-strength with increasing colour, together with the relative enhancement of Ba II in the redder stars.

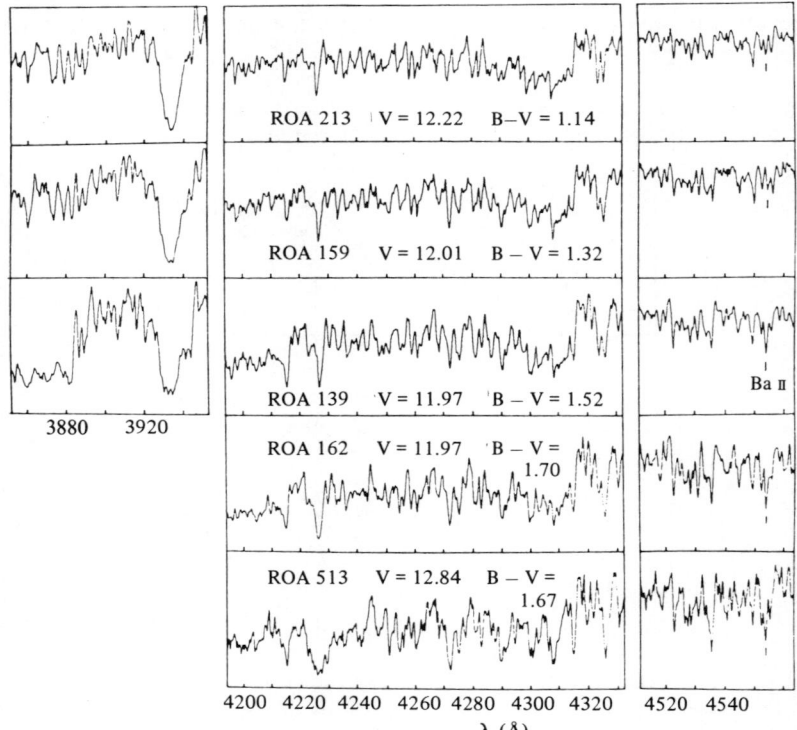

standard models can be used to represent the stars in ω Cen. Details of the choice of atmospheric parameters, and the model calculations will be given elsewhere; suffice it here to say that the model atmospheres used were those of Bell, Eriksson, Gustafsson & Nordlund (1976), the synthetic spectrum code was written by Cottrell (see Cottrell & Norris 1978), and the effective temperatures are based in part on infrared colours kindly made available by A. R. Hyland. The point that is clear from fig. 6.7 is that when standard model atmosphere techniques are used there is an increase in all elements as one moves redward across the giant branch of ω Cen. There is one final comment that should be made concerning figs. 6.6 and 6.7. While all lines appear to strengthen as one moves redward, the Ba II $\lambda 4554$ line increases at a faster rate than those of the other elements. It will be important to see if more detailed investigation of a larger sample of stars confirms this behaviour.

6.5 Discussion

The basic result of this investigation is the strong correlation between the enhancement of cyanogen and of the heavy elements, and the implied connection between the indicators of primordial abundance variations and

Fig. 6.7. A comparison between observed (——) and synthetic (——) spectra for 3 giants. Note the apparent Ba II enhancement in the cooler stars.

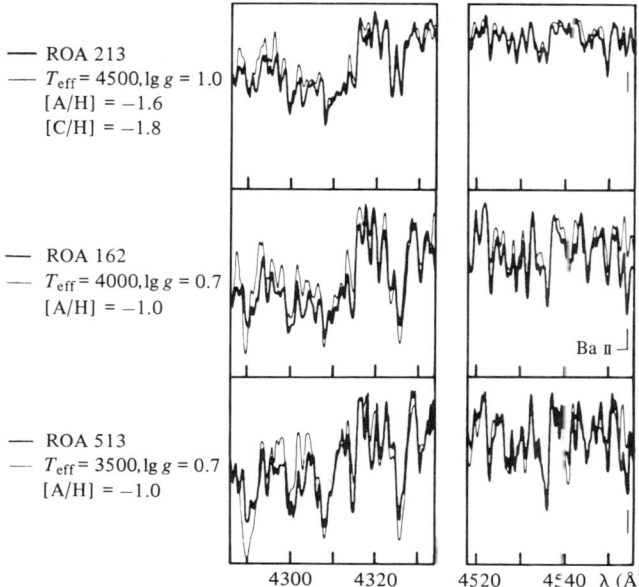

of mixing. How can these facts be reconciled? Three alternatives present themselves. First, as suggested by Norris & Bessell (1977), large enhancements of cyanogen and/or carbon monoxide might cause changes in atmospheric structure (cooling in the outer layers) which would lead to a stronger-lined spectrum. Second, the cyanogen and s-process element variations may also be primordial. Third, objects which have primordial enhancements of the heavy elements may preferentially mix during their giant branch phase. Since we see no evidence for such preferential mixing (in the sense of its being a universal phenomenon) in the more metal rich clusters (such as M5 (Zinn 1977) and 47 Tuc (Feast & Thackeray 1960; Norris 1978; Hesser 1978)) we regard this final suggestion as unlikely and will not discuss it further. Let us address the other possibilities.

6.5.1 Atmospheric effects

The model calculations of Gustafsson, Bell, Eriksson & Nordlund (1975, their figures 8 and 9) show that the inclusion of the effects of CN and CO in the infrared leads to changes in the temperature structure in the atmospheres of red giants: at $T_{\rm eff} = 4000$ K, $\lg g = 2.25$, $[A/H] = 0.0$, for example, models which include the effects of CO and CN are some 100–200 K cooler in the outer layers (due primarily to the effects of CO). This leads

Fig. 6.8. A comparison of two synthetic spectra in the region of Ca II H and K. The model atmospheres differ only in their nitrogen abundance, which leads to the temperature difference (as a function of optical depth τ) seen in the inset. The narrower line corresponds to the cooler atmosphere.

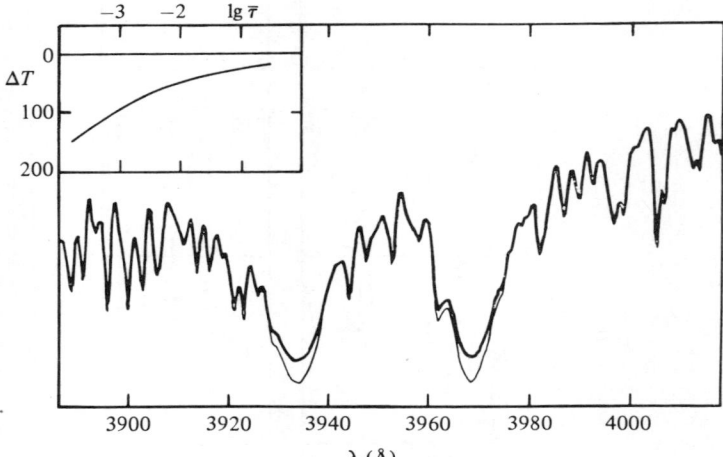

one to ask if anomalous molecular abundances in the ω Cen giants can lead to spurious line strengthening due to changes in atmospheric structure rather than increased heavy element abundance. Calculations which have been performed at Mount Stromlo lead me to believe that this is unlikely to be the case. Fig. 6.8 shows synthetic spectra in the region of Ca II H and K for two models having $T_{\text{eff}} = 4000$, $\lg g = 1.0$, $[A/H] = -1.8$, but which have different temperature structures in the outer layers due to one having a large (30 times) nitrogen enhancement. The difference in the outer layers is ~ 150 K in the sense that the nitrogen-rich atmosphere is cooler. The main effect is to deepen the cores of H and K, while the wings are unaffected; at the resolution of the present observations the other features appear not to be affected. This will not explain the current observations.

Some success in producing anomalously strong-lined objects was achieved with models of very high helium abundance. This seems a rather artificial solution. Increased helium would lead to a bluer giant branch (through increased atomic weight in the envelope), though this might be offset by increased blocking, and higher molecular opacity. It would lead also to a spread in a brightness on the horizontal branch which, on the average, is not observed (Butler, Dickens & Epps 1978).

6.5.2 Primordial cyanogen and s-process variations?

If we suppose that half the stars in ω Cen have primordial cyanogen enhancements, and that this is due to a mean enhancement of nitrogen by a factor 10, it follows that we are required to produce ~ 1000 $M_\odot$ of nitrogen during the primordial stages (assuming a cluster mass $\sim 5 \times 10^6$ $M_\odot$, and that initially $[N/H] = -1.5$). We must also produce some material with $[N/Fe] \sim +1.6$ (see Dickens & Bell 1976). The s-process elements are also to be enhanced, presumably at a faster rate than the other elements. Whether a first generation of high-mass stars of low abundance can fulfil these needs is at present unclear. It is interesting, however, that Iben (1975, 1976) finds that a 7 $M_\odot$ star of solar composition dredges up significant amounts of ^{12}C (produced during partial helium burning) and s-process material during its thermal pulsations on the asymptotic giant branch. Whether the ^{12}C is converted into ^{14}N during the interpulse phases depends on the temperature at the base of the convective envelope and hence on the mixing length adopted: for $l/H_p = 1$, ^{12}C is destroyed, while for $l/H_p = 0.7$ it is not. If we suppose, however, that conversion is complete; if we adopt 0.3 $M_\odot$ and 0.25 as the mass of material and the fraction of ^{12}C in this material dredged up during the thermal pulsation phases (Iben

& Truran 1978; Iben 1977); and if we suppose that this 7 $M_\odot$ star is in some way typical of the massive first generation stars in ω Cen in the range $\sim$ 5–15 $M_\odot$ of which there were $\sim$ 10 000 (a number which follows if one assumes an initial mass function $dN/d \lg M \propto M^{-1.5}$, and that the first generation of stars in the mass range 0.2–0.8 $M_\odot$ make up $\sim 10^6$ $M_\odot$ of the cluster, and have mean mass ~ 0.5 $M_\odot$); then we find that ~ 750 $M_\odot$ of ^{14}N could be produced. Clearly there are too many 'ifs' in this chain for the result to be taken too seriously; the estimate does show, however, that a primordial origin of cyanogen enhancement is not totally implausible. What are most needed at this point are calculations relevant to the postulated first generation of stars in ω Cen, to ascertain if nitrogen can be synthesized as a primary element in these objects. A similar need exists in respect of s-process enhancement, since the degree of enrichment will presumably depend on the abundances in the postulated first generation of stars.

Clearly more work remains to be done, both observationally and theoretically, before the role of mixing and primordial variations can be fully understood. Observationally we must determine the distribution of nitrogen and the s-process elements relative to calcium and iron. Theoretically it is crucial to know whether large relative nitrogen enhancements can be produced primordially. Finally the role of the weak-G-band and the CH stars, which have not been considered here, needs to be understood. To me, however, it appears that a primordial origin for the bulk of the enhancements seen in ω Cen is currently a most attractive hypothesis.

References

Bell, R. A., Eriksson, K., Gustafsson, B. & Nordlund, Å. (1976). *Astron. Astrophys. Suppl.* **23**, 37.
Bessell, M. S. & Norris, J. (1976). *Astrophys. J.* **208**, 369.
Butler, D., Dickens, R. J. & Epps, E. (1978). *Astrophys. J.* **225**, 148.
Cottrell, P. L. & Norris, J. (1978). *Astrophys. J.* **221**, 893.
Dickens, R. J. & Bell, R. A. (1976). *Astrophys. J.* **207**, 506.
Feast, M. W. & Thackeray, A. D. (1960). *Mon. Not. Roy. astron. Soc.* **120**, 463.
Freeman, K. C. & Rodgers, A. W. (1975). *Astrophys. J. Lett.* **201**, L71.
Gustafsson, B., Bell, R. A., Eriksson, K. & Nordlund, Å. (1975). *Astron. Astrophys.* **42**, 407.
Hesser, J. E. (1978). *Astrophys. J. Lett.* **223**, L117.
Hesser, J. E., Hartwick, F. D. A. & McClure, R. D. (1977). *Astrophys. J. Suppl.* **33**, 471.
Iben, I., Jr (1975). *Astrophys. J.* **196**, 525.
Iben, I., Jr (1976). *Astrophys. J.* **208**, 165.
Iben, I., Jr (1977). *Astrophys. J.* **217**, 788.
Iben, I., Jr & Truran, J. W. (1978). *Astrophys. J.* **220**, 980.
Lloyd Evans, T. (1977a). *Mon. Not. Roy. astron. Soc.* **178**, 345.

Lloyd Evans, T. (1977b). *Mon. Not. Roy. astron. Soc.* **181**, 591.
Mallia, E. A. (1976). *Astron. Astrophys.* **48**, 129.
Norris, J. (1978). In *The HR Diagram*, IAU Symposium no. 80, ed. A. G. Davis Philip & D. S. Hayes, p. 195. Dordrecht: D. Reidel.
Norris, J. & Bessell, M. S. (1975). *Astrophys. J. Lett.* **201**, L75.
Norris, J. & Bessell, M. S. (1977). *Astrophys. J. Lett.* **211**, L91.
Searle, L. (1977). In *The Evolution of Galaxies and Stellar Populations*, ed. B. M. Tinsley & R. B. Larson, p. 219. New Haven: Yale University Observatory.
Searle, L. (1978). Results reported at NATO Advanced Study Institute on Globular Clusters, Cambridge.
Zinn, R. (1977). *Astrophys. J.* **218**, 96.

7
Nucleosynthesis in evolved globular clusters†

ICKO IBEN, Jr

7.1 Introduction

It is well established that galactic globular clusters are among the oldest objects in the universe and that the most evolved single stars in globular clusters that are still burning nuclear fuel began life on the main sequence over 10^{10} years ago with masses of about $0.8\,M_\odot$. In such stars, the nucleosynthesis that is of special interest here is that which is ultimately manifested by a change in the surface composition during the lifetime of the star.

The standard theory of the evolution of low-mass stars (standard meaning no rotation, no magnetic fields, simple-minded treatment of convection, etc.) suggests that, at most, two instances of a change in surface composition will occur.

The first instance occurs when, following the exhaustion of hydrogen over the inner 10–13 per cent of its mass, the low-mass star evolves along the giant branch and envelope convection reaches downward into the star to bring up products of hydrogen burning that have been formed in the interior during the earlier main-sequence phase. The net result is a small decrease in the surface abundance of ^{12}C and an increase in the ^{14}N abundance by exactly the same amount that ^{12}C is decreased (Iben 1964).

Since ^{14}N is a 'second generation' element (requiring the prior presence of ^{12}C, which is a 'first generation' element), it might be expected that the initial abundance of ^{14}N relative to the abundance of ^{12}C is much less in a globular cluster star of Population II composition than it is in the Sun. Hence, the change in the ratio $^{14}N/^{12}C$ along the giant branch might be expected to be much more dramatic in a globular cluster star than the change experienced by a typical Population I star.

The second instance of a change in the surface composition that is suggested by standard theory occurs after the star has exhausted helium

† Supported in part by NSF grant AST 75-21801.

in its core and has developed an electron-degenerate carbon–oxygen core above which hydrogen and helium burning take place (Schwarzschild & Härm 1967). Once all of the matter initially in the hydrogen-rich envelope has been processed through this double-shell burning phase, the surface should exhibit the final products of complete hydrogen and helium burning at low temperatures; namely, ^{12}C, ^{16}O and ^{22}Ne. By then, however, the remnant star may have contracted to white dwarf dimensions and dimmed to such an extent that its spectrum is inaccessible to ground-based observations.

There would be no more to relate if it were not for the existence in globular clusters (and in the field) of stars whose surface compositions differ radically from that of the average cluster star in a way that cannot be achieved within the framework of the standard theory (assuming that each star began life with a distribution of elements similar to that now found at the surface of the *average* unevolved star); see Kraft, p. 87, this volume.

For example, there are approximately eight known CH stars in galactic globular clusters (Bond 1975, and private communication). These stars are giants and lie in the HR diagram somewhat to the red of the main giant branch. At their surfaces carbon appears to be more abundant than oxygen (in contrast to the normal case) and several s-process elements are present both (i) at an abundance significantly greater than normal and (ii) in a distribution that is different from the solar-system distribution of s-process elements.

Recently, Carbon, Kraft & Langer (1979; see Kraft, p. 95, this volume) have shown that the *average* carbon abundance decreases monotonically with increasing luminosity in stars along the giant branch in the cluster M92. While this is qualitatively in accord with the predictions of the standard theory, the total decrease in the average carbon abundance is by a factor of five, or by over three times the predicted value.

Whereas the CH stars found in several galactic globular clusters are only somewhat redder than the average giant-branch star of similar absolute brightness, there exists in Magellanic Cloud globular clusters a class of extremely red G stars (Feast & Lloyd Evans 1973; Catchpole & Feast 1973). The occurrence of such stars averages about one per cluster and their extreme redness suggests a very large overabundance of carbon relative to the milder overabundances in the CH stars found in galactic globular clusters.

The incidence of field analogues to the 'weird' stars found in globular clusters provides additional evidence that the standard theory must perhaps be modified in some way. The Ba stars exhibit many of the properties of CH stars, in particular showing over-abundances of s-process elements and a wide variety of s-process distributions (Warner 1965; Danziger 1966). The carbon stars (N stars, S stars, and the like) in the field are quite probably in the same evolutionary stage as are the asymptotic giant branch stars in globular clusters (but are on the average slightly more massive and hence evolve to higher luminosities before completing the double-shell burning stage). Some of these exhibit surface abundance characteristics that cannot be accounted for by the standard theory of the evolution of low-mass stars (see e.g. Scalo 1976; Iben & Truran 1978).

One of the most fascinating developments of the past three years has been the demonstration of large variations in the abundances of Fe (Freeman & Rodgers 1975; Rodgers 1978; Norris, p. 117, this volume; Freeman 1978), Ca (see Norris, p. 114, this volume), and CN (see Norris, p. 114, this volume) among giants within the cluster ω Cen and of large variations in Fe (Butler, Dickens & Epps 1978) among RR Lyrae stars in this same cluster. If these variations were due to evolutionary processes, then one might consider every star in ω Cen to be a weird star, deviating from the norm. However, the variations in CN, in Ca and in Fe appear to be tightly correlated (Norris, p. 117, this volume). Since neither Fe nor Ca are *made* in the interior of a standard low-mass model star, let alone carried to the surface by some mixing process, the observed variations are probably inherited by the stars at birth.

7.2 External sources of pollution

Before inventing exotic phenomena to account for the weird stars in globular clusters, let us examine the possibility that the surface composition characteristics of (at least some of) the weird stars are primordial, that is, due to nucleosynthesis in an earlier generation of stars. The main conclusions will be that the large random variations in the abundance of an element such as Fe among stars of a given cluster can be readily understood as being primordial (see also Norris, p. 121, this volume) and that at least *some* of the weird stars might owe their abundance peculiarities to accidents of birth.

One may define three classes of primordial pollution. The first and most obvious source of external pollution is the injection into the globular cluster of chemically evolved matter from outside the cluster, at a time before or during the interval of rapid star formation in the cluster. If the 'foreign' or chemically-evolved matter is not uniformly mixed into the pre-existing cluster gas prior to the formation of cluster stars, one would quite naturally expect a variation from cluster star to cluster star in the fraction of chemically evolved matter at the stellar surface.

In particular, suppose that the protocluster gas is bombarded by a stream of chemically evolved matter which penetrates a thin shell near the surface of the cluster. Cooling accelerated by the introduction of the coolants in the evolved matter, plus shock compression occurring at the interface between the original cluster matter and the incident stream could initiate star formation that works its way toward the centre of the cluster. Then, those stars formed nearer the periphery of the cluster would contain a higher fraction of chemically evolved matter than those near the centre. In time, the migration of stars through the cluster would engender a more or less random distribution of the more metal-rich stars.

A second form of primordial pollution will be due to nucleosynthesis *within* the cluster as a consequence of the existence of intermediate- and high-mass stars that are presumably formed during the period of active star formation within the cluster. This source of pollution might account for abundances in a few weird cluster stars as well as contribute to the overall spread in the heavy element abundances among stars in the cluster. For example, the existence of isolated stars with weird abundances such as the CH stars in galactic globular clusters and the extreme C stars in the Magellanic Cloud globular clusters might be due to the formation of such stars out of gas clouds that were in the vicinity of an intermediate-mass star that has produced an excess of carbon and s-process elements and injected this matter into the surrounding gaseous medium via a wind or planetary nebula. Similarly, a star formed out of matter that has just been shocked and polluted by chemically evolved matter from a nearby supernova might be expected to exhibit unusual characteristics relative to those of stars born out of material containing well-mixed products of nucleosynthesis from a large number of sources.

A final form of primordial pollution will affect the surface composition of the initial secondary in a close binary system in which the initial primary is of intermediate or high mass. If the initial primary expands to fill its Roche lobe after it has inserted into its envelope products of interior nucleosynthesis, these products will ultimately appear at the surface of the initial secondary which is currently experiencing standard low-mass evolutionary processes in its interior.

I am persuaded that the decrease in the cluster-average iron-to-hydrogen ratio with mean cluster distance from the galactic centre (see van den Bergh, p. 176, this volume) can best be explained as a consequence of a decrease in the extent of pollution from *outside* the cluster, rather than as the consequence of the increase in a mysterious 'mass-loss term' plus the assumption that the primary source of pollution is nucleosynthesis *within* the cluster. The most obvious argument against this latter picture is that there exists no straightforward physical process which will achieve an increase in the amount of mass lost by a cluster (during its period of active star formation) with increasing average cluster distance from the galactic centre. In fact, one might even expect that the degree of interchange between a cluster and the ambient medium would be larger for those clusters with the smaller orbital semi-major axes simply because the density of the ambient medium increases as one approaches the galactic centre.

To my mind, the most reasonable picture of the chemical evolution of matter that ultimately finds its way primordially into globular-cluster stars begins with the assumption that the average rate of star formation, whether within a potential globular cluster or outside it, increases with propinquity to the galactic centre. This follows from the observed increase in mean matter density (averaged over dimensions large compared to the size of a typical protocluster) with propinquity to the centre. One might suppose that the period of most rapid star formation in the Galaxy coincides with the last few galactic 'e-folding' times as the overall collapse of the Galaxy is stabilized by balance between centrifugal forces and gravitational forces. The last e-folding time is of the order of the current period of galactic rotation (few $\times 10^8$ yr) and is short compared to the time for an outlying protocluster to negotiate its entire orbit (about 10^9 yr).

Near the galactic centre and in what is now called the galactic disc, one might expect early on in galactic history a tremendous jumble of gas clouds and stars, with the frequency of cloud–cloud and cloud–star encounters being sufficiently high to prevent the survival of many star clusters (Fall & Rees 1977). But those that do survive must have exchanged (in both directions) considerable matter with their surroundings and therefore should now contain some stars of high metal content.

Consider the possible fate of a protocluster with large orbital diameter. Chances are that it will be closer to its 'apogalacton' than to its 'perigalacton' during the period of most intense chemical evolution of matter near the galactic centre. By the time this protocluster passes nearest the galactic centre, the density of its as yet chemically unevolved gas may well be greater than that of the density of the chemically evolved matter through which it passes (much of the original gas in the disc having been turned irreversibly into low-mass stars). Instead of being *swept clean of* gas as it passes through the disc, the protocluster will *sweep up* and retain the chemically evolved gas in the disc that blocks its path and thus experience *external* pollution, the degree of pollution being greater, the more closely the cluster approaches the galactic centre before the most active period of star formation *within the cluster* begins.

One might in fact suppose that the addition of externally formed pollutants is the trigger that sets off the most active period of star formation within the cluster. That is, the chemically evolved matter will contain coolants to a much greater extent than does chemically unevolved matter. The enhanced cooling rates thus achieved and the shocks that appear as cluster gas and disc gas collide should speed up the rate of protostar condensation within the cluster. One now imagines the matter in the protocluster rapidly turning into stars and further chemical evolution due to internal 'astration' taking place within the cluster as it moves outward toward apogalacton.

With much of the original cluster gas transformed into stars, one might expect that, during subsequent passages of the cluster through the galactic disc, gas is swept from the cluster by gas in the disc. Thus, disc gas experiences external pollution from nucleosynthesis originating in globular clusters, but further star formation in the globular cluster is dramatically curtailed and all traces of chemical evolution become fossilized in old stars.

In this way we have accounted for the general run of cluster [Fe/H] with distance from the galactic centre: (i) the central plateau of high [Fe/H] for clusters confined to within about 3 kpc of the galactic centre may be attributed to a high degree of interaction between cluster matter and ambient matter (high degree of participation of original cluster matter in chemical evolution in the collapsing disc); (ii) the plateau of low [Fe/H] in clusters with orbital diameter larger than about 10 kpc is due to the fact that chemical evolution is dominated primarily by internal sources of pollution which, due to the uniformity of overall cluster properties, should produce, in first approximation, a uniformity in net internal pollution (for another point of view, see Truran & Cameron 1971); and (iii) the decrease in average [Fe/H] with average cluster orbital diameter (for apogalacton between about 3 kpc and 10 kpc) may be ascribed to the fact that the closer the cluster approaches the galactic centre, the greater is the pollution it has experienced as a consequence of encountering and incorporating highly chemically evolved matter in the disc.

An important fact which must also be accommodated by this general picture is the variation in cluster-average [Fe/H] among clusters of roughly the same orbital diameter. A large number of effects must be involved. For example, for a given orbital diameter, the degree of external pollution will be larger for that cluster which is furthest from the galactic centre during the period of most active star formation in regions near the centre.

As another example, the mass of the protocluster and the density of matter within the cluster must be involved in an important way. All other things being equal (e.g. fixed mean density), the greater the mass of the cluster, the larger is its volume per unit surface area and hence the smaller is the overall external pollution to be expected on passage through the galactic disc. On the other hand, the greater the mass of the cluster, the more extensive might be the degree of internal pollution, since the more massive cluster might be expected to have a larger 'optical depth' inhibiting the loss of its internally manufactured pollutants.

Another effect might be the variation in mass spectrum of stars formed in both the cluster and in the environment through which it passes when experiencing its greatest rate of external pollution. We know that an element such as ^{3}He is produced primarily only by low-mass stars; s-process elements are produced primarily by stars of intermediate mass; ^{16}O, ^{20}Ne, ^{24}Mg, ^{32}S, ^{36}A, ^{40}Ca, ^{56}Fe (maybe) are produced only by high-mass stars; ^{12}C is not produced by most low-mass stars, but it is produced by stars of intermediate mass and perhaps also by massive stars. An element such as ^{14}N is produced by all stars that contain some ^{12}C at birth, but the amount of ^{12}C converted into ^{14}N is a function of initial stellar mass. Thus, the relative proportions of the elements that appear in chemically evolved matter clearly depend on the rate of star formation as a function of mass (the mass spectrum) and, since the mass spectrum is undoubtedly a very sensitive function of environment, this effect will contribute to a variation of, say, [Fe/H] for clusters of even the same orbital dimensions.

Still another factor that may influence the type and degree of internal pollution experienced by a cluster is the escape velocity from the cluster relative to the speed with which highly processed matter is ejected from its stellar sources. We believe that, during their asymptotic giant-branch phases, low- and intermediate-mass stars eject matter via low-velocity winds. Thus ^{12}C and s-process elements are inserted at low velocity into the interstellar medium. On the other hand, a high-mass star will presumably become a supernova and eject ^{12}C, ^{16}O, ^{40}Ca, ^{56}Fe, etc., at high velocity (10^4 km s^{-1}). Thus, much of the ^{16}O, ^{40}Ca and ^{56}Fe produced in supernovae that appear near the surface of a cluster may be lost from the cluster, whereas ^{12}C and s-process elements produced in stars of intermediate mass may be retained. Also, all other things being equal, the more massive the cluster, the less pronounced will this mass selectivity be.

One could continue indefinitely in this vein, but we have already strayed far beyond the primary objective of simply making plausible the contentions that: (1) most of the variations or *spreads* in composition variables that are observed among currently evolving cluster stars are primordial; that is, variations in such elements as Fe and Ca are incorporated in the stars at birth (this must surely be the explanation of the large spreads in Ca, [Fe/H] and CN found in ω Cen); and (ii) at least some of the composition 'weirdos' may simply be accidents of birth (that is, weirdness can be on occasion primordial).

7.3 Relevant features of the standard evolution of low-mass stars

Since weird stars are, by definition, of infrequent occurrence, it is indeed fortunate that the standard theory of the evolution of low-mass stars (no rotation, no magnetic fields, etc.) does not lead naturally to the production of weird abundances. The results of this standard theory are well documented (see, e.g., reviews by Iben 1971, 1974; Renzini 1977; Castellani, p. 65, this volume) and need not be repeated here. For our purposes, only a few salient features require emphasis.

During the main-sequence phase, when the conversion of hydrogen into helium constitutes the major stellar energy source, almost all of the primordial ^{12}C within the inner approximately 35 per cent of the star is converted into ^{14}N. When hydrogen becomes exhausted over the inner 13 per cent of the star's mass, its envelope swells to red giant proportions, simultaneously developing a deep conventive envelope. The depth (in mass) of the convective envelope increases as the star brightens along the giant branch. When the base of the envelope reaches the mass within which ^{12}C has been converted into ^{14}N during the main-sequence phase, the surface abundance of ^{14}N begins to rise and the surface abundance of ^{12}C begins to drop. This alteration in surface abundances ceases once the newly formed hydrogen-burning shell, on moving outward in mass, encounters the base of the inward moving convective envelope and drives this base back outward again.

Thereafter, the star continues to brighten as a red giant while the growing electron-degenerate helium core becomes more dense and hotter. Neutrino losses predicted by both the charge-current and the neutral-current theories of weak interactions maintain the central temperature lower than the temperature in the middle of the core. The increase in the *average* temperature in the core is due to the conversion of gravitational potential energy released by the ashes of hydrogen burning as these are deposited into an ever more massive helium core of essentially constant volume. When the mass in the helium core reaches approximately 0.5 $M_\odot$, the triple-α reaction ($3\alpha \to {}^{12}$C) initiates a thermal runaway at the location of the temperature maximum in the core; nuclear energy is liberated more rapidly than it can be carried outward by electron conduction and the temperature in the burning region rises rapidly. In fact, the intensity of the nuclear reactions produces a flux of outward flowing energy so high that turbulent convective motions are generated; a convective shell grows outward from the temperature maximum in the core.

In standard models, the outer edge of the convective shell does not reach the location of the hydrogen–helium discontinuity before expansion engendered by the overpressure developed in the shell terminates the runaway and damps out the thermal pulse. Within the region occupied by the convective shell at its maximum extent is a residue of ^{12}C, the initial product of helium burning. Temperatures in the convective shell at no time become high enough for the release of neutrons via the $^{22}Ne(\alpha,n)$ ^{25}Mg reaction (Schwarzschild & Härm 1967; Demarque & Mengel 1971).

During the power-down of the helium-shell flash the base of the convective envelope, which is confined to the hydrogen-rich region above the hydrogen–helium discontinuity, moves outward in mass and then back inward, returning approximately to the position it occupied before the flash (but ever so slightly further in). It is important to note that the base of the convective envelope in standard models does not reach inward beyond the original location of the hydrogen–helium discontinuity and that, therefore, no fresh ^{12}C is dredged up into the envelope.

A sequence of inward thermal flashes follows, each flash lifting the degeneracy in its vicinity, until the entire core is again maintained in balance with gravitational forces by the kinetic motions of non-degenerate electrons and nuclei. No products of nucleosynthesis freshly made in the core make their way to the surface. In the HR diagram, the star has become a horizontal-branch resident (if its surface component of heavy elements is low enough) or it takes up a position far down on the giant branch as a 'clump' star (if its surface is rich in heavy elements). The horizontal-branch (or clump) phase is uneventful except for possible semi-convective 'burps' (A. V. Sweigart & A. Renzini, preprint) and occasional RR Lyrae oscillations.

After helium is exhausted at its centre, the star develops a helium-burning shell that eats its way outward toward the hydrogen-burning shell. Then, as the star climbs upward along the asymptotic branch, a sequence of thermal pulses ensues. These pulses are similar in character to the thermal flashes that have occurred earlier in the core except that now electrons in the unstable region are not degenerate. Again, a convective shell forms and extends outward from the temperature maximum where the rate of nuclear energy generation is greatest and the thermal runaway is most pronounced.

And again, during the course of the thermal pulse, the base of the convective envelope in the hydrogen-rich region at first recedes outward and then, during pulse power-down, moves back inward (in mass).

As the mass of the inert carbon–oxygen core below the thermally pulsing region increases, the amplitude of the thermal pulse grows. The maximum temperature reached in the transient convective shell increases with each pulse, as do the maximum rate of energy generation in the shell, the closest proximity of the outer edge of the convective shell to the hydrogen–helium discontinuity, and the furthest inward (in mass) extent of the base of the convective envelope achieved during pulse power-down.

However, in stars initially as light as those now evolving along the asymptotic giant branch in globular clusters, the amplitude of even the final pulses is not large enough to produce the interesting effects that stars in the initial mass range 3–8 $M_\odot$ experience (Gingold 1974; Sweigart 1973, 1974). In particular, at no time during the pulse does the outer edge of the convective shell reach the hydrogen–helium discontinuity with the consequent injection of hydrogen into the helium- and carbon-rich convective region; the temperatures in the shell never become high enough for the ^{22}Ne$(\alpha,n)^{25}$Mg reaction to provide an interestingly large flux of neutrons; and the base of the convective envelope does not reach inward during pulse power-down far enough to dredge up the matter containing fresh ^{12}C made in the convective shell during the peak of the pulse.

Once most of the matter in the star has been processed through both hydrogen and helium burning, the hydrogen-rich envelope suffers a dynamic instability and is ejected as a planetary nebula. The remnant carbon–oxygen core cools off along the white-dwarf sequence (Wood 1974; Härm & Schwarzschild 1975).

7.4 The binary hypothesis

Having determined that the standard theory of single-star evolution does not produce an exotic surface composition in models of low mass, it is worthwhile to explore more fully other schemes for a possible explanation of the stars in globular clusters of weird surface composition. We have already remarked in a general way on the possibility of primordial pollution providing an explanation for at least some weirdos. An interesting example of primordial pollution involves membership in a binary system, the initial primary being a star of intermediate mass (3–8 $M_\odot$).

The choice of an intermediate-mass primary is due to the rather strong theoretical evidence (Iben 1975a, b, 1976, 1977) that stars of intermediate mass (i) produce ^{12}C and s-process elements quite naturally in the intermittently appearing convective shell during the asymptotic giant-branch phase, *and* (ii) dredge up into the envelope convective region large quantities of these freshly made elements.

The production of s-process elements is made possible by the fact that the carbon–oxygen core can grow (in principle) as large as the Chandrasekhar mass (although mass loss from the surface restricts stars which reach this limit to those with initial mass greater than about 5 $M_\odot$). The dredge-up of freshly made ^{12}C begins when the mass of the carbon–oxygen core exceeds about 0.8 $M_\odot$. Once the mass of the carbon–oxygen core exceeds about 1 $M_\odot$, temperatures in the convective shell become large enough for the production of a large flux of neutrons by means of the ^{22}Ne$(\alpha, n)^{25}$Mg reaction. Thereupon, s-process elements are formed and dredged to the surface along with ^{12}C.

The sequence of events that may be imagined to give rise to a currently observed weirdo among globular cluster stars goes like this (see also Renzini, Mengel & Sweigart 1977; K. Smith & P. Demarque, preprint). We assume that the initial secondary has a mass less than 0.8 $M_\odot$ (the initial main-sequence mass of a typical cluster giant) and adopt an initial orbital separation larger than the combined radii of the secondary and of the primary at a time when the primary first becomes an asymptotic giant-branch star. By choosing the initial orbital radius properly (remembering that, since the primary and also the whole system is losing mass via a stellar wind from the primary, the orbital characteristics of the system are changing), we can arrange for the primary to overflow its Roche lobe *after* it has built up sizeable surface overabundances of ^{12}C and s-process elements. In this way we deposit onto the secondary a surface layer of highly processed matter. If we have arranged initial orbital characteristics even more carefully, we might be able to wind up with an 0.8 $M_\odot$ star burning hydrogen at its centre even while sporting a highly processed surface. After 10^{10} yr this star with an 'icing' will then become a red giant with a weird composition.

Among the arguments that can be raised against the binary hypothesis is the probability that the occurrence of binaries with 'just the right characteristics' is small compared to the frequency of occurrence of weird stars. Another argument is based on the details of surface composition. At least one Ba star (the population I field-star analogue of a CH star) exhibits magnesium isotopes in a solar-system distribution, while those s-process elements which are overabundant are not present in a solar-system distribution among themselves (Tomkin & Lambert, 1979). One of the major characteristics of the s-process distribution formed when the reaction ^{22}Ne(α, n)^{25}Mg provides the necessary neutrons is that, because ^{25}Mg and its progeny capture precisely the correct fraction of emitted neutrons, the s-process elements in the atomic mass range $A = 70$–204 are formed precisely in the solar-system distribution (Truran & Iben 1977). The penalty that one pays for this result is that the final distribution of magnesium isotopes is far from solar. Thus, the observations of Tomkin & Lambert rule out the binary hypothesis for at least one star of weird composition.

Still another argument against the binary hypothesis is the fact that the distributions of s-process elements in weird stars are quite variable from one weird star to another; we have already pointed out that one of the most remarkable features of the intermediate-mass environment is that it produces the s-process elements essentially uniquely in a solar-system distribution.

The properties of S stars in the field are a further argument against the binary hypothesis as an explanation for all weird surface compositions. The S stars, which were first discovered by Merrill (1952) to exhibit absorption lines due to the ephemeral technetium nucleus (2×10^5 yr half-life), contain carbon and oxygen in roughly equal amounts (it is the fact that the carbon and oxygen are locked into carbon monoxide that simplifies the spectrum and allows the technetium absorption lines to be seen). The existence of technetium implies that s-process elements are *currently* being made in the S stars and are therefore not at the surface as a consequence of deposition from some more massive companion. However, the fact that carbon is not far more abundant than oxygen in those S stars in which s-process elements are as much as a factor of 20 overabundant relative to iron (Boesgaard 1970) demonstrates that the highly overabundant s-process elements do not owe their existence to the processes that produce these elements in intermediate-mass stars, wherein the dredge-up of s-process elements is in lock step with the dredge-up of ^{12}C.

It is this last line of evidence which is perhaps the most direct argument for the existence of a non-standard process that occurs in some single stars before they reach the asymptotic giant branch phase. Granted that the S stars are now making s-process elements (appearance of technetium) and were hence probably of intermediate mass when on the main sequence, the large overabundance of the stable s-process elements had nevertheless to be produced prior to the double-shell source phase since, during this phase, the rate at which ^{12}C is enhanced is comparable to the rate at which s-process elements are enhanced. Even if the major abundance characteristics of the

S stars were primordial, these characteristics had to be achieved in an earlier generation of stars undergoing non-standard evolution. We are therefore forced to look for some non-standard process that occurs in only a few stars as a consequence of some physically realizable perturbation whose magnitude might be expected to be variable from one star to another.

7.5 On the establishment of peculiarities during the helium flash

Perhaps the most promising situation where perturbations might lead to dramatic variations in behaviour is that phase in the evolution of a low-mass star that immediately follows the ignition of helium in the electron-degenerate hydrogen-exhausted core.

A tantalizing clue is provided by the investigation by Schwarzschild & Härm (1967) of the properties of thermally pulsing models and the investigation by Sanders (1967) of possible associated nucleosynthesis. This latter work has been elaborated beautifully by Ulrich (1973). If one can somehow inject small quantities of hydrogen into a region containing ^{12}C and ^{4}He at high temperatures, then the ultimate result is the production of one neutron for every proton injected and the subsequent capture of this neutron by ^{56}Fe or by one of its neutron-rich progeny, thereby contributing to the buildup of s-process elements. If the number of injected protons is a function of the strength of some perturbation and if the extent to which freshly produced s-process elements are brought to the surface is also a function of the strength of this same perturbation, then one can possibly account both for the high variability in the distribution of s-process elements found in stars such as the Ba and CH stars and for the high variability in the total overabundance of s-process elements among S stars.

The possibility of the injection of protons into the helium core is made plausible by the fact that, during the helium flash in standard models, convection extends from the region of maximum energy-generation rate up to within a very few scale heights of the hydrogen–helium discontinuity. Several studies indicate that, if convection is formally made very inefficient (Sugimoto 1964; Edwards 1969), the outer edge of the convection zone generated in the helium core can be brought arbitrarily close to the hydrogen–helium discontinuity. Several other numerical experiments show that, if the strength of the flash is artificially enhanced (Thomas 1967; Paczyński & Tremaine 1977), then the extent of dredge-up during flash power-down is also enhanced to the extent that the ^{12}C produced during the flash is brought to the surface.

The most obvious perturbation that might increase the strength of the flash is the distortion associated with rapid rotation. That some stars in globular clusters are affected by rotation is possibly indicated by the fact that the total reduction in the average carbon abundance along the giant branch in at least one cluster (M92) appears to be over three times that predicted by standard theory without rotation (Carbon *et al.* 1979; Kraft, p. 95, this volume). Sweigart & Mengel (1979) have recently shown that, for 'reasonable' rotation rates, meridional currents might well bring into the convective envelope sufficient products of hydrogen burning that are formed just outside the hydrogen-burning shell during the first red giant branch to account for the observations in M92.

If one accepts this evidence for 'reasonable' rotation rates in the interiors of red giants, one must then ask how rotation might affect the development of the helium flash. One sees intuitively that by countering inward gravitational forces, rotation reduces the pressure necessary for quasistatic equilibrium and hence, for a given core mass, induces a lower temperature. In order to attain the temperature necessary for igniting helium, the core must therefore grow to a larger mass than in the absence of rotation. The effect is much the same as if one were to increase artificially the neutrino-cooling rates in standard models of non-rotating stars. The effect of such an increase can be seen by comparing the models of Thomas (1967), in which the neutrino rates were chosen to be four times larger than is suggested by the charge-current theory of the weak interactions, with the results of standard models which use the smaller rates (Demarque & Mengel 1971). The helium-shell flash obtained by Thomas is more intense than that obtained with standard rates and a significant dredge-up of processed matter accompanying pulse power-down occurs only in the Thomas model. This effect is demonstrated also by Paczyński & Tremaine (1977). Note, however, that although dredge-up occurs for sufficiently large cooling rates, the edge of the convective shell formed in the helium-burning region *does not reach* the hydrogen–helium discontinuity; s-process elements are therefore *not* formed during the flash and are *not* available for dredging up into the envelope. In any case, Różyczka (1978) estimates quantitatively the degree by which 'reasonable' rotation rates will enhance the flash and concludes that the enhancement is insufficient to produce detectable effects.

Another obvious perturbation is the presence of internal magnetic fields. Just how internal magnetic fields of a 'reasonable' magnitude and configuration might affect the course of the flash has not been explored quantitatively, but one can imagine several possibilities. One effect might be the inhibition of convection. In order for this effect to be important, it is necessary for the energy density in the magnetic field to be comparable with the energy density associated with convective motions. That is, one requires that $H^2/8\pi \gtrsim f(\frac{1}{2}\rho v_c^2)$, where H is the strength of the magnetic field, ρ is the mass density, v_c is a typical speed of convective currents, and f is a factor of the order of, say, 0.1. The inequality may be rewritten as $(v_c/v_s) < (10^{-8} (H/10^6 \text{ gauss}) (10^6 \text{ g cm}^{-3}/\rho) (c/v_s)$, where c is the speed of light in vacuum and v_s is the sound speed in an electron-degenerate medium. In such a medium, $v_s/c \approx (5/3 \mu_e)^{\frac{1}{2}} \epsilon_F/M_H c^2$, where μ_e = electron molecular weight, ϵ_F = electron Fermi energy, and M_H = mass of the proton. Just prior to the helium flash, $v_s/c \sim 1.7 \times 10^{-2}$, $\rho \sim 10^6$ g cm^{-3}, so that our requirement becomes $v_c \gtrsim 2 \times 10^{-6} v_s (H/10^6 \text{ gauss})$.

During the flash, convective speeds are typically about three orders of magnitude smaller than sound speed (Edwards 1969). The apparent presence of magnetic fields of strength $\sim 10^{12}$ gauss at the surface of some neutron stars implies the existence of fields on the order of 10^6 gauss in the hydrogen-exhausted core of at least some red giants that are approaching the helium-flash phase $(H_{RG}/H_{NS} \sim (10^6 \text{ cm}/10^9 \text{ cm})^2)$. It would thus appear that a magnetic field of 'reasonable' strength (10^6 gauss) is about three orders of magnitude too small to inhibit convective motions.

However, since the weird stars we wish to understand are of low frequency relative to normal stars, perhaps it is actually 'reasonable' to suppose that, in a few stars, magnetic field strengths can reach 10^9 gauss in the core. Further, convective motions build up over a finite interval of time; magnetic 'bottling up' of energy could therefore be effective near the outer edge of the growing convective core, where convective motions are being established, even for magnetic field strengths on the order of only 10^5–10^6 gauss.

Although the numerics just presented do not look especially promising, the possibility remains that a sufficiently strong and properly geometrized magnetic field configuration should be able to operate *in the same sense as decreasing the convective efficiency*. Decreasing the convective efficiency has been demonstrated by Sugimoto (1964) and by Edwards (1969) to increase the intensity of the helium flash and thereby to increase the likelihood that core convection can reach and extend somewhat beyond the location of the hydrogen–helium discontinuity during the flash. That is, by bottling up convection, the flash proceeds further and leads to an enlargement of the maximum size of the convective region.

As a final remark it should be emphasized that the standard treatment of the helium flash could well be inappropriate for all but the most extreme metal-deficient stars. At one time (Iben 1967) it was thought that the $^{14}\text{N}(\alpha,\gamma)^{18}\text{F}$ reaction might provide a trigger for the helium flash that would explain what was then thought to be an observational fact: that the absolute luminosity at the red-giant tip in a globular cluster decreases with the average value of [Fe/H] characterizing the cluster. Since then it has been shown that the cross-section for capture on ^{14}N is far too small for this process to play an important role in the development of the helium flash (Parker 1968) and that the magnitude at the red-giant tip is essentially independent of [Fe/H] (Cohen, Frogel & Persson, preprint; see also p. 146, this volume). Since Cohen, Frogel & Persson estimated cluster distances by using horizontal-branch fitting and since the luminosity of the horizontal branch is expected to decrease with increasing Z, it appears that the red-giant tip may in fact be dimmer in clusters with larger average [Fe/H]. This recent observational result is to be contrasted with the theoretical prediction that $\text{d}\lg L/\text{d}\lg Z \sim +0.09$ (Rood 1973), when it is solely the $3\alpha \rightarrow {}^{12}\text{C}$ reactions that trigger the helium flash.

Several years ago, Mitalas (1974) pointed out that, at densities similar to those near the centre of a red-giant core, the Fermi energy of the electrons is quite close to the threshold energy for electron capture on ^{14}N and that there will therefore be a finite abundance of ^{14}C in the core. As a consequence, the reaction $^{14}\text{C}(\alpha,\gamma)^{18}\text{O}$ must be considered as a possible trigger for the helium flash, especially in stars of Population I composition. Mitalas concluded that the rate of the $^{14}\text{C}(\alpha,\gamma)^{18}\text{O}$ reaction is probably too small for the reaction to be of any importance, but recently, Kaminisi, Arai & Yoshinaga (1975) and Kaminisi, Keisuke & Arai (1975) have shown that Mitalas may have underestimated the rate by almost a factor of 10^6! The N–C–O sequence of reactions may therefore be the trigger for the helium flash in stars of sufficiently large heavy element abundances.

An important feature of the $^{14}\text{N}(e^-,\nu)\,^{14}\text{C}(\alpha,\gamma)^{18}\text{O}$ sequence of reactions is that it is *pycnonuclear* as well as *thermonuclear*, due to the strong dependence on density of the abundance of ^{14}C. Whereas the triple-α rate goes like

$$\text{Rate}\,(3\alpha) \propto Y_{\text{He}}^3\, \rho^2\, T^{-3/2} \exp(-\alpha/T),$$

the $^{14}\text{N} \rightarrow {}^{18}\text{O}$ rate goes like

$$\text{Rate}\,(\text{N} \rightarrow \text{O}) \propto Y_{\text{N}}\, Y_{\text{He}}\, \rho\, T^3 \exp[-|(\epsilon_{\text{th}}-\epsilon_{\text{F}})|/kT] \times T^{-3/2} \exp(-\beta/T),$$

where $\epsilon_{\text{F}} \propto \rho^{2/3}$. Here Y_{He} is the abundance by number of ^4He, Y_{N} is the abundance by number of ^{14}N, ρ is the density, T is the temperature, and ϵ_{F} is the electron Fermi energy.

Because of its strong density dependence, the N $\rightarrow$ O set of reactions will achieve its maximum value much closer to the centre of the star than will the 3α reaction. The net result may be a weaker flash in stars with large enough initial CNO abundances for the N–O reactions to trigger the flash.

If the N → O reactions can trigger the flash, one also has a natural explanation of the apparent observational result that the luminosity at the red-giant tip in a cluster is smaller in a cluster with higher metallicity. For, assuming a roughly monotonic correlation between [Fe/H] and the abundance of CNO elements, the stars with larger [Fe/H] will in general contain more CNO to start with and will therefore experience the helium flash for smaller core masses and hence smaller surface luminosities.

A final feature of the N → O induced flash is that it also provides a way of accounting for the difference between the general characteristics of weird stars in the Galaxy and in the Magellanic Clouds. One might expect that the stronger the flash, the more easily perturbations can induce mixing. Hence, if a smaller abundance of the CNO elements means a stronger flash, weirdness should be more pronounced in stars of low heavy-element abundances. This may provide the explanation for the presence of extreme C stars in the Magellanic Cloud globular clusters and the total absence of such stars in galactic globular clusters. It may also partially explain why the incidence of carbon stars in the Galaxy increases with distance from the galactic centre (Blanco 1965; Blanco, Blanco & McCarthy 1978) although it is just as likely that this trend in the frequency of carbon stars is due to a monotonic variation in the primordial abundances of carbon and oxygen.

It is a pleasure to thank Moshe Elitzur, Jay Gallagher, Telemachos Mouschovias, Bob Spulak, Jim Truran, Ron Webbink and Stan Wyatt for helpful conversations and for comments on the manuscript.

Note added in proof. Recently, R. McClure (preprint) has compiled evidence that most Ba stars may be binaries and Spulak (1980) has shown that, *whatever* the rate of the $^{14}C(\alpha,\gamma)^{18}O$ reaction, the sequence of reactions transforming ^{14}N into ^{18}O does not initiate the helium flash. Thus, perhaps as many as 1500 words in this chapter have become obsolete since their utterance.

References

Blanco, V. M. (1965). In *Galactic Structure*, ed. A. Blaauw & M. Schmidt, vol. 5, p. 241. University of Chicago Press.
Blanco, B. M., Blanco, V. M. & McCarthy, M. F. (1978). *Nature*, **271**, 638.
Boesgaard, A. M. (1970). *Astrophys. J.* **161**, 163.
Bond, H. E. (1975). *Astrophys. J. Lett.* **202**, L47.
Butler, D., Dickens, R. J. & Epps, E. (1978). *Astrophys. J.* **225**, 148.
Carbon, D., Kraft, R. P. & Langer, E. (1979). *Astrophys. J.* (in press).
Catchpole, R. M. & Feast, M. W. (1973). *Mon. Not. Roy. astron. Soc.* **164**, 11.

Danziger, I. J. (1966). *Astrophys. J.* **143**, 527.
Demarque, P. & Mengel, J. G. (1971). *Astrophys. J.* **164**, 317.
Edwards, A. C. (1969). *Mon. Not. Roy. astron. Soc.* **146**, 445.
Fall, S. M. & Rees, M. J. (1966). *Mon. Not. Roy. astron. Soc.* **181**, 37P.
Feast, M. W. & Lloyd Evans, T. (1973). *Mon. Not. Roy. astron. Soc.* **164**, 15P.
Freeman, K. C. (1978). Results reported at NATO Advanced Study Institute on Globular Clusters, Cambridge.
Freeman, K. C. & Rodgers, A. W. (1975). *Astrophys. J. Lett.* **201**, L71.
Gingold, R. A. (1974). *Astrophys. J.* **193**, 177.
Härm, R. & Schwarzschild, M. (1975). *Astrophys. J.* **200**, 324.
Iben, I., Jr (1964). *Astrophys. J.* **140**, 1631.
Iben, I., Jr (1967). *Astrophys. J.* **147**, 650.
Iben, I., Jr (1971). *Publ. astron. Soc. Pacific*, **83**, 697.
Iben, I., Jr (1974). *Ann. Rev. Astron. Astrophys.* **12**, 215.
Iben, I., Jr (1975a). *Astrophys. J.* **196**, 525.
Iben, I., Jr (1975b). *Astrophys. J.* **196**, 549.
Iben, I., Jr (1976). *Astrophys. J.* **208**, 165.
Iben, I., Jr (1977). *Astrophys. J.* **217**, 788.
Iben, I., Jr & Truran, J. W. (1978). *Astrophys. J.* **220**, 980.
Kaminisi, K. Arai, K. & Yoshinaga, K. (1975). *Prog. theor. Phys.* **53**, 1853.
Kaminisi, K., Keisuke & Arai, K. (1975). *Physics Reports of the Kumamoto University*, vol. 2, no. 1, 19.
Merrill, P. W. (1952). *Science*, **115**, 484.
Mitalas, R. (1974). *Astrophys. J.* **187**, 155.
Paczyński, B. & Tremaine, S. D. (1977). *Astrophys. J.* **216**, 57.
Parker, P. D. (1968). *Phys. Rev.* **173**, 1021.
Renzini, A. (1977). In *Advanced Stages in Stellar Evolution*, 7th Course of the Swiss Society of Astronomy and Astrophysics, Saas-Fee, ed. P. Bouvier & A. Maeder, p. 149. Geneva Observatory Publ.
Renzini, A., Mengel, J. G. & Sweigart, A. V. (1977). *Astron. Astrophys.* **56**, 369.
Rodgers, A. W. (1978). Results reported at NATO Advanced Study Institute on Globular Clusters, Cambridge.
Rood, R. T. (1973). *Astrophys. J.* **184**, 815.
Różyczka, M. (1978). *Acta Astron.* **28**, 19.
Sanders, R. H. (1967). *Astrophys. J.* **150**, 971.
Scalo, J. M. (1976). *Astrophys. J.* **206**, 474.
Schwarzschild, M. & Härm, R. (1967). *Astrophys. J.* **150**, 961.
Spulak, R. G. (1980). *Astrophys. J.* in press.
Sugimoto, D. (1964). *Prog. theor. Phys.* **32**, 703.
Sweigart, A. V. (1973). *Astron. Astrophys.* **24**, 459.
Sweigart, A. V. (1974). *Astrophys. J.* **189**, 289.
Sweigart, A. V. & Mengel, J. G. (1979). *Astrophys. J.* **229**, 624.
Thomas, H.-C. (1967). *Z. Astrophys.* **67**, 420.
Tomkin, J. & Lambert, D. L. (1979). *Astrophys. J.* **227**, 209.
Truran, J. W. & Cameron, A. G. W. (1971). *Astrophys. Space Sci.* **14**, 179.
Truran, J. W. & Iben, I., Jr (1977). *Astrophys. J.* **216**, 797.
Ulrich, R. K. (1973). In *Explosive Nucleosynthesis*, ed. D. N. Schramm & W. D. Arnett, p. 139. Austin: University of Texas Press.
Warner, B. (1965). *Mon. Not. Roy. astron. Soc.* **129**, 263.
Wood, P. R. (1974). *Astrophys. J.* **190**, 609.

8
Infrared observations of red giants in globular clusters

S. E. PERSSON AND JAY A. FROGEL†

8.1 Introduction

Infrared observations of globular cluster red giants provide a valuable addition to studies of these stars at optical wavelengths (e.g. Glass & Feast 1973a, b, 1977; Cohen, Frogel & Persson 1978; Pilachowski 1978). Bolometric luminosities and accurate effective temperatures can be obtained from broad-band photometric data at J, H, K and L (1.25 μm, 1.65 μm, 2.2 μm, and 3.5 μm). Estimates of the abundances of the important molecules CO and H_2O can also be obtained by making photometric measurements of the strengths of the 1.9 μm H_2O band and the 2.3 μm CO band. Recent advances in InSb detector systems have made stars in the brightest 3 or 4 magnitudes of many clusters accessible to these types of observation. Those clusters which have already been observed include 47 Tuc (Glass & Feast 1973a, b; Frogel & Persson, unpublished data), ω Cen (Glass & Feast 1973a, b, 1977; and this discussion), M3, M13, M71 and M92 (Cohen et al. 1978; Pilachowski 1978; Frogel, Persson & Cohen 1979), M5 (Pilachowski 1978; Frogel & Persson, unpublished data), M10 and NGC 6171 (Pilachowski 1978) and M22 (Frogel & Persson, unpublished data).

In sections 8.2 and 8.3 we discuss the dependence of the observed colours and indices on various physical parameters and review some of the work in the references cited above. In section 8.4 we summarize our results on the red giants in ω Cen.

8.2 Results from broad-band photometry

The first near-infrared JHKL work on globular cluster stars was that of Glass & Feast (1973a, b) who observed a number of giants and red variables in ω Cen and 47 Tuc. Their two-colour plots, particularly J−H

† Guest Investigator at the Hale Observatories.

versus H − K, showed that the energy distributions of the cluster stars depart from those of black bodies, as is the situation for Population I M giants. The explanation for this effect is that the H⁻ opacity, which is dominant at these wavelengths, has a minimum near 1.65 μm, so that we are seeing an 'excess' flux in the H band. Their observations also showed that two of the TiO variables in ω Cen have temperatures considerably cooler than that of the ordinary giant branch stars at the same luminosity.

More recent work by Glass & Feast (1977) has shown that the giant branch of ω Cen is wide in a colour–magnitude diagram of K versus J − K, as is also seen in the (V, B − V) plane. Our observations of ω Cen, discussed in section 8.4 of this chapter, agree with this result.

Recently, several groups have extended near-infrared observations to a larger number of fainter stars in several clusters. Two important parameters obtainable from this type of data are effective temperatures (T_{eff}) and bolometric corrections (BC).

8.2.1 Effective temperatures

Quite accurate values of T_{eff} for cool stars of different metallicities and surface gravities can be obtained from V − K colours corrected for reddening. This is because H⁻ is the dominant opacity source at both V and K, and thus variations in [Fe/H] and lg g largely cancel out. Atmospheric models for metal abundances ranging from solar to 10^{-2} times solar and values of lg g appropriate to globular cluster giants were calculated to calibrate the (V − K)/T_{eff}, (J − K)/T_{eff} and BC/T_{eff} relationships; detailed results are presented in Cohen et al. (1978). Fig. 8.1 shows the calibration for V − K and for R − I. With the measurement accuracy now obtainable, namely ±0.03 mag in V − K, and because of the large wavelength baseline, one can obtain temperatures to better than ±50 K from V − K. Fig. 8.1 shows that for a given [Fe/H] and lg g, a spread of ±0.10 mag in V − K corresponds to ±60 K at 4200 K; a similar spread in R − I corresponds to ±200 K. The dependence of the calibration on metal abundance and gravity is sufficiently weak that negligible additional uncertainty is incurred if these quantities are not well known. In practice the V − K, J − K and BC calibrations for T_{eff} agree to within 50–100 K. The calibration also agrees fairly well with that of Johnson (1966) but disagrees with Osborn's (1973) scale based on DDO photometry. Details are discussed in Cohen et al. (1978).

Since Rayleigh scattering is an important source of opacity in the blue, B − V colours cannot be used to obtain reliable temperatures. If the reddening is uncertain, J−K colours can be used to obtain T_{eff} since $E(J-K) = 0.17E(V-K) = 0.48E(B-V)$, but this determination will suffer from the short baseline of J−K as compared to that of V−K.

8.2.2 Bolometric luminosities

Bolometric corrections and total luminosities are found by simply integrating the observed (reddening-corrected) energy distributions from U to K or L (3600 Å to 2.2 μm or 3.5 μm). For typical giant stars, extrapolation to wavelengths shorter than the U band or longward of K or L does not add appreciable error, nor does the inevitable smoothing over all the absorption features. Because we are interested in the differences between stars in a given cluster, or between clusters having metal abundances down from solar by a factor of 3 or more, these approximations are not critical, and the relative bolometric luminosities are accurate to ±0.05 mag. The situation for the coolest stars, for example the red variables in 47 Tuc and ω Cen, is not so clear-cut, as they may have excess radiation at the longer

Fig. 8.1. Colours versus effective temperature for the model atmosphere results given in Cohen et al. (1978). The two curves in each case show the maximum range in colour for the parameters $-2.0 < $ [Fe/H] $ < 0.0$ and $0.5 < \lg g < 2.0$. Note the scale change between R−I and V−K.

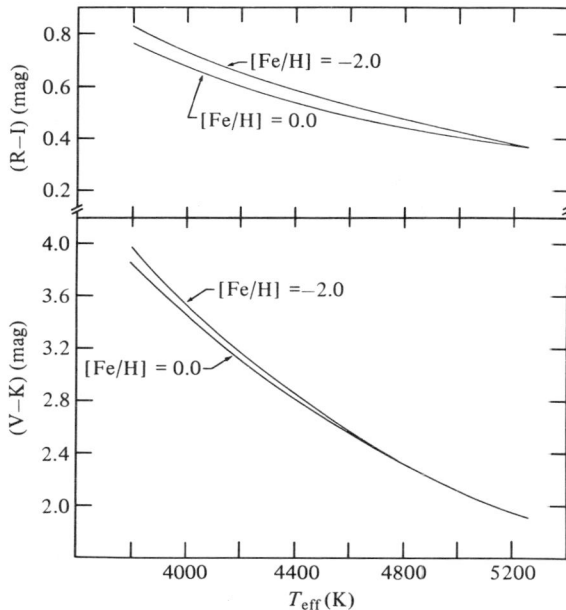

wavelengths arising from circumstellar shells (Glass & Feast 1973a, b, 1977).

8.2.3 The physical HR diagram

Fig. 8.2 presents the $\lg L$ versus $\lg T_{\text{eff}}$ diagram for four clusters, derived directly from the observational data and published distance moduli, while fig. 8.3 adds a number of new stars recently measured in M71 (Frogel *et al.* 1979). Several theoretical tracks due to Rood (1972) are shown for comparison. A number of conclusions can be drawn from these figures. First, the well-known correlation between increasing B−V colour of the giant branch and cluster metallicity holds when the physical parameters, T_{eff} and M_{bol}, are examined. An exception to this is that the locus of the M3 stars lies to the right of the M13 locus, though both clusters have nearly identical abundances. (For M3, [Fe/H] = −1.8, while for M13, [Fe/H] = −1.6 (Cohen 1978). These values agree with the results of Searle & Zinn (1978).) Secondly, the luminosities of the brightest giants in all of the clusters are nearly the same. This is quite different from the situation in V versus B−V diagrams and arises because of the increasing redness

Fig. 8.2. The theoretical HR diagram for four clusters (Cohen *et al.* 1978) and two evolutionary tracks from Rood (1972). The Rood tracks correspond to $M = 0.8\,M_\odot$, $X = 0.7$, $Y = 0.3$, and $Z/Z_\odot = 0.005$ (track 1) and 0.05 (track 2). ○, M67; △, M3; ●, M13; +, M92.

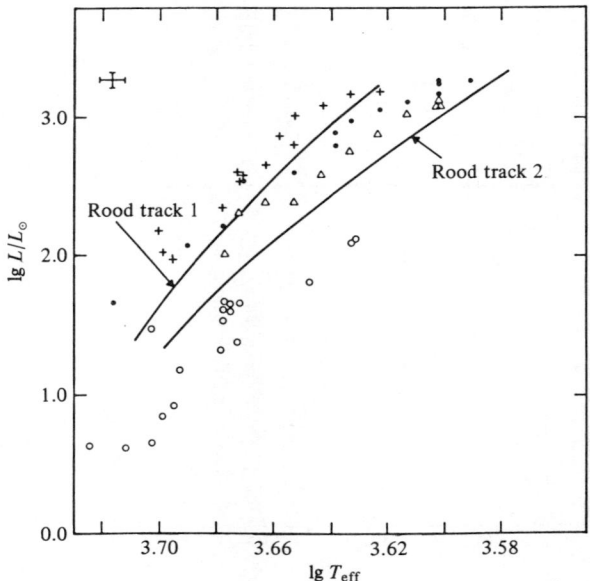

and BCs for the coolest stars in the more metal-rich clusters. Thus, while the V magnitudes of cluster giants may not be useful as distance indicators, their bolometric magnitudes (or their K magnitudes) probably are. Thirdly, in comparing the observational data with theory, we note that the agreement in the general location of the tracks shows that Rood's choice of mixing length to set the zero-point of the $T_{\rm eff}$ axis was reasonable. The Yale tracks (Ciardullo & Demarque 1977) do not fit quite as well along the $T_{\rm eff}$ axis; this is discussed by Cohen et al. (1978). Also, the observed luminosities at the red-giant tip agree well with the termination of Rood's tracks.

An apparent disagreement between the observations and the tracks in both fig. 8.2 and 8.3 is a systematic deviation in the sense that the coolest stars are observed to be cooler than the tracks, while the lower luminosity, hotter stars are hotter than the tracks. This effect was also noted in 47 Tuc by Lloyd Evans (1974) who transformed Rood's theoretical tracks to an ordinary colour–magnitude diagram by means of Johnson's (1966) calibrations. Figs. 8.2 and 8.3 show that this difference between theory and observation is present in all of the clusters observed.

8.3 Results from narrow-band observations

Two narrow-band indices have been measured in a number of stars in different globular clusters (Cohen et al. 1978; Frogel et al. 1979; Pilachowski 1978). These are the CO index (a 2.20–2.36 μm colour; Baldwin,

Fig. 8.3. Same as fig. 8.2, but with the addition of the M71 stars (Frogel et al. 1979), and a metal-rich track, denoted RG5, with $Z/Z_\odot = 0.5$, from Rood (1972). The tracks denoted RG1 and RG2 correspond to tracks 1 and 2 in fig. 8.2. The small enclosed area is that of the M67 clump.

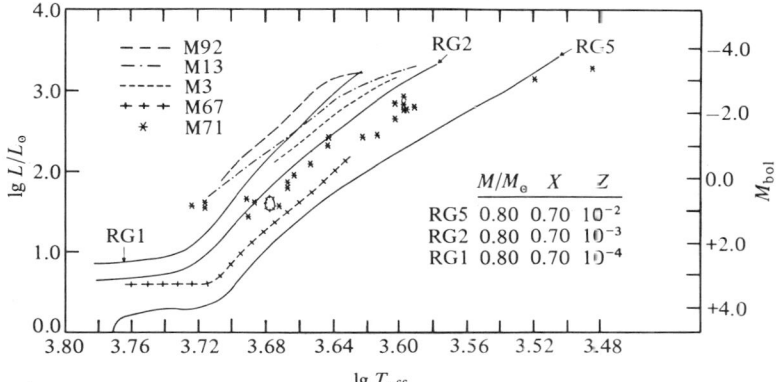

Frogel & Persson 1973) and an H_2O index (a 2.20–2.00 μm colour; Aaronson, Frogel & Persson 1978). Here we will consider the former. The CO index measures the strength of the absorption due to the first overtone band of CO. For Population I G, K and M stars the CO index increases with luminosity for giants and supergiants of the same effective temperature and generally increases with decreasing temperature for stars of the same luminosity class. The CO index is near zero for dwarf stars of all effective temperatures (Frogel 1971; Baldwin et al. 1973; Persson, Aaronson & Frogel 1977). In metal-poor stars, the CO index also varies with metal abundance at a given T_{eff}, as discussed below. These various dependences arise from several effects: saturation of the bands in metal-rich stars; a faster than linear change in CO column density with abundance in metal-poor stars; and turbulent broadening of the bands in giants and

Fig. 8.4. (a) The observed CO index plotted as a function of $(V-K)_o$ colour (corrected for reddening) for four clusters in Cohen et al. (1978). The mean dwarf and giant lines are from observations discussed by Persson et al. (1977) and Frogel et al. (1978). The single error bar refers to all the measurements in this and subsequent plots. (b) The behaviour of the H_2O index, not further discussed. ○, M67; △, M3; ●, M13; +, M92.

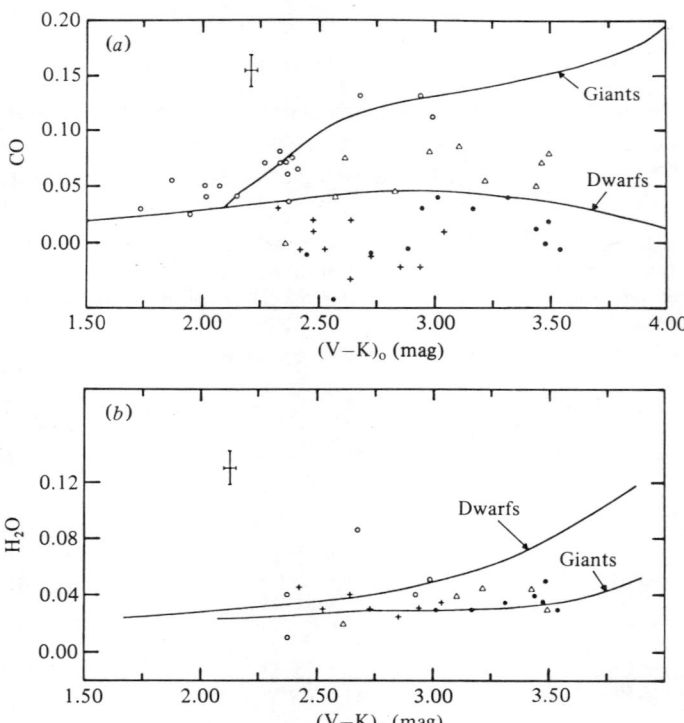

supergiants. The H_2O indices of metal-poor stars are no larger than would be expected from the slope of the energy distributions near 2 μm, and we shall not discuss them further.

Figs. 8.4 and 8.5 display the CO indices for a number of cluster stars, plotted as a function of V−K colour. Two of the important conclusions of Cohen et al. (1978), Frogel et al. (1979) and Pilachowski (1978), are evident from these figures. First, the distribution of CO index within a cluster is fairly narrow, and the mean index increases with metal abundance going from M92 to M71, M67 or the field giants. The M71 stars have CO indices intermediate between those of the M67 stars and the metal-poor clusters, which is consistent with M71 having a metal abundance *less* than that of M67.

Secondly, the individual M3 stars have consistently larger CO indices than do those of M13. These two clusters have very nearly the same metal abundance (Cohen 1978), and thus they clearly depart from the general and expected strengthening of the CO bands with metal abundance. A similar effect is seen in a comparison of M5 and M10 (Pilachowski 1978), two other clusters with similar [Fe/H]. The M5 giants have CO strengths similar to those of the M3 stars while the M10 giants are like those in M13. As mentioned in Cohen et al. (1978), and discussed in detail by Pilachowski

Fig. 8.5. CO versus $(V-K)_o$ for the envelopes of the distributions of the individual stars in clusters studied by Cohen et al. (1978), and for individual M71 stars from Frogel et al. (1979). The T_{eff} scale is from fig. 8.1. The effect of a ± 0.05 mag uncertainty in $E(B-V)$ for M71 is indicated by the arrow. The numbers near the M71 stars 29 and B are their $(V-K)_o$ colours.

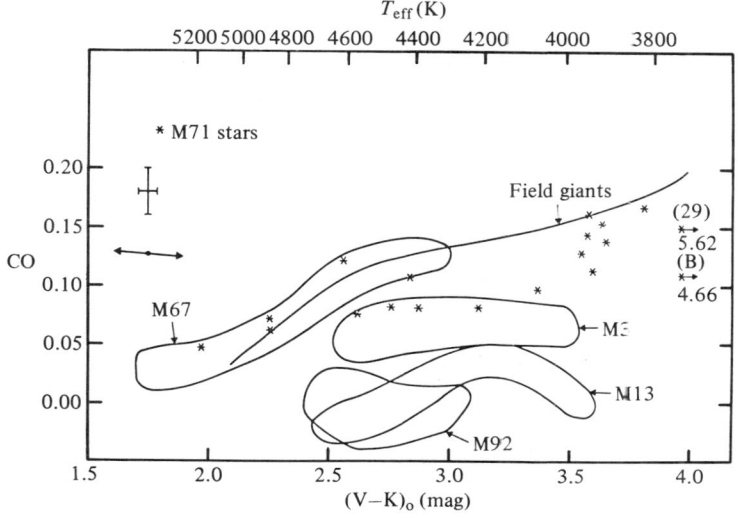

(1978), this anomaly could be related to the differences in horizontal branch morphology – M5 and M3 have both red and blue horizontal branch stars and M13 and M10 have only blue. Since differences in CO index strengths are probably directly related to a difference in [CNO/Fe] in these clusters, the data agree with the idea that CNO is the cause of second-parameter effects in globular clusters (Castellani & Tornambè 1977; Hartwick & McClure 1972; the review by Renzini 1977; and Kraft, p. 92, this volume). Alternatively, the CO strengths and relative populations of the red and blue sides of the horizontal branches could be symptomatic of a more subtle common mechanism. We note that CNO variations do not explain the shifts in giant branch locations of M3 and M13 in figs. 8.2 and 8.3 (cf. Renzini 1977).

8.4 Infrared observations of giants in ω Cen

One of the most peculiar globular clusters known is ω Cen (NGC 5139). Its distinctive characteristics include a broad giant branch (in a V versus $B-V$ diagram), evidence for a wide range in Ca II abundance among the RR Lyrae stars, many stars with both weak and strong spectral features due to CN, CH, and s-process elements, and the presence of a number of red variables (Dickens & Woolley 1967; Cannon & Stobie 1973; Dickens, Feast & Lloyd Evans 1972; Eggen 1972; Freeman & Rodgers 1975; Dickens & Bell 1976; Bessell & Norris 1976a, b; Norris, p. 113, this volume; Freeman, p. 105, this volume). Attempted explanations of these phenomena by the authors cited above and by others have generally centred around a combination of (i) primordial abundance variations among the stars, and (ii) mixing of the products of various nuclear reactions to the surfaces of the giants (see for example, Norris, p. 119, this volume). The more fundamental problem of why ω Cen is unique among galactic globular clusters has so far been approached only in a qualitative fashion with explanations usually involving its large mass, its ellipticity, and possible high stellar rotational velocities (e.g. Bessell & Norris 1976a, b).

In sections 8.2 and 8.3 we discussed the contributions that infrared photometry can make to the study of ordinary globular clusters. Here we shall present preliminary results of an infrared study of the giant stars in ω Cen. Fifty-five stars on the upper giant branch were observed with the 2.5-metre telescope at Las Campanas Observatory. The selection of stars and UBV photometry were taken from Cannon & Stobie (1973). No known field stars were included. All of the observed colours and indices were corrected for a reddening corresponding to $E(B-V) = 0.11$.

Fig. 8.6. CO versus $(V-K)_o$ for individual stars in ω Cen. Some stars are identified with their ROA numbers in this and subsequent plots.

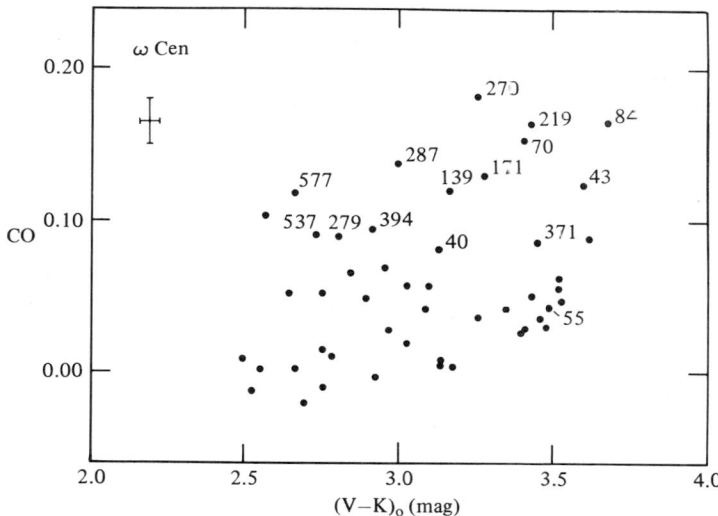

Fig. 8.7. CO versus $(V-K)_o$ for the envelope of the distribution of the ω Cen stars compared to those for the other clusters (see also fig. 8.5).

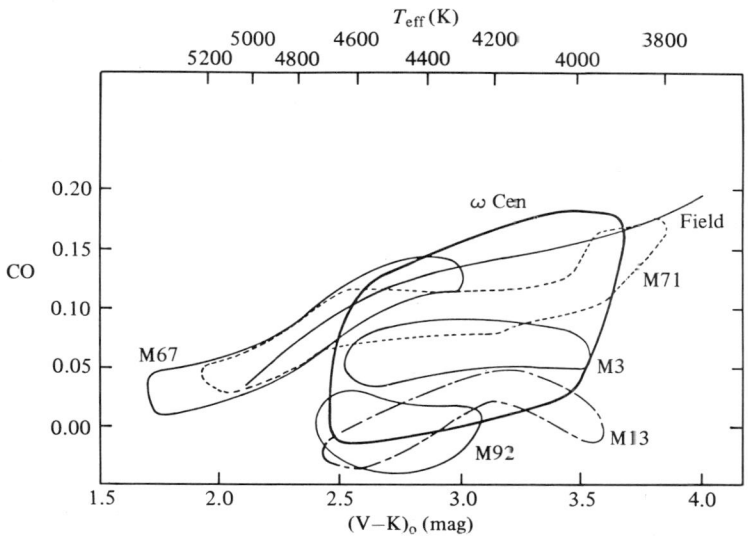

8.4.1 The CO indices

Figs. 8.6 and 8.7 illustrate the first set of results. These figures show the dependence of CO on $(V-K)_o$ for the stars in ω Cen and a comparison of this dependence with that obtained for the other clusters. For the ω Cen giants we see a uniform spread of CO index upwards from a well-defined lower bound for the entire range of $(V-K)_o$ sampled by our data. This is in marked contrast to the giants in the other clusters studied. In fact, at a given $(V-K)_o$, the ω Cen stars display CO indices which range from those characteristic of the extremely metal-poor cluster M92 to greater than those characteristic of the slightly metal-poor cluster M71 or of field stars. Several stars having strong CO, for example ROA 40 and ROA 70, also have strong CN and CH, while the most extreme carbon star ROA 55 does not have particularly strong CO. There is thus no one-to-one correspon-

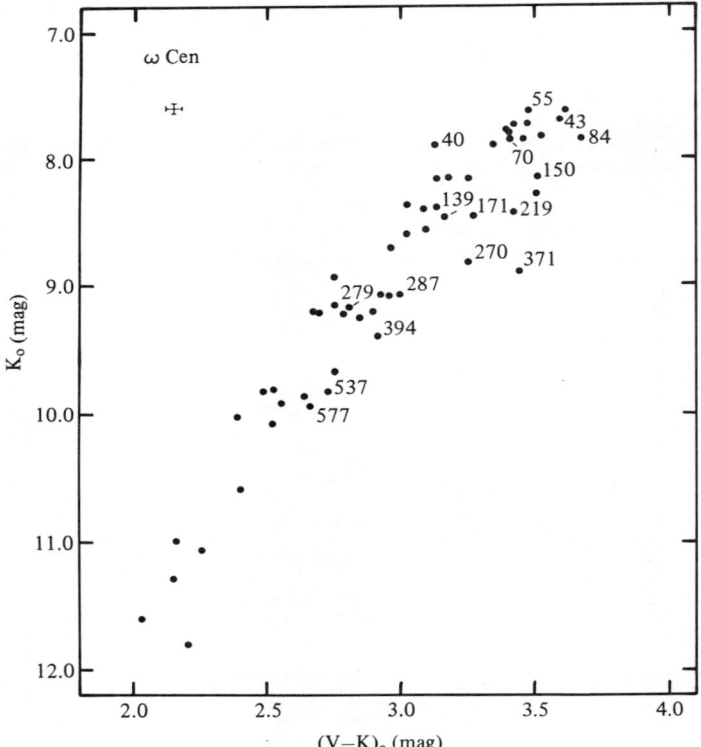

Fig. 8.8. An infrared colour–magnitude diagram for the red giants in ω Cen.

dence between the presence of s-process material on the surface, or strong CH, CN or C_2 bands, and strong CO absorption.

8.4.2 Colour–magnitude diagrams

The second set of results is illustrated in figs. 8.8 and 8.9. Fig. 8.8 presents a K_o versus $(V-K)_o$ colour-magnitude diagram for the ω Cen giants. At a given luminosity there is a spread in $(V-K)_o$ of about 0.4 mag which corresponds to 300 K in T_{eff}. (For the other clusters, the maximum spread is only 0.1 mag in $(V-K)_o$.) This result is consistent with that of Glass & Feast (1973a, b, 1977). Furthermore, the spread in $(V-K)_o$ occurs at all luminosities on the giant branch, as does the spread in $B-V$. The dispersion in ω Cen exceeds the range in colour shown by M92, M3 and M13.

In order to construct fig. 8.9, we integrated the energy distributions of the ω Cen stars to yield bolometric corrections and luminosities, as described in section 8.3. Effective temperatures were derived from the

Fig. 8.9. The physical HR diagram for the ω Cen giants compared to the mean lines for the clusters in Cohen *et al.* (1978) and Frogel *et al.* (1979). The individual stars in M3, M13, M92 and M71 are plotted in fig. 8.2 and 8.3. The derivation of bolometric luminosities and effective temperatures is discussed in the text.

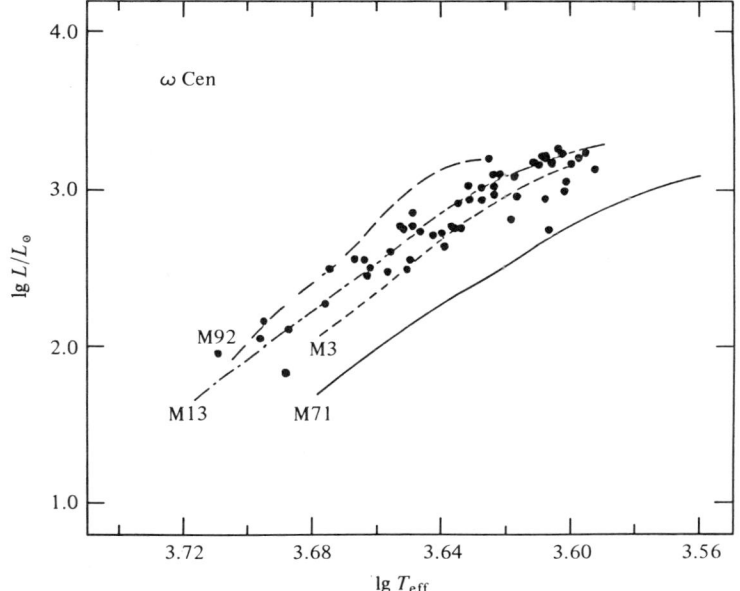

calibration given in fig. 8.1. Because the model atmosphere results show that reliable temperatures can be obtained even for stars of solar composition, we do not expect that blanketing, backwarming, or flux redistribution effects should be very serious for the non-extreme ω Cen stars. However, the marked spectral peculiarities present in some of the ω Cen stars, for example ROA 55 and ROA 70, may introduce errors into the values of T_{eff} derived from $(V-K)_0$. The (true) distance modulus of ω Cen was taken to be 13.6 mag (Harris 1976).

The blue edge of the distribution of the ω Cen stars in fig. 8.9 is close to the distribution for M92, as expected for the low metal abundance deduced from spectra (Freeman & Rodgers 1975), from periods of RR Lyrae variables, and from the height of the giant branch tip above the horizontal branch (Dickens & Woolley 1967). This agrees with the conclusions of Bessell & Norris (1976a, b) who compared the two clusters in V versus B−V diagram. The spread in $(V-K)_0$ at a given K_0 is, of course, preserved in fig. 8.9. A plausible explanation of the observed spread is that it is due to a range in heavy metal abundance. The ω Cen giants span the region occupied by M92, M3, and M13 in fig. 8.9; a few of the

Fig. 8.10. Excess CO absorption ΔCO versus excess $(V-K)_0$ colour $\Delta(V-K)$. The calculation of these quantities is described in the text. The two smooth curves give the expected change of mean CO index with colour in going from M92 to M71 via M3 (upper curve) or M13 (lower curve). The tick marks on the curves locate the mean CO indices for the M3 and M13 stars read at a $(V-K)_0$ where $\lg L/L_\odot = 2.8$ in a K_0 versus $(V-K)_0$ diagram.

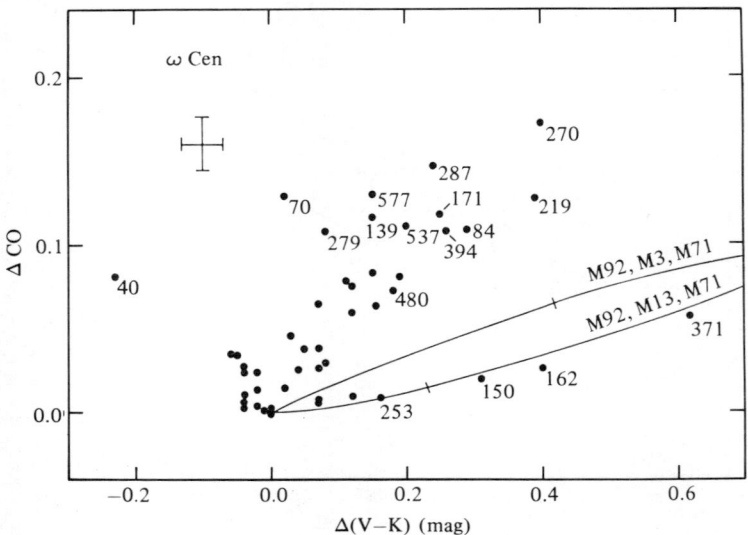

stars in our sample are nearly as red as the M71 locus. We may thus conclude that [Fe/H] probably has a range exceeding unity in ω Cen. This is consistent with evidence presented by Norris (p. 116, this volume) and Freeman (1978).

8.4.3 Comparison of the spread in different colours and indices

To compare the spread in colours on the giant branch, let us first define ΔCO as the excess CO absorption above the lower boundary of the distribution in fig. 8.7, and $\Delta(V-K)$ as the excess colour measured from the blue edge of the distribution in a K_o versus $(V-K)_c$ diagram, which is close to the locus of the M92 stars. Fig. 8.10 shows the relationship between ΔCO and $\Delta(V-K)$.

It is clear from this plot that most of the stars having strong CO absorption compared to M92 lie on the redward side of the giant branch in a K_o versus $(V-K)_o$ diagram. How much of this effect is due *just* to

Fig. 8.11. Excess $(B-V)_o$ colour, $\Delta(B-V)$, versus $\Delta(V-K)$. The calculation of $\Delta(B-V)$ values is described in the text. The smooth curve connects the mean location of the giant branch in K_o versus $(B-V)_o$ and K_o versus $(V-K)_o$ diagrams for the other clusters, as described in the text. The tick marks on the smooth curve locate the giant branches of M3 and M13 in these colour–magnitude diagrams, with the colours read off at $\lg L/L_\odot = 2.8$.

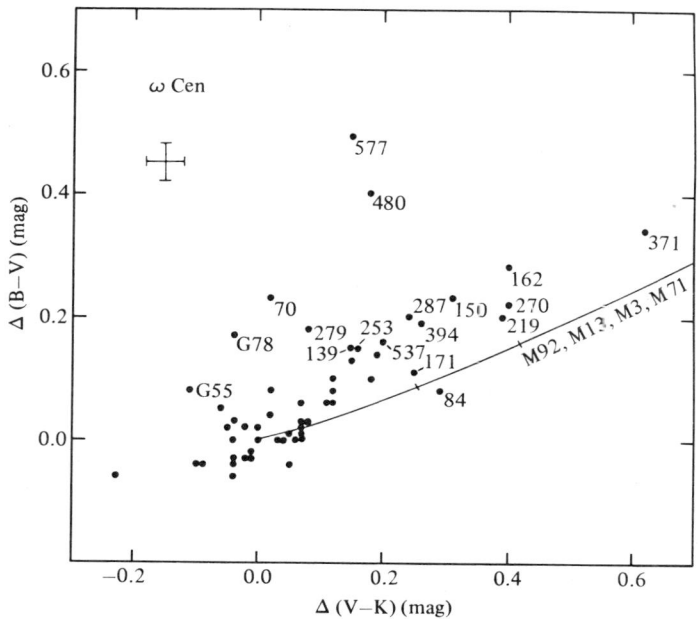

the expected increase of CO strength with decreasing temperature and increasing metallicity? The two lines in the diagram represent the limits of the average enhancement of CO with $V-K$ in going from M92 through M3, M13 and M71. Thus, the increase of excess CO strength with excess $V-K$ colour is substantially steeper for the ω Cen giants than we find in going from M92 to M71, and we may say that half of the stars observed have evidence for CO strength *in excess* of that expected for their temperature and metallicity. The lack of a unique correspondence between the presence of s-process elements and excess CO absorption shows up again in this plot: the stars ROA 371 (s-process plus strong CN) and ROA 162 (s-process) have very weak CO, and happen to lie close to the mean relationship for the cluster-to-cluster variation. Also, several stars having s-process enhancements lie within the area occupied by most of the stars.

To compare the spread in $(V-K)_0$ with that in $(B-V)_0$, we take the excess $(B-V)_0$ colour from the blue edge of the distribution in a K_0 versus $(B-V)_0$ diagram. The resulting $\Delta(B-V)$ excesses are plotted against the $\Delta(V-K)$ excesses in fig. 8.11. We expect a correlation here, simply because the cooler effective temperatures of the stars with large $\Delta(V-K)$ should produce a reddening of $(B-V)_0$. The lines on fig. 8.11 represent the observed dependence of the shift in giant branch location in $V-K$ and $B-V$ going from M92 to M3 and M13. We conclude that much of the $B-V$ spread in a V versus $B-V$ diagram is produced by the spread in T_{eff} at a given luminosity caused by variations in [Fe/H].

Nevertheless, there are several stars which have a *further* excess $B-V$ colour. These stars may be affected by blanketing in B due to CH, CN or the continuous opacity source found by Bond & Neff (1969) in Ba II stars (see also Bond 1974) and advocated by Bessell & Norris (1976a, b) as a partial explanation for the spread in $B-V$.

8.4.4 Conclusions

We conclude that the infrared observations of the ω Cen giants can be most easily explained if there is a range in primordial [Fe/H]. The CO index displays a wide variation at all $V-K$ which is due either to variable mixing of triple-α carbon to the surfaces of the stars, or to a primordial variation in CNO/Fe. If the variation is primordial, then CNO/Fe increases with Fe, because the variation of CO with [Fe/H] in ω Cen is steeper than found for the cluster-to-cluster variation. This goes in the opposite sense to that found for CO and [Fe/H] in M3 and M13. Some of the spread in $B-V$ can be explained by a spread in T_{eff} at a given luminosity; however, some stars have a further excess in $B-V$ colour.

Finally, we might have expected to see some red horizontal branch stars in ω Cen because of the range in [Fe/H], and yet none are seen. If CNO is indeed the 'second parameter', then the range in CO index in ω Cen might also lead us to expect some red horizontal branch stars.

We wish to thank our collaborators M. Aaronson, J. Cohen and K. Matthews for help in various phases of this research. This work was supported in part by NSF grant AST 76-22676.

References

Aaronson, M., Frogel, J. A. & Persson, S. E. (1978). *Astrophys. J.* **220**, 442.
Baldwin, J. R., Frogel, J. A. & Persson, S. E. (1973). *Astrophys. J.* **184**, 427.
Bessell, M. S. & Norris, J. (1976a). *Astrophys. J.* **208**, 369.
Bessell, M. S. & Norris, J. (1976b). *Astrophys. J.* **210**, 618.
Bond, H. E. (1974). *Astrophys. J.* **194**, 95.
Bond, H. E. & Neff, J. S. (1969). *Astrophys. J.* **158**, 1235.
Cannon, R. D. & Stobie, R. S. (1973). *Mon. Not. Roy. astron. Soc.* **162**, 207.
Castellani, V. & Tornambè, A. (1977). *Astron. Astrophys.* **61**, 427.
Ciardullo, R. B. & Demarque, P. (1977). *Trans. astron. Obs. Yale Univ.* **35**.
Cohen, J. G. (1978). *Astrophys. J.* **223**, 487.
Cohen, J. G., Frogel, J. A. & Persson, S. E. (1978). *Astrophys. J.* **222** 165.
Dickens, R. J. & Woolley, R. v.d. R. (1967). *Roy. Obs. Bull.* no. 128.
Dickens, R. J., Feast, M. W. & Lloyd Evans, T. (1972). *Mon. Not. Roy. astron. Soc.* **159**, 337.
Dickens, R. J. & Bell, R. A. (1976). *Astrophys. J.* **207**, 506.
Eggen, O. J. (1972). *Astrophys. J.* **172**, 639.
Freeman, K. C. (1978). Results reported at NATO Advanced Study Institute on Globular Clusters, Cambridge.
Freeman, K. C. & Rodgers, A. W. (1975). *Astrophys. J. Lett.* **201**, L71.
Frogel, J. A. (1971). *Ph.D. thesis*, California Institute of Technology.
Frogel, J. A., Persson, S. E., Aaronson, M. & Matthews, K. (1978). *Astrophys. J.* **220**, 75.
Frogel, J. A., Persson, S. E. & Cohen, J. G. (1979). *Astrophys J.* **227**, 499.
Glass, I. S. & Feast, M. W. (1973a). *Mon. Not. Roy. astron. Soc.* **163**, 245.
Glass, I. S. & Feast, M. W. (1973b). *Mon. Not. Roy. astron. Soc.* **164**, 424.
Glass, I. S. & Feast, M. W. (1977). *Mon. Not. Roy. astron. Soc.* **181**, 509.
Harris, W. E. (1976). *Astron. J.* **81**, 1095.
Hartwick, F. D. A. & McClure, R. D. (1972). *Astrophys. J. Lett.* **176**, L57.
Johnson, H. L. (1966). *Ann. Rev. Astron. Astrophys.* **4**, 193.
Lloyd Evans, T. (1974). *Mon. Not. Roy. astron. Soc.* **167**, 393.
Osborn, W. (1973). *Astrophys. J.* **186**, 725.
Pilachowski, C. A. (1978). *Astrophys. J.* **224**, 412.
Persson, S. E., Aaronson, M. & Frogel, J. A. (1977). *Astron. J.* **82**, 729.
Renzini, A. (1977). In *Advanced Stages in Stellar Evolution*, 7th Course of the Swiss Society of Astronomy and Astrophysics, Saas-Fee, ed. P. Bouvier & A. Maeder, p. 149. Geneva Observatory Publ.
Rood, R. T. (1972). *Astrophys. J.* **177**, 681.
Searle, L. & Zinn, R. (1978). *Astrophys. J.* **225**, 357.

9
Infrared observations of globular clusters in M31 and a comparison with galactic globulars and elliptical galaxies

JAY A. FROGEL,[†] S. E. PERSSON AND JUDITH G. COHEN

9.1 Introduction

The usefulness of near infrared photometric data for the study of elliptical galaxies and of globular clusters in our galaxy has been demonstrated (Frogel, Persson, Aaronson & Matthews 1978; Aaronson, Frogel & Persson 1978b; Aaronson, Cohen, Mould & Malkan 1978a; Cohen, Frogel & Persson 1978; Strom *et al.* 1976). The basic parameters – broad band $V-K$ colours and CO and H_2O absorption indices – are sensitive to or can set constraints on bolometric luminosities, mean metal abundances, relative numbers of dwarf and giant stars, and giant branch luminosity functions.

A comparison of the photometric properties of globular clusters with those of elliptical galaxies is of interest because of the possibility of synthesizing galaxian light from that of a variety of individual clusters. This is particularly true for the Milky Way globulars since their integrated light can presumably be understood by a study of their individual stars. The globular clusters associated with M31, however, have two properties which make them easier to study than the Milky Way clusters: they are all at the same distance; and their integrated light can be measured quite easily. An examination of the M31 globulars is also valuable in its own right since there are few extra-galactic systems whose associated globular clusters can be easily studied. Yet, since the detailed study of them by van den Bergh (1969), little more has been done.

After a brief review of some of the conclusions reached in the references cited above, we present preliminary results of a study of the infrared photometric properties of the brightest M31 clusters. This work is being carried out with the 5-metre Hale telescope. We will compare these clusters with galactic globulars (Aaronson *et al.* 1978a) and with early-type

[†] Guest Investigator at the Hale Observatories.

galaxies (Frogel et al. 1978) to investigate what new conclusions can be drawn from the infrared data.

9.2 The infrared indices

9.2.1 CO and H_2O indices

As discussed by Persson & Frogel (p. 147, this volume), G, K and M giants display absorption near 2.4 μm due to CO. This absorption increases with luminosity for galactic giants and supergiants of the same effective

Fig. 9.1. The relationships between $(V-K)_0$, CO and [Fe/H] for those galactic globulars which have abundance determinations based on observations of individual stars. The $(V-K)_0$ and CO values are adapted from Aaronson et al. (1978a), while the [Fe/H] values are from Searle & Zinn (1978) and from Butler (1975).

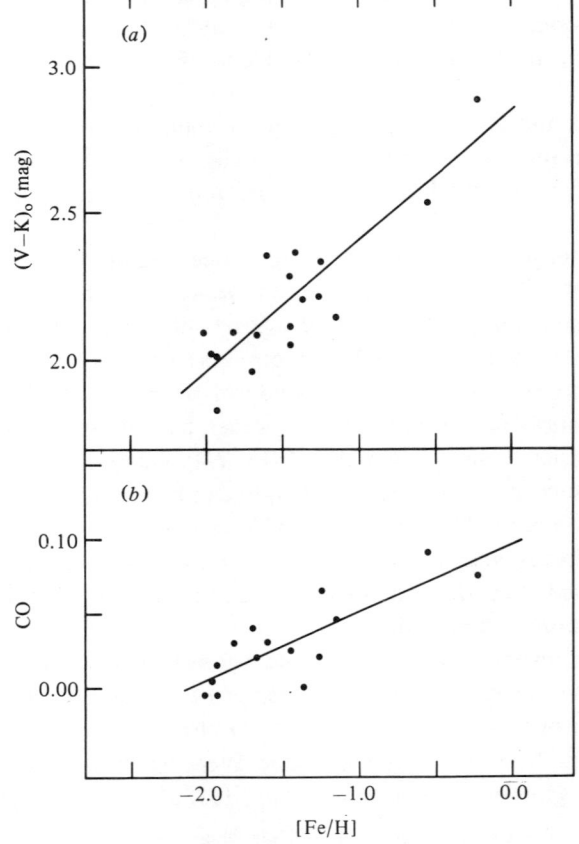

temperature and increases with decreasing temperature for stars of the same luminosity class. The CO absorption is much weaker in dwarfs than in giants for all effective temperatures (Frogel 1971; Baldwin, Frogel & Persson 1973; Persson, Aaronson & Frogel 1977). In globular cluster stars, the CO index is additionally dependent on metal abundance at a given $T_{\rm eff}$ (Cohen et al. 1978; Frogel, Persson & Cohen 1979; Pilachowski 1978).

CO indices measured in the integrated light of elliptical galaxies are found to be strong and help to quantify the dominance of giant light over dwarf light in these systems. Furthermore, the CO index is only weakly dependent on galaxian luminosity over a range of 6 mag in $M_{\rm V}$ (Frogel et al. 1978, and in preparation). For globular clusters in our galaxy, the CO index varies strongly with [Fe/H] over the range $-2.0 <$ [Fe/H] < -0.3. Fig. 9.1(b) shows the dependence of CO index on [Fe/H] for those galactic globular clusters observed by Aaronson et al. (1978a) and with metallicities determined by Searle & Zinn (1978) or by Butler (1975). The best linear fit to the data gives

$$\text{CO} = 0.096 + 0.045 \text{ [Fe/H]}, \qquad (1)$$

which is close to that derived by Aaronson et al. (1978a) from somewhat different sources of metallicity determinations.

Measurements of the 1.9 μm H_2O absorption band complement the CO band data since H_2O is quite strong in M dwarfs but weak in K and early M giants. Only in giants later than M4 does this band become strong, and it is then extremely sensitive to effective temperature. For example, for giants between M2 and M7, the H_2O index increases from 0.06 to 0.36 mag, while CO increases only from 0.20 to 0.24 mag (Aaronson et al. 1978b; Persson et al. 1977). Absorption due to H_2O is only weakly evident in the integrated light of Milky Way globulars (Aaronson et al. 1978a), but is quite strong in early-type galaxies (Aaronson et al. 1978b). In fact it is so strong that about 40 per cent of the 2 μm light in these galaxies has to arise from M4–M6 giants.

9.2.2 $V-K$ colour

The $V-K$ colours of giant stars are primarily sensitive to effective temperature and independent of metallicity for a range in metallicity from solar to 0.01 times solar or lower (Cohen et al. 1978). Since the K light, and to a lesser extent the V light of a cluster is dominated by stars on the giant branch, we expect that the $V-K$ colours of clusters with similar ages should correlate well with metal abundance since the giant branch shifts to hotter temperatures in metal-poor clusters. (Older clusters are also expected to have redder V−K colours because the relative number of red giant stars increases with age, and the main sequence turnoff is redder.) This is indeed the case for globular clusters in our galaxy (Frogel et al. 1978; Aaronson et al. 1978a). Fig. 9.1(a) displays the dependence of $(V-K)_o$, the de-reddened colour, on [Fe/H] for galactic globulars. Again, the photometry is from Aaronson et al. (1978a) and the metallicity determinations are from Searle & Zinn (1978) and Butler (1975). This correlation differs from the finding of Grasdalen (1974) as discussed by Aaronson et al. (1978a) and by Pritchet (1977). The best linear fit to these data is

$$(V-K)_o = 2.84 + 0.44 \,[\text{Fe/H}], \qquad (2)$$

which is not significantly different from that derived by Aaronson et al. (1978a). This result will be used later to derive metallicities for the M31 clusters.

The V−K colours of E and S0 galaxies depend on metal abundance as well (e.g. Strom et al. 1976; Frogel et al. 1978). Because of the dependence of metal abundance on M_V for these galaxies (e.g. Faber 1973), one expects a correlation between V−K and M_V. That this is indeed the case can be seen from observations of E and S0 galaxies in the Virgo and Coma clusters

Fig. 9.2. An example of the dependence of integrated $(V-K)_o$ colours of E and S0 galaxies on M_V. The data plotted are for the Virgo and Coma galaxies from Persson et al. (1979).

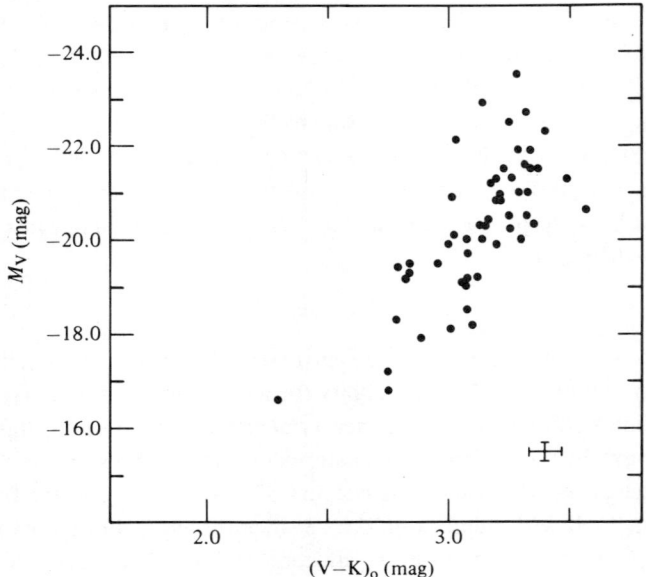

(Persson, Frogel & Aaronson, 1979, and in preparation) presented in fig. 9.2. The use of these correlations to obtain mean metallicities for galaxies has been discussed by Aaronson et al. (1978a) and by Tinsley (1978).

9.3 The M31 globulars

9.3.1 Reddening corrections

Many of the clusters in M31 that we have observed lie in regions of that galaxy where there is evidence for a considerable amount of extinction. A proper quantitative analysis of the infrared photometry requires that this extinction (and reddening) be taken into account for each cluster separately. From multichannel spectrophotometric data, Searle (1978) has computed a number of reddening-free parameters which are sensitive to the intrinsic colours and metallicities of the M31 globulars. His reddening-free calcium index Q_K correlates well with the 5000–8000 Å continuum slope for unreddened clusters. At a given Q_K, comparison of this continuum slope with the zero-reddening slope gives a value for the extinction. If, for the present, we assume that van de Hulst's curve No. 15 (Johnson 1966) holds in M31, then we find that $E(V-K)$ is typically less than 0.35, although some clusters have values as much as three times greater. Our subsequent discussion is based on individually corrected cluster colours and magnitudes.

9.3.2 A metallicity calibration for the M31 globulars

We will assume that Searle's (1978) reddening-free calcium absorption strength indicator Q_K is proportional to [Fe/H]. Fig. 9.3 shows Q_K as a function of $(V-K)_o$. This relationship is tight with the scatter of points consistent with that expected from observational errors alone. A linear fit to these data gives
$$(V-K)_o = 1.20 + 1.32\, Q_K. \tag{3}$$

If now we assume that the same relation between $(V-K)_o$ and [Fe/H] holds for the M31 globulars as was found for the galactic ones (eq. (2)), we can obtain [Fe/H] directly from the $(V-K)_o$ colours of the M31 clusters and provide a calibration for Searle's Q_K values. First, for the calibration we obtain
$$Q_K = 1.24 + 0.33\, [\text{Fe/H}]. \tag{4}$$

Secondly, from the $(V-K)_o$ colours, we conclude (fig. 9.3) that the metal abundance of the M31 globulars ranges from about twice solar to about a factor of 100 down from solar, in qualitative agreement with the results of van den Bergh (1969) and Spinrad & Schweizer (1972). The six M31 clusters measured by Harris & Canterna (1977) are among the ones we have measured. Their values for [Fe/H] do not differ systematically from ours, but the scatter is nearly ±0.4 in [Fe/H].

We emphasize that the validity of any determination of [Fe/H] for the M31 globulars via a transferral of an empirically derived $(V-K)$ versus [Fe/H] relation for the galactic globulars is dependent on at least two assumptions: (i) the M31 clusters have the same stellar luminosity function as the galactic ones; and (ii) both cluster systems have the same age. An age difference of 5×10^9 yr, however, would cause a difference of only 0.1 mag in $(V-K)_o$ (Struck-Marcell & Tinsley 1978).

Fig. 9.3. Searle's (1978) reddening-free metallicity parameter Q_K plotted against reddening-corrected $(V-K)_o$ colour. The best linear fit to these data is displayed. The dependence of [Fe/H] on $(V-K)_o$ is that derived from the galactic globulars in fig. 9.1(a). Typical one sigma standard error bars are shown for the observed quantities.

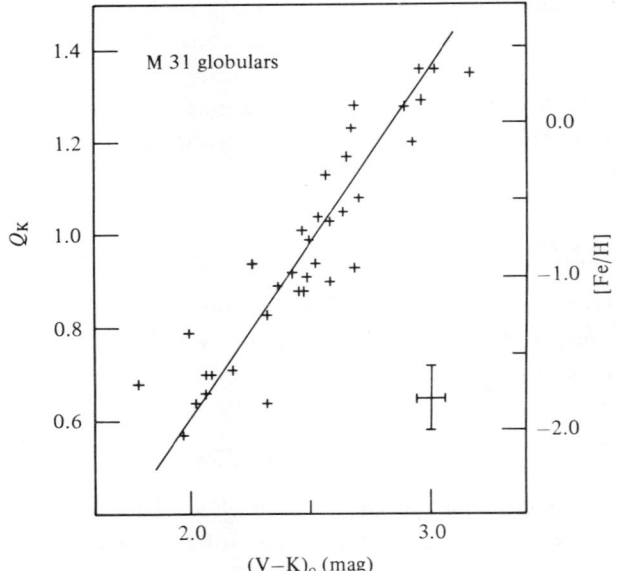

9.4 A comparison of the photometric properties of globular clusters and early-type galaxies

9.4.1 Line-strengths and $(V-K)_o$

Van den Bergh's (1969) line-strength parameter L is the one metallicity parameter presently available for both the M31 and galactic globular cluster systems. Fig. 9.4 displays this parameter as a function of $(V-K)_o$ for the two cluster systems. The relation between [Fe/H] and $(V-K)_o$ is from eq. (2). Although there are few metal-rich galactic globulars having infrared photometry, we see that the two cluster systems are indistinguishable. This does not imply that there are no differences in the distribution

Fig. 9.4. Van den Bergh's (1969) L indices plotted against $(V-K)_o$ colours for M31 clusters (this work) and galactic globular clusters (Aaronson et al. 1978a). The enclosed area contains 18 galactic globulars. A reddening vector is displayed. Typical one sigma standard error bars are indicated. The smaller of the two $(V-K)_o$ errors applies to the M31 clusters. The metallicity calibration is from eq. (2).

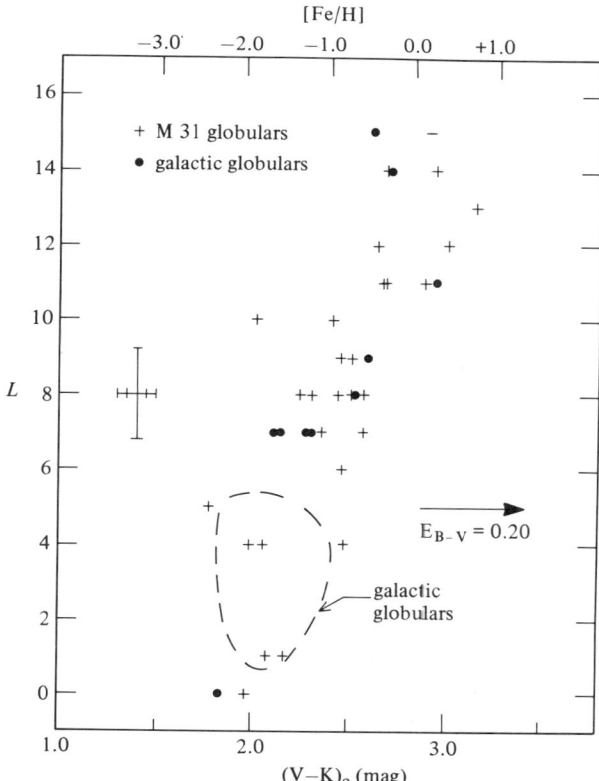

over metallicity between the M31 globulars and the galactic ones (van den Bergh 1967, 1969, and p. 183, this volume; Hanes 1977, and p. 233, this volume; and Searle 1978).

9.4.2 $(U-V)_o$ versus $(V-K)_o$

Fig. 9.5 contains all of the M31 globulars we have observed for which there are photoelectric UBV data available (van den Bergh 1969). Also shown are the areas occupied by the early-type galaxies (from Persson *et al.* 1979) and the galactic globulars (from Aaronson *et al.* 1978a). For these two types of objects, data points which lie outside of the enclosed areas are also displayed.

The M31 clusters clearly overlap the region occupied by the galactic ones. Differences in the relative numbers of the Milky Way and M31 clusters as a function of colour are in part due to observational bias since Aaronson *et al.* (1978a) did not observe many of the metal-rich clusters near the galactic centre.

The two cluster systems together overlap the early-type galaxies; however, there is a tendency for galaxies with the same $(U-V)_o$ as the globulars to be redder in $(V-K)_o$ than the latter. This is particularly evident for the E-type galaxies. (We will show in S. E. Persson, J. A. Frogel & M. Aaronson (in preparation) that faint ellipticals are systematically

Fig. 9.5. Reddening-corrected broad-band colours. The enclosed areas contain large numbers of galactic globulars and early-type galaxies. The data are primarily from Aaronson *et al.* (1978a), Frogel *et al.* (1978) and Persson *et al.* (1979). +, M31 globulars; ⊗, galactic globulars; ●, E galaxies; ○, S0 galaxies.

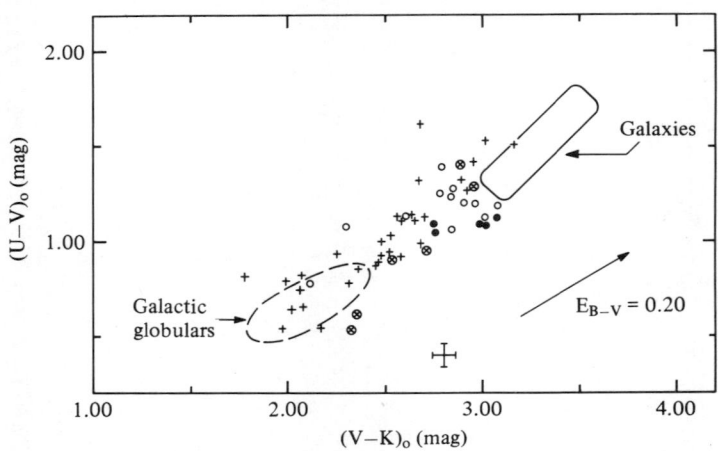

redder in $(V-K)_o$ than the faint S0s at a given $(U-V)_o$. Furthermore, the range of colours for galaxies extends to much redder colours in both $(U-V)_o$ and $(V-K)_o$ than the reddest globulars in either system.

9.4.3 CO versus $(V-K)_o$

Fig. 9.6 shows the CO index plotted versus $(V-K)_o$ for the eight M31 globulars with measured CO indices. Also indicated are the regions occupied by early-type galaxies (Frogel *et al.* 1978; Frogel & Persson, unpublished data) and the galactic globulars (Aaronson *et al.* 1978a). Two conclusions can be drawn from this figure: first, as in the $(U-V)_o$ versus $(V-K)_o$ plot, the two cluster systems mingle; secondly, taken together, the globular clusters appear to have weaker CO indices than the galaxies for a given $(V-K)_o$, a point first noted in Frogel *et al.* (1978). Observations of the LMC globulars (unpublished data) show that for a given $(V-K)_o$ they too have CO indices as much as 0.1 mag weaker than those of early-type galaxies.

9.4.4 Discussion

The data we have presented show qualitative overlap in the infrared photometric characteristics of the M31 and galactic globulars. However, the clusters differ in several ways from early-type galaxies. Most striking

Fig. 9.6. Reddening-corrected $(V-K)_o$ colours and CO indices. The enclosed areas contain large numbers of galactic globulars from Aaronson *et al.* (1978a) and early-type galaxies, mostly from Frogel *et al.* (1978). The reddest galaxy shown individually is NGC 4486B. +, M31 globulars; ⊗, galactic globulars; ●, galaxies.

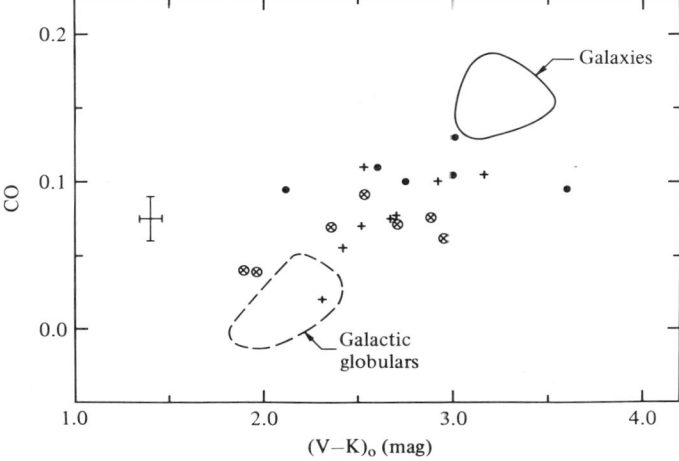

is the fact that the galaxies extend to considerably redder $(U-V)_o$ and $(V-K)_o$ colours and to larger values of the CO index. Recall that the M31 clusters which are reddest in $(V-K)_o$ and also have the strongest CO absorption were found to have [Fe/H] about *twice solar*.

Taken together with the observation that early-type galaxies display strong H_2O absorption (Aaronson *et al.* 1978b), these results lead us to conclude that a population of very late-type stars is present in the galaxies but not in the clusters, even the most metal-rich ones. For example in Aaronson *et al.* (1978b) it was shown that in O'Connell's (1976) models which best fit the integrated light of luminous galaxies, nearly 40 per cent of the light at 2.2 μm came from the M6 III bin $(V-K = 7.2)$. Even metal-rich galactic globulars such as M71 and M69 (Frogel, Persson & Cohen 1979; and Frogel & Persson, unpublished data) are not known to have any stars redder than $V-K = 5.6$.

What is the nature of these late-type stars? One possibility is that they are representatives of a population possessing several times solar metal abundance and that they are present only in galaxies (even low-mass dwarfs) perhaps because of a greater number of metal enhancement events. The stellar synthesis models of Aaronson *et al.* (1978a), based on the Ciardullo & Demarque (1977) evolutionary tracks, include stars with up to twice solar metal abundance but with no stars beyond the first red giant tip. While these models adequately reproduce the globular cluster data, they fail to predict the $U-V$, $V-K$ relation for bright galaxies and the CO indices for faint galaxies. In order to fit the data by the techniques of Aaronson *et al.* (1978a), it would be necessary to combine models of different metal abundance, the most metal-rich component having 5 or 6 times the solar abundance or more. That such a high value of metal abundance would be required can be seen directly from the data presented here. A twice-solar metal abundance globular cluster in M31 has a $(V-K)_o$ of only 3.0 and a CO index of 0.10, while a typical E galaxy has $(V-K)_o$ ~ 3.3 and a CO index ~ 0.15.

A second possibility, which does not require such a high-metal-abundance component, is that these stars lie beyond the tip of the first red giant branch and are of approximately solar metal abundance. For example, they could be highly evolved asymptotic giant branch stars with masses of about 0.6 $M_\odot$ or greater. Such stars, while not significantly redder than the first giant branch, can be 10 times more luminous than the red giant tip (Gingold 1976). There could also be a contribution from Mira-type stars which can have $V-K$ of 10 or greater. The tracks computed by Ciardullo & Demarque (1977) do not include such exotic stellar species (Mould 1978;

Tinsley 1978), but Tinsley (1978) has included them in her models in a semi-empirical way.

A combination of age and small metallicity differences between the globular cluster systems and elliptical galaxies could also act to produce the observed photometric differences. As noted earlier, the integrated $V-K$ colour of a stellar system reddens with increasing age, and absorption due to CO and H_2O would also increase. The addition of a relatively metal-poor population would tend to make $U-V$ bluer because of the increased contribution of horizontal branch stars to the U light. There is, of course, no evidence in support of significant age differences between galaxies and globular clusters. In any case, if additional infrared observations of globular clusters continue to show differences between clusters and galaxies, it will complicate the problem of synthesizing the integrated light of galaxies from that of globular clusters.

9.5 The colour–magnitude relation

Fig. 9.7 presents a reddening-corrected colour–magnitude diagram for the M31 globulars which we have observed. A comparison with the list of Vetešník (1962) and the observations of Searle (1978) indicates that, while

Fig. 9.7. The colour–magnitude diagram for the observed M31 globulars. A reddening vector and the approximate completeness limit in V of our survey are shown.

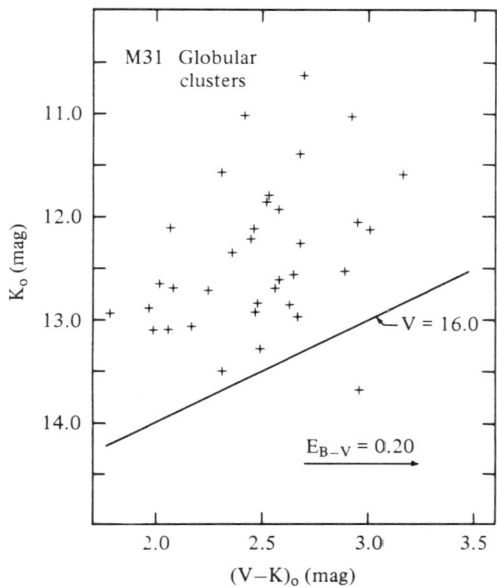

our sample is reasonably complete down to an *apparent* V magnitude of 16.0, we may be missing about half a dozen clusters with intrinsic $(V-K)_0$ between 3.0 and 3.5 and $K_0 \approx 12.0 \pm 0.5$. These limitations should be born in mind during the following discussion.

Fig. 9.8. is a similar colour–magnitude diagram for the galactic globulars, based in part on the data of Aaronson *et al.* (1978a). For those clusters for which no infrared photometry is available we predicted $(V-K)_0$ colours by using the $U-V$ colours of Harris & van den Bergh (1974), reddening values from Harris (1976), and the relation between $(U-V)_0$ and $(V-K)_0$ of Aaronson *et al.* (1978a). M_{K_0} values were obtained from the distance moduli tabulated by Harris (1976) and the integrated V magnitudes collected by Peterson & King (1975). Thus nearly all known galactic globulars are included in fig. 9.8.

Both fig. 9.7 and 9.8 display a weak tendency for the brightest clusters in K_0 to be redder than the fainter clusters. To investigate this further, we

Fig. 9.8. A colour–magnitude diagram for Milky Way globulars. The observed values of $(V-K)_0$ are from Aaronson *et al.* (1978a). For clusters with no infrared photometry $(V-K)_0$ values were predicted from their $(U-V)_0$ colours as described in the text.

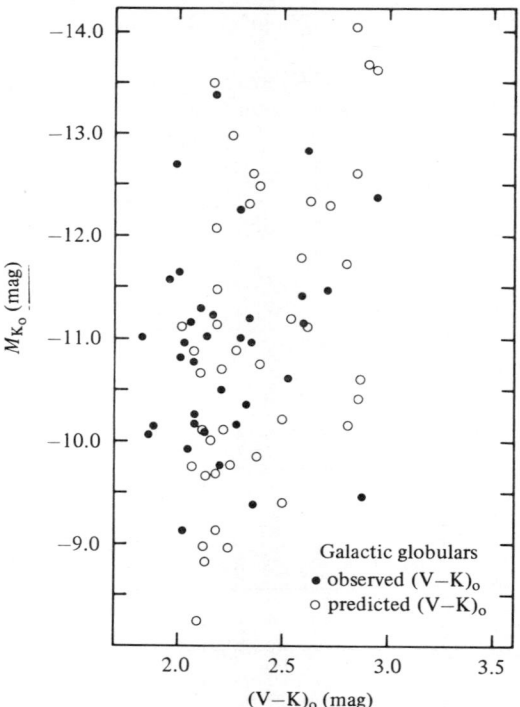

have binned the clusters in $(V-K)_o$ intervals. Fig. 9.9 shows the results of this binning. We first note that again there is no significant difference between the two cluster systems. Secondly, there does appear to be a trend of increasing luminosity with increasing $(V-K)_o$. Similar trends appear in plots of K_o against Q_K, and with the binning in Q_K, as expected from the tight correlation of $(V-K)_o$ with Q_K.

In view of the lack of infrared data for the galactic globulars, and the selection effects for the M31 globulars mentioned above, more data are required to confirm or deny the apparent correlation between colour and luminosity in globular clusters. Nevertheless, we shall suggest some mechanisms which could lead to such a relationship.

The first effect, which should operate in all clusters, concerns the evolutionary lifetimes of red giant stars as a function of metallicity. With the use of the Ciardullo & Demarque tracks we have calculated how the integrated light of a cluster depends on metallicity and initial mass

Fig. 9.9. Mean absolute K_o magnitudes (formed by averaging fluxes) for globular clusters in $(V-K)_o$ bins 0.25 mag wide beginning at $(V-K)_o = 1.75$. The last M31 bin is 0.50 mag wide, extending from 2.75 to 3.25. The error bars represent the standard deviations of the mean values. The values for the galactic globulars include all of those with predicted $(V-K)_o$ colours as discussed in the text. The means for this complete sample do not differ significantly from the values that are obtained by considering only those clusters with observed $(V-K)_o$ colours. ×, M31 globulars; ●, galactic globulars.

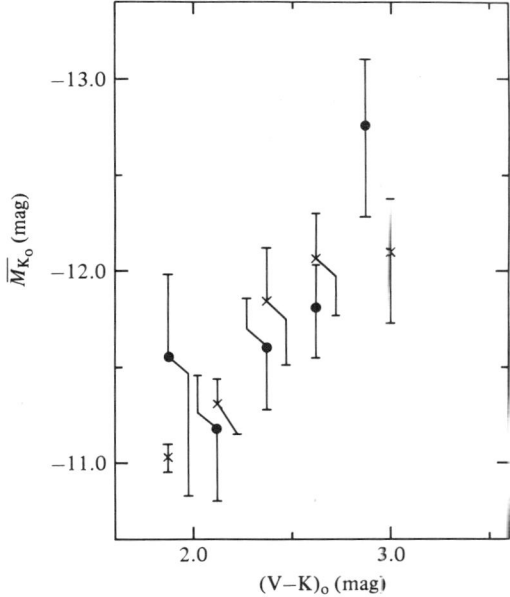

function. If we start out with two clusters, each having the same number of stars, and a slope of $x = 0$ for the initial mass function (the Salpeter value is $x = 1.35$), but one cluster having [Fe/H] $= +0.3$ and the other [Fe/H] $= -2.3$ (about the range inferred for the M31 globulars), then after 13×10^9 yr the metal-rich cluster will be nearly 1.2 mag brighter at K than the metal-poor one. The colours will be $(V-K)_o = 3.4$ and 1.9 respectively. For $x = 2.35$ the difference in the K magnitude is reduced to 0.61 and the $(V-K)_o$ colours become 3.3 and 2.1 respectively. An intermediate value of x would result in differences of K_o and $(V-K)_o$ in qualitative agreement with fig. 9.9. There are several effects which contribute to making a metal-rich cluster brighter than a metal-poor one. As [Fe/H] increases, stars evolve off the main sequence more slowly so that there will be more of them at any given time. Also, the helium flash occurs at a higher luminosity in stars of higher Z. The luminosity dependence observed at K will not be as apparent at V partly because of the increasing contribution to the V light by horizontal branch stars in the metal-poor clusters.

A second possibility is that of metal-enhanced star formation (see e.g. Talbot & Arnett 1973) which would give rise to variations in both the mass function slope and the absolute numbers of stars at the present turnoff. If all other parameters are more or less equal in the primordial gas globules that give birth to globular clusters, those globules that are more metal-rich may be able to convert their mass into stars more efficiently than those that are less metal-rich. Da Costa's (1977) work seems to indicate that this is a possible mechanism; however, there is no other firm observational evidence in support of metal-enhanced star formation.

Finally, we have the (converse) possibility – that of self-enrichment, as is generally believed to occur in elliptical galaxies. Perhaps the most massive clusters have been able to retain some enriched gas ejected by the first generations of stars (see for instance, Iben, p. 127, this volume).

JAF is grateful to the Director of the Hale Observatories for Guest Investigator privileges and for the hospitality of a month's stay during which time this chapter was prepared. We thank L. Searle and R. Zinn for many conversations and are grateful to Searle for giving us access to his unpublished data and to M. Aaronson for a tabulation of his galactic globular cluster data. O. Eggen, J. Elias and F. Schweizer commented on an early version of this chapter. We are very grateful to S. Beckwith for help in making some of the observations.

This work was supported in part by NSF grant AST 74-18555A2 and NASA grant NGL 05-002-207.

Note added in proof. We have found that a numerical error was made in the calculation of the M_{K_o} magnitudes for the galactic globular clusters only in fig. 9.8 and 9.9. The large apparent slope of magnitude versus colour in fig. 9.9 is actually much smaller for the galactic globular clusters. None of the arguments or conclusions of this chapter are affected.

J.A.F.   S.E.P.   J.G.C.

References

Aaronson, M., Cohen, J. G., Mould, J. & Malkan, M. (1978a). *Astrophys. J.* **223**, 824.
Aaronson, M., Frogel, J. A. & Persson, S. E. (1978b). *Astrophys. J.* **220**, 442.
Baldwin, J. R., Frogel, J. A. & Persson, S. E. (1973). *Astrophys. J.* **184**, 427.
Butler, D. (1975). *Astrophys. J.* **200**, 68.
Ciardullo, R. B. & Demarque, P. (1977). *Trans. astron. Obs. Yale Univ.* **35**.
Cohen, J. G., Frogel, J. A. & Persson, S. E. (1978). *Astrophys. J.* **222**, 165.
Da Costa, G. (1977). *Ph.D. thesis*, Australian National University.
Faber, S. M. (1973). *Astrophys. J.* **179**, 731.
Frogel, J. A. (1971). *Ph.D. thesis*, California Institute of Technology.
Frogel, J. A., Persson, S. E., Aaronson, M. & Matthews, K. (1978). *Astrophys. J.* **220**, 75.
Frogel, J. A., Persson, S. E. & Cohen, J. G. (1979). *Astrophys. J.* **227**, 499.
Gingold, R. A. (1976). *Astrophys. J.* **204**, 116.
Grasdalen, G. L. (1974). *Astron. J.* **79**, 1047.
Hanes, D. A. (1977). *Mon. Not. Roy. astron. Soc.* **179**, 331.
Harris, H. C. & Canterna, R. (1977). *Astron. J.* **82**, 798.
Harris, W. E. (1976). *Astron. J.* **81**, 1095.
Harris, W. E. & van den Bergh, S. (1974). *Astron. J.* **79**, 31.
Johnson, H. L. (1966). In *Nebulae and Interstellar Matter*, ed. B. M. Middlehurst & L. H. Aller, p. 167. University of Chicago Press.
Mould, J. R. (1978). *Astrophys. J.* **220**, 434.
O'Connell, R. W. (1976). *Astrophys. J.* **206**, 370.
Persson, S. E., Aaronson, M. & Frogel, J. A. (1977). *Astron. J.* **82**, 729.
Persson, S. E., Frogel, J. A. & Aaronson, M. (1979). *Astrophys. J. Suppl.* **39**, 61.
Peterson, C. J. & King, I. R. (1975). *Astron. J.* **80**, 427.
Pilachowski, C. A. (1978). *Astrophys. J.* **224**, 412.
Pritchet, C. (1977). *Astron. J.* **82**, 471.
Searle, L. (1978). Results reported at NATO Advanced Study Institute on Globular Clusters, Cambridge.
Searle, L. & Zinn, R. (1978). *Astrophys. J.* **225**, 357.
Spinrad, H. & Schweizer, F. (1972). *Astrophys. J.* **171**, 403.
Strom, S. E., Strom, K. M., Goad, J. W., Vrba, F. J. & Rice, W. (1976). *Astrophys. J.* **204**, 684.
Struck-Marcell, C. & Tinsley, B. M. (1978). *Astrophys. J.* **221**, 562.
Talbot, R. J., Jr & Arnett, W. D. (1973). *Astrophys. J.* **186**, 51.
Tinsley, B. M. (1978). *Astrophys. J.* **222**, 14.
van den Bergh, S. (1967). *Astron. J.* **72**, 70.
van den Bergh, S. (1969). *Astrophys. J. Suppl.* **19**, 145.
Vetešník, M. (1962). *Bull. astron. Inst. Czechoslovakia*, **13**, 180.

10
Globular clusters and galaxy evolution†

SIDNEY VAN DEN BERGH

10.1 Introduction

Globular clusters are the oldest known objects in the universe. The first globulars probably formed some time before the darkness of intergalactic space began to be illuminated by the first generation of quasars. Circumstantial evidence in favour of this view is provided by the observation that many globular clusters exhibit a very low metal abundance whereas there is no *strong* evidence that quasars are particularly metal-poor (Strittmatter & Williams 1976). The observation that the heavy-element abundance *ratios* in quasars are similar to those in the Sun lends additional support to the view that the quasar phenomenon occurs in regions in which interstellar material has already been polluted by at least one generation of evolving stars.

10.2 The early evolution of galaxies

Because of their great ages globular clusters should be able to provide us with significant information on the earliest phases of galaxy evolution. In particular we have every reason to believe that globulars (which are quite tightly bound gravitationally) are able to survive the coalescence of ancestral protogalaxies which is postulated in some theories of galaxy formation.

In sections 10.2.1–10.2.4 below we confront the entire spectrum of presently fashionable ideas on galaxy formation with available evidence from globular cluster observations. These data appear to rule out models in which galaxies form from globular clusters and models in which galaxy formation takes place by mergers of ancestral dwarf systems. The globular cluster data seem to favour models in which galaxies evolve from large (possibly lumpy) protogalaxies.

† Dominion Astrophysical Observatory Contribution No. 383, NRC No. 16865.

The discussion of evolutionary models A to D presented below is an extension of one originally given in van den Bergh (1975). It should, perhaps, be emphasized that models A to D differ physically only if star formation begins before sub-units begin to merge.

10.2.1 Galaxy formation from globular clusters (model A)

Peebles & Dicke (1968) have suggested that the first bound systems to have formed in the expanding universe were gas glouds with masses similar to those of globular clusters. They argue that such protoclusters subsequently banded together to form galaxies. Difficulties with this model, some of which were pointed out by Peebles & Dicke themselves, are the following:

(i) The Peebles & Dicke picture (model A) does not account for the observation (see Harris 1976, and fig. 10.1) that the heavy element content of globular clusters correlates with position in the Galaxy, i.e. model A does not explain how all metal-rich clusters knew that they should migrate towards the galactic nuclear bulge. A stratification of globulars similar to that observed in the Galaxy is also suggested by the analysis of Hartwick & Sargent (1974) who find metal-poor clusters in M31 to have a higher velocity dispersion than do metal-rich globulars.

(ii) Model A does not explain why (van den Bergh 1974) the average

Fig. 10.1. Spectral type versus distance from the galactic centre. The figure shows that metal-rich (G-type) globulars have $R < 10$ kpc, whereas metal-poor (F-type) clusters occur at all R values.

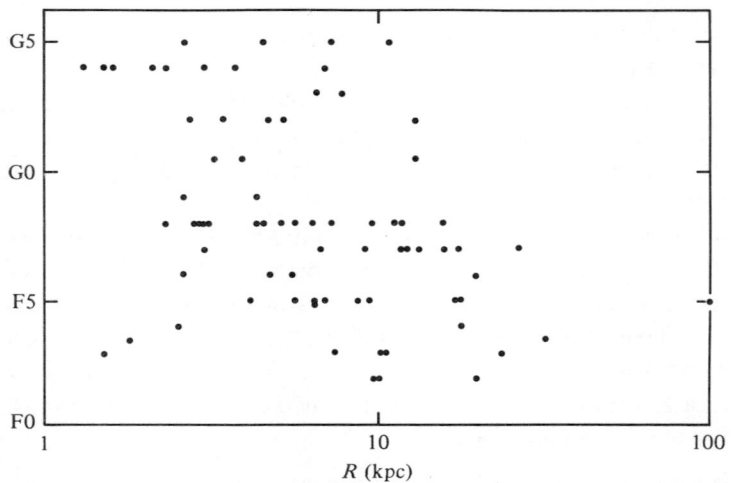

metallicity of globular cluster families depends on the luminosity (or mass) of their parent galaxies.

(iii) Model A does not account in a natural way for the observation (cf. Hanes 1977b) that such a small (10^{-2}–10^{-3}) fraction of the original globular cluster-sized gas clouds in the galactic halo ended up as globular clusters. In the case of very low-density galaxies (such as the Fornax system) it is difficult to argue that most of the originally existing globulars were disrupted by tidal forces.

(iv) On the basis of the model proposed by Peebles & Dicke one might have expected to find large numbers of intergalactic globular clusters which did not manage to find a home within individual galaxies. At the present time there is no evidence for the existence of large numbers of such 'intergalactic tramps'.

(v) Finally model A throws no light on the nature and origin of the differences between globular clusters on the one hand and the faintest dwarf spheroidal systems on the other (see Zinn, p. 191, this volume). Even though such objects have comparable masses their volumes differ by factors of $\sim 10^3$.

10.2.2 Galaxy formation by merger of dwarf galaxies (model B)

It has recently been proposed by Doroshkevich, Saar & Shandarin (preprint) that 'galaxies are formed by the clustering of small structural units – gaseous (or partly stellar) clouds similar to dwarf galaxies'. In the same vein Toomre (1977) writes 'It seems inconceivable that there wasn't a great deal of merging of sizable bits and pieces (including quite a few lesser galaxies) early in the career of every major galaxy.' It is difficult to see how this hypothesis can be reconciled with the observations.

The following arguments suggest that the Galaxy was not built up by the coalescence of dwarf spheroidal galaxies and (or) objects similar to the Magellanic Clouds:

(i) The majority of globular clusters in the SMC and a few of the LMC globulars seem to contain some red giant stars with $B-V > 2.0$. No galactic globular cluster is known to contain such very red stars. Stars with $B-V > 2.0$ have also been observed in the Sculptor dwarf spheroidal system (Hodge 1965).

(ii) Van Agt (1967) has shown that the short period W Vir (BL Her) stars in dwarf spheroidal systems obey a different period–luminosity relation than do those that occur in galactic globular clusters (see Zinn, p. 203, this volume).

(iii) W Vir stars with periods longer than 10 days are quite common in the Galaxy and in M31 but are very rare in the Magellanic Clouds and absent in those dwarf spheroidal galaxies which have so far been studied (see Zinn, p. 202, this volume).

(iv) No galactic globular cluster is as highly flattened as are NGC 121 (Tifft 1963) in the SMC and NGC 1978 (Hodge 1960) in the LMC.

(v) It is difficult to see how model B could account for the fact that relatively metal-rich (G-type) globulars only occur near the centre of the Galaxy. The assumption that the metals in the G-type globulars were swept up by protoclusters as they passed through enriched gas near the centre of the Galaxy meets with the difficulty (see fig. 10.2) that the metallicity of galactic globulars does not appear to correlate with cluster luminosity. Such a correlation might have been expected since dense massive clusters would sweep up less enriched gas *per unit mass* than low-density clusters. For the M31 globulars (van den Bergh 1969) plotted in fig. 10.2 there is some indication that the four clusters brighter than $M_V = -10$ have above-average metallicity. This *might* indicate that the most massive

Fig. 10.2. Cluster luminosity versus spectral type for the Galaxy (lower panel) and for M31 (upper panel). The lack of correlation between mass and metallicity of galactic globulars argues against the hypothesis that protoglobulars swept up heavy elements as they passed through metal-rich interstellar clouds.

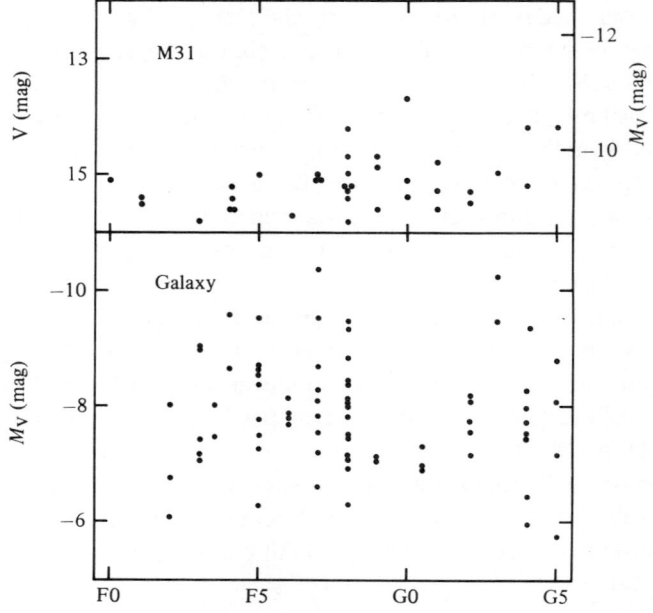

globulars are able to retain some of the heavy elements formed by the first generation of stars. Some support for this speculation is provided by the observation (Butler, Bell, Dickens & Epps 1978) that the giants and the RR Lyrae stars in ω Cen (which is the most luminous galactic globular) exhibit a range in heavy element abundances. It would be interesting to search for similar effects in NGC 6388, which is the second brightest ($M_V = -10.2$) known galactic globular cluster. Furthermore, NGC 6388 is more tightly bound than is ω Cen.

The observations discussed in points (i)–(iv) above show that the globular clusters associated with the Galaxy have 'family traits' which differ systematically from those of the globular clusters in the Magellanic Clouds and from those of the Population II stars in dwarf spheroidal galaxies. Presumably such 'family traits' of globular cluster systems are determined by the locus of protoclusters in Y, Z, t (or perhaps Y, Fe/CNO, t) space. This indicates that the chemical evolution of the gas clouds from which galactic globular clusters formed had a different history from that of those clouds which produced globular clusters in the present satellite galaxies of our Milky Way system. This conclusion is inconsistent with the hypothesis that the Galaxy formed from the merger of dwarf galaxies such as the Magellanic Clouds.

10.2.3 Galaxy formation from lumpy protogalaxies (model C)

Larson (1978) and Tinsley (1979) point out that the Jeans mass within a massive collapsing protogalaxy may be only $\sim 10^8$ $M_\odot$ so that matter might be highly clumped on scales that are much smaller than an entire protogalaxy. On this picture most star formation will take place in the densest subsystems that later merge as the collapse proceeds. Alternatively high-velocity collisions between dense clumps of matter might trigger bursts of star formation.

Recently Searle (1977) has calculated the frequency distribution of globular cluster abundances that is predicted by a model in which one has complete mixing within each clump but no mixing between clumps. He finds good agreement between theory and observations of halo globular clusters for a model in which each clump suffers ten 'enrichment events'. On the other hand, Hartwick (1976) finds that available observations can be represented by using a homogeneous protogalaxy (model D) in which gas and stars collapse together in free fall. The main uncertainty in both the Searle and Hartwick analyses is that the fraction of all matter that goes into globular clusters may itself be dependent on time and/or metallicity.

10.2.4 Galaxy formation by collapse of a single protogalaxy (model D)

Early ideas on the collapse of our Galaxy from a large protogalaxy are discussed in papers by von Weizsäcker (1955) and Oort (1958). A detailed discussion of this model is contained in the classical paper of Eggen, Lynden-Bell & Sandage (1962). Observations of globular clusters suggest no straightforward way of distinguishing the Eggen, Lynden-Bell & Sandage picture (model D) from that in which the Galaxy evolved from a clumpy protogalaxy (model C).

10.3 The radial density distribution of clusters

In first approximation a spiral galaxy consists of an exponential disc that is embedded within a spheroidal (core + halo) component. From the point of view of galaxy evolution it is of some interest to enquire whether the radial density distributions within the spheroidal components of spirals are similar to those in ellipticals which are observed to obey the 'Hubble law', i.e. if σ, the observed surface density, and ρ, the true space density, fall off as $\sigma \propto R^{-2}$ and hence $\rho \propto R^{-3}$ respectively, where R is the galactocentric distance.

Fig. 10. 3. N is the number of globulars *on the near side of the Galaxy* for which the distance to the galactic nucleus is larger than R. Dots and scale on left refer to faint clusters with $M_V \geqslant -7.5$. Solid line and scale on right refer to bright clusters with $M_V < -7.5$. The dashed line represents the relation $\rho \propto R^{-3}$.

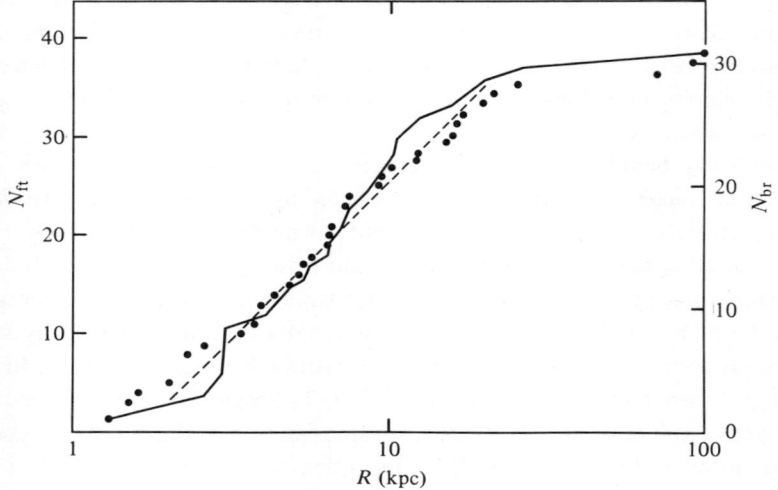

Fig. 10.3 shows a plot of the radial distribution of galactic globular clusters, based mainly on the cluster distances given by Harris (1976). In order to minimize the effects of incompleteness only those clusters on the near side of the Galaxy are plotted. Fig. 10.3 shows that both the bright ($M_V < -7.5$) and the faint ($M_V \geq -7.5$) globulars have a radial density distribution that is consistent with $\rho \propto R^{-3}$ over the range 2 kpc $< R <$ 20 kpc. For $R >$ 30 kpc the density of both bright and faint galactic globulars falls well below that expected from a Hubble law. Oort & Plaut (1975) find that a relation of the form

$$\rho = \alpha R^{-3} \tag{1}$$

also gives a satisfactory fit to their observations of RR Lyrae stars in the range 1 kpc $< R <$ 5 kpc. The fact that both RR Lyrae stars and globular clusters follow the same radial density relation is consistent with the notion that these objects are representatives of the same stellar population.

Fitting eq. (1) to the data in fig. 10.3 yields $\alpha = 4.8$ for globular clusters.

Fig. 10.4. Plot of the central relaxation time (Lightman, Press & Odenwald 1978) versus distance from the centre of the Galaxy (Harris 1976). The relaxation times for clusters near the centre of the Galaxy are seen to be $\sim 10^2$ times shorter than are those for the clusters in the galactic halo.

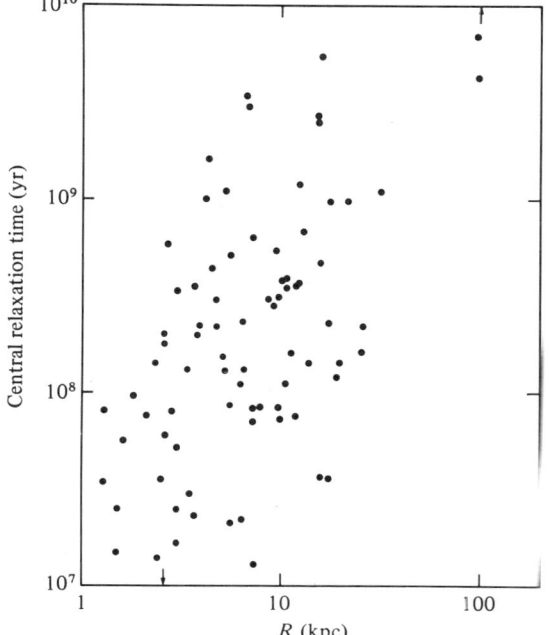

This may be compared to the Oort & Plaut value $\alpha = 880$ for field RR Lyrae stars. It should, of course, be emphasized that the ratio α_{RR}/α_{GC} might itself be a function of R because the population density of cluster-type variables in the instability strip is dependent on metallicity.

In a recent paper Lightman, Press & Odenwald (1978) have suggested that large numbers of globular clusters have disappeared as a result of core collapse. They estimate that the original number of galactic globulars was between ~ 2 and ~ 1000 times larger than it is at present. Inspection of fig. 10.4 shows that the central relaxation time of globulars in the nuclear bulge of the Galaxy is $\sim 10^2$ times shorter than that for halo clusters. It follows that the population of globular clusters near the galactic centre would have been much more severely depleted by core collapse deaths than the population of halo clusters. This effect would have produced a flattening of the radial density gradient for galactic globular clusters. The fact that both RR Lyrae stars and globular clusters (which belong to similar populations) are presently observed to have the same radial density gradient $\rho \propto R^{-3}$ therefore militates strongly against the suggestion that a significant fraction of all galactic globulars have disappeared as a result of core collapse. The same point has also been made by Harris & Petrie (1978) who note that the globular clusters and the surface brightness of M49 and M87 exhibit the same $\sigma \propto R^{-2}$ dependence. The similarity between the radial density distribution of galactic globular clusters and the density gradient within elliptical galaxies has also been emphasized by de Vaucouleurs (1977).

If globulars represent a fair sample of the galactic halo then the halo has a structure comparable to that of an elliptical galaxy. A possible difference between the halos of spirals and the envelopes of ellipticals is, however,

Table 10.1. *Spatial distribution of galactic globular clusters of known spectral type[a]*

Spectrum	F2.5	F4.5	F6.5	F8.5	G0.5	G2.5	G4.5
$\langle \lg |Z| \rangle$	0.76	0.66	0.60	0.17	0.56	0.22	0.00
	±0.13	±0.14	±0.09	±0.11	±0.20	±0.13	±0.10
$\langle \lg R \rangle$	0.97	0.96	0.96	0.72	0.74	0.74	0.50
	±0.16	±0.12	±0.09	±0.04	±0.25	±0.08	±0.08
n	8	14	13	18	3	7	12

[a] Distances in kpc, with $|Z|$ = distance above or below galactic plane and R = galactocentric distance.

that they may exhibit different enrichment gradients. According to Strom *et al.* (1976) the nuclear bulges of the edge-on S0 galaxies NGC 3115 and NGC 4762 show large abundance gradients extending over scales of a few kpc whereas Strom & Strom (1978a, b) find only small halo abundance gradients in the large majority of elliptical galaxies. Additional observations of the metallicity gradients in elliptical galaxies and in the halos of spirals would clearly be very desirable. Possible structural differences between the halos of ellipticals and those of some spirals have recently been discussed by Kormendy & Bruzual (1978).

Galactic globular cluster data (which are plotted in fig. 10.1) show that relatively metal-rich G-type clusters are confined to a region with a radius $R \lesssim 10$ kpc whereas metal-poor globulars with F-type spectra are found at all distances from the galactic centre.

Information on the spatial distribution of galactic globulars of known spectral type is summarized in table 10.1. Taken at face value these data suggest that the subsystem of metal-rich G-type clusters is more highly flattened than is the subsystem of metal-poor F-type globulars. This conclusion should, however, be regarded with caution because errors in the distances to G-type clusters near the galactic centre will tend to increase $\langle R \rangle$, the average distance of clusters from the galactic nucleus. This bias will, however, be counteracted by the fact that many of the G-type clusters close to the galactic plane are hidden behind absorbing clouds, so that the true mean value of $\langle |Z| \rangle$ is smaller than the observed value of this parameter.

Unfortunately, the globular clusters with $R > 20$ kpc are not sufficiently numerous to draw any strong conclusions about the existence of an abundance gradient in the outer halo of the Galaxy. Kraft *et al.* (1979) also emphasize the fact that available data on RR Lyrae stars at high latitudes are still insufficient to provide firm information on the abundance gradient in the outer part of the galactic halo.

10.4 The metallicity of globular cluster families

Some years ago I used the Hale 5-metre telescope to obtain photometry and spectra of the brightest globular clusters in M31. These observations (van den Bergh 1969) showed that (i) the average line-strength of the globular clusters associated with the Andromeda Nebula is greater than that observed in galactic globulars, and (ii) M31 contains a larger fraction of intrinsically red globulars than does the Galaxy. The latter conclusion has been challenged by Hanes (1977a, and p. 233, this volume) who concluded that 'the suggestion that the relative numbers of globular clusters of different metallicities in M31 and the Galaxy are significantly different (van den Bergh) is not substantiated'. It is shown below that this conclusion is incorrect.

10.4.1 The colours of globular clusters

According to Racine (1973) the photometric parameter

$$Q = (U-B) - 0.72(B-V), \quad (2)$$

which was used by van den Bergh (1969), is not entirely independent of reddening for late-type globulars. He therefore introduced the parameter

$$\mathscr{R} = \frac{(U-B) + 0.10[1.00-(B-V)]}{2(B-V)-1.00}, \quad (3)$$

which appears to be essentially reddening-free for globular clusters of all spectral types. A comparison of the frequency distributions of $\mathscr{R}$ values for

Fig. 10.5. Frequency distribution of the parameter $\mathscr{R}$, which measures the intrinsic colour of globular clusters in the Galaxy and in M31. Photoelectric observations of clusters in the Andromeda Nebula are indicated by the solid histogram, photographic data are shown by lighter shading.

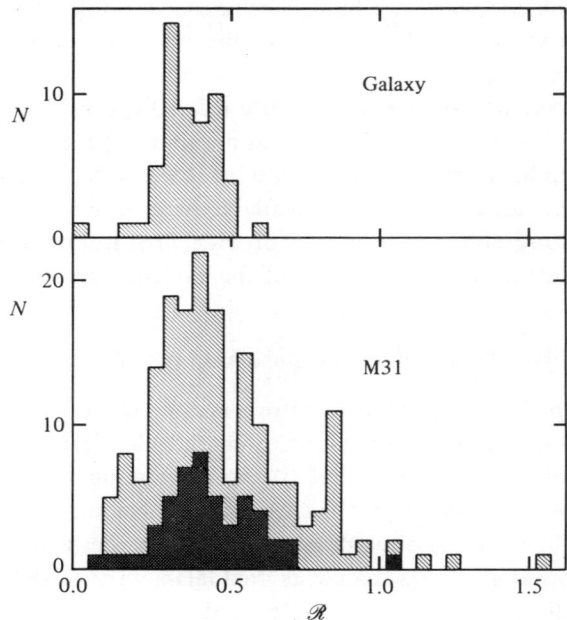

55 photoelectrically observed galactic globular clusters and for 180 M31 globulars, for which either photoelectric or photographic UBV colours are available (Harris 1974), is shown in fig. 10.5. This figure shows that the M31 cluster sample contains many more intrinsically red globulars (which have large $\mathscr{R}$ values) than does the galactic cluster sample.

Application of the Kolmogorov–Smirnov two-sample test (see Beyer 1966) shows that the hypothesis that the M31 and galactic globular cluster $\mathscr{R}$ values are drawn from the same population can be rejected at the 0.1 per cent confidence level.

Due to contamination by the background light of the Andromeda Nebula itself the photographically measured colours of many of the fainter clusters in M31 are probably unreliable. It might therefore be more meaningful to compare only the 49 photoelectrically observed clusters in M31 (see the solid histogram in fig. 10.5) with the 55 galactic globular clusters for which Harris (1974) lists intrinsic $U-B$ and $B-V$ colours. Application of the Kolmogorov–Smirnov test to these two samples of photoelectric data shows that the hypothesis that they were drawn from the same population can be rejected at the 5 per cent level. Again the difference between these two samples is in the sense that M31 contains a larger fraction of intrinsically red clusters (which have large $\mathscr{R}$ values) than does the Galaxy.

The M31 cluster sample is essentially unbiased with respect to metallicity because the clusters were selected on the basis of their apparent luminosity. (A few bright clusters near the nucleus of M31 were not observed because of bright background nebulosity. Furthermore the search for bright halo clusters with $R > 20$ kpc is incomplete.) The absence of bias in the sample of galactic globular clusters for which UBV colours are available is more difficult to establish. In general $U-B$ colours are not available for those galactic globular clusters that are either (i) highly reddened, or (ii) of very low surface brightness. Most of the highly reddened clusters lie in the general direction towards the galactic centre and might therefore be expected to have larger than average $\mathscr{R}$ values. On the other hand scattered clusters with a low surface brightness usually lie far out in the galactic halo and are therefore probably bluer than average.

10.4.2 Line-strengths and spectral types of clusters

Application of the Komogorov–Smirnov two-sample test to the spectroscopic line-strength data on 42 galactic globulars and 37 globular clusters in M31 (van den Bergh 1969) shows that the hypothesis that these two samples are drawn from the same population can be rejected at the 0.1 per cent confidence level. As expected from the colorimetric data the spectra indicate that M31 contains a larger fraction of strong-lined clusters than does the Galaxy. The sample of M31 clusters observed spectroscopically was entirely unbiased since selection was based only on apparent magnitude. It is, however, quite possible that the rather small sample of galactic globulars for which spectroscopic line-strength (L) classes were obtained is not representative of the entire galactic globular cluster population. Nevertheless it seems highly unlikely that selection effects could account entirely for the striking differences that are exhibited by the samples that were actually chosen for observation. Of the galactic globular cluster sample almost half turned out to be weak-lined ($L \leq 2$) whereas such metal-poor clusters made up only 8 per cent of the M31 sample. (Some of the metal-poor globulars near M31 might actually be interlopers that have been tidally stripped from M32).

A larger sample of 74 L values (representing almost two-thirds of all known galactic globular clusters) may be constructed by converting the integrated spectral types given by Burstein & McDonald (1975) to L values. The Kolmogorov–Smirnov test shows that the hypothesis that this enlarged sample of 74 galactic L values is drawn from the same population as the homogeneous sample of 37 L values of clusters in M31 can also be rejected at the 0.1 per cent confidence level.

10.4.3 Metallicities in other globular cluster families

Intercomparison of available spectroscopic line-strength classifications shows that the percentages of metal-poor ($L \leq 2$) globular clusters in M31, the Galaxy and the Fornax dwarf spheroidal system are 8 per cent, 48 per cent and 100 per cent respectively. These data suggest that metal-poor globulars are preferentially associated with low-mass galaxies. Additional support for the suggestion that the mean metallicity and mean intrinsic colour of globular cluster families is correlated with luminosity (or mass) of their parent galaxy is provided by the following observations:

(i) Danziger's (1973) photometry and the morphology of cluster red giant branches (Gascoigne 1966; Hesser, Hartwick & Ugarte 1976) suggest that none of the true globular clusters in the Magellanic Clouds are of the strong-lined (47 Tuc-like) variety. A similar result is obtained from the integrated cluster colour data of Bernard & Bigay (1974). This conclusion has more recently been confirmed by spectra of individual cloud cluster stars obtained by Cowley & Hartwick (in preparation).

(ii) The globulars in and near the dwarf galaxy NGC 205 (Hodge 1973) are, with one exception, quite blue and hence presumably metal-poor. The exception is NGC 205: III which, on the basis of its radial velocity (van den Bergh 1969), is a member of the M31 cluster system that happens to be projected on NGC 205.

It is not yet clear if the correlation between mean metallicity of cluster families and luminosity of parent galaxy can be extrapolated to objects more massive than the Andromeda Nebula. For 3 globulars associated with M87, Racine, Oke & Searle (1978) find a mean metallicity similar to that for M31 clusters. It should, however, be emphasized that the clusters studied by Racine, Oke & Searle are located between 20 kpc and 30 kpc from the nucleus of M87. They are therefore extreme halo objects which might have below average metallicity.

10.5 The nuclei of galaxies

Spiral and elliptical galaxies that are brighter than $M_v = -15$ usually contain a small semi-stellar nucleus. Tremaine, Ostriker & Spitzer (1975) and Tremaine (1976) have made the ingenious suggestion that such nuclei might be formed by the accumulation of globular clusters that are dragged into the centres of galaxies by dynamical friction. Observational evidence on the nuclei of M31, M32 and M33, which militates against this suggestion, is presented below.

10.5.1 The nucleus of M31

Of 37 M31 globular clusters (van den Bergh 1969) for which spectra are available only No. 282 has a line-strength that is comparable to that observed in the spectrum of the nucleus of the Andromeda Nebula. UBV photometry of cluster No. 282 yields $B-V = 0.92$ and $U-B = 0.47$. These colours are significantly bluer than are those of the nucleus of M31 itself for which Sandage, Becklin & Neugebauer (1969) obtain $B-V = 1.08$ and $U-B = 0.80$. The fact that even the strongest-lined globular known in M31 is much bluer than the nucleus argues strongly against the hypothesis that globular cluster-like stars contribute significantly to the total light of the nucleus of the Andromeda Nebula.

An additional argument against the idea that the nucleus of M31 is composed of disintegrated globulars is provided by the observation (Morton & Thuan 1973) that the nucleus has $M/L_V = 13$ compared to $M/L_V \sim 2$ (Illingworth 1976) for typical globulars.

10.5.2 The nucleus of M32

The nucleus of M32 is brighter than that of M31. Its total luminosity corresponds to that of dozens of typical globulars. It seems unlikely that a small galaxy such as M32 ever contained such a large number of globulars. (NGC 205, which is approximately as bright as M32, contains only seven globulars.) Furthermore, the mass-to-light ratio for the centre of M32 (Richstone & Sargent 1972) is larger than that for globulars (Illingworth 1976).

10.5.3 The nucleus of M33

A spectrogram of the semi-stellar nucleus of the Triangulum Nebula has been described by van den Bergh (1976). This nucleus is found to have a late A-type spectrum from $K/H + H\epsilon$, an F2–F4 spectrum from $\lambda 4226/H\delta$ and a type F3–F4 from $CH/H\gamma$. Taken at face value these data suggest that relatively young stars, similar to those in the disc of M33, contribute most of the light of the nucleus of the Triangulum Nebula. Alternatively the early integrated spectral type of the nucleus of M33 might be produced by *very* metal-poor old stars. Inspection of the spectra of galactic globulars shows that only stars as metal-poor as those in M15 and M92 could mimic the observed spectrum of the nucleus of M33. It appears rather implausible that the nucleus of M33 consists of only the most metal-poor globulars while that of M31 is built up from clusters that are metal-richer than any that have survived to the present time.

In summary the data seem to suggest that most of the mass and most of the light of the nuclei of M31, M32 and M33 is probably *not* derived from disintegrated globular clusters. This indicates that the formation of the nuclei of galaxies is not directly related to the existence of globular clusters. In any case it would have been difficult to understand how a nucleus such as that in M87, which is believed to have a mass of $\sim 5 \times 10^9$ $M_\odot$ (Sargent *et al.* 1978), could have been formed by the accumulation of globular clusters.

References

Bernard, A. & Bigay, J. H. (1974). *Astron. Astrophys.* **33**, 123.
Beyer, W. H. (1966). *Handbook of Tables for Probability and Statistics*, pp. 323–5. Cleveland: The Chemical Rubber Co.
Burstein, D. & McDonald, L. H. (1975). *Astron. J.* **80**, 17.
Butler, D., Bell, R. A., Dickens, R. J. & Epps, E. (1978). In *The HR Diagram*, IAU Symposium no. 80, ed. A. G. Davis Philip & D. S. Hayes, p. 183. Dordrecht: D. Reidel.
Danziger, I. J. (1973). *Astrophys. J.* **181**, 641.
de Vaucouleurs. G. (1977). *Astron. J.* **82**, 456.

Eggen, O. J., Lynden-Bell, D. & Sandage, A. R. (1962). *Astrophys. J.* **136**, 748.
Gascoigne, S. C. B. (1966). *Mon. Not. Roy. astron. Soc.* **134**, 59.
Hanes, D. A. (1977a). *Mon. Not. Roy. astron. Soc.* **79**, 331.
Hanes, D. A. (1977b). *Mon. Not. Roy. astron. Soc.* **180**, 309.
Harris, W. E. (1974). Unpublished Ph.D. thesis, University of Toronto
Harris, W. E. (1976). *Astron. J.* **81**, 1095.
Harris, W. E. & Petrie, P. L. (1978). *Astrophys. J.* **223**, 88.
Hartwick, F. D. A. (1976). *Astrophys. J.* **209**, 418.
Hartwick, F. D. A. & Sargent, W. L. W. (1974). *Astrophys. J.* **190**, 283
Hesser, J. E., Hartwick, F. D. A. & Ugarte, P. (1976). *Astrophys. J. Suppl.* **32**, 283.
Hodge, P. W. (1960). *Astrophys. J.* **132**, 346.
Hodge, P. W. (1965). *Astrophys. J.* **142**, 1390.
Hodge, P. W. (1973). *Astrophys. J.* **182**, 671.
Illingworth, G. (1976). *Astrophys. J.* **204**, 73.
Kormendy, J. & Bruzual, G. (1978). *Astrophys. J. Lett.* **223**, L63.
Kraft, R. P., Trefzger, C. & Suntzeff, N. (1979). In *The Large-Scale Characteristics of the Galaxy*, IAU Symposium no. 84, ed. W. B. Burton, p. 463. Dordrecht: D. Reidel.
Larson, R. B. (1978). In *Chemical and Dynamical Evolution of our Galaxy*, IAU Colloquium no. 45, ed. E. Basinska-Grezesik & M. Mayor, p. 3. Geneva Observatory Publ.
Lightman, A. P., Press, W. H. & Odenwald, S. F. (1978). *Astrophys. J.* **219**, 629.
Morton, D. C. & Thuan, T. X. (1973). *Astrophys. J.* **180**, 705.
Oort, J. H. (1958). In *Stellar Populations*, ed. D. J. K. O'Connell, p. 415. Amsterdam: North-Holland.
Oort, J. H. & Plaut, L. (1975). *Astron. Astrophys.* **41**, 71.
Peebles, P. J. E. & Dicke, R. H. (1968). *Astrophys. J.* **154**, 891.
Racine, R. (1973). *Astron. J.* **78**, 180.
Racine, R., Oke, J. B. & Searle, L. (1978). *Astrophys. J.* **223**, 82.
Richstone, D. & Sargent, W. L. W. (1972). *Astrophys. J.* **176**, 91.
Sandage, A. R., Becklin, E. E. & Neugebauer, G. (1969). *Astrophys. J.* **157**, 55.
Sargent, W. L. W., Young, P. J., Boksenberg, A., Shortridge, K., Lynds, C. R. & Hartwick, F. D. A. (1978). *Astrophys. J.* **221**, 731.
Searle, L. (1977). In *The Evolution of Galaxies and Stellar Populations*, ed. B. M. Tinsley & R. B. Larson, p. 219. New Haven: Yale University Observatory.
Strittmatter, P. A. & Williams, R. E. (1976). *Ann. Rev. Astron. Astrophys.* **14**, 307.
Strom, K. M. & Strom, S. E. (1978a). *Astron. J.* **83**, 73.
Strom, K. M. & Strom, S. E. (1978b). *Astron. J.* **83**, 1293.
Strom, S. E., Strom, K. M., Goad, J. W., Vrba, F. J. & Rice, W. (1976). *Astrophys. J.* **204**, 684.
Tifft, W. G. (1963). *Mon. Not. Roy. astron. Soc.* **125**, 199.
Tinsley, B. M. (1979). In *The Large-Scale Characteristics of the Galaxy*, IAU Symposium no. 84, ed. W. B. Burton, p. 431. Dordrecht: D. Reidel.
Toomre, A. (1977). In *The Evolution of Galaxies and Stellar Populations*, ed. B. M. Tinsley & R. B. Larson, p. 420. New Haven: Yale University Observatory.
Tremaine, S. D. (1976). *Astrophys. J.* **203**, 345.
Tramaine, S. D., Ostriker, J. P. & Spitzer, L. (1975). *Astrophys. J.* **196**, 407.
van Agt, S. L. T. J. (1967). *Bull. astron. Inst. Netherlands*, **19**, 275.
van den Bergh, S. (1969). *Astrophys. J. Suppl.* **19**, 145.
van den Bergh, S. (1974). In *The Formation and Dynamics of Galaxies*, IAU Symposium no. 58, ed. J. R. Shakeshaft, p. 157. Dordrecht: D. Reidel.
van den Bergh, S. (1975). *Astron. Astrophys.* **44**, 231.
van den Bergh, S. (1976). *Astrophys. J.* **203**, 764.
von Weizsäcker, C. F. (1955). *Z. Astrophys.* **35**, 252.

11
The dwarf spheroidal galaxies

ROBERT ZINN

11.1 Introduction

The Galaxy is surrounded by a retinue of smaller galaxies which are almost certainly satellites. The largest of these are, of course, the irregular galaxies, the Large and Small Magellanic Clouds. The other galaxies are much less massive and appear to be pure Population II systems. These objects are often called the dwarf spheroidal galaxies even though some of them are elliptical in shape.

If the dwarf spheroidal galaxies are not round and their stellar populations are Population II, why not call them dwarf elliptical galaxies? There is little reason why not to do this since the only differences between the dwarf spheroidal galaxies and the dwarf elliptical galaxies (e.g. M32; NGCs 205, 185, 147) appear to be that the dwarf spheroidal galaxies are less massive and have lower surface brightnesses. The dwarf spheroidal galaxies probably represent the low-mass end of the mass distribution of elliptical

Table 11.1. *Properties of the dwarf spheroidal galaxies*

	Absolute magnitude (M_V)	Radius (kpc)	Mass ($M_\odot$)	Central density ($M_\odot/pc^3$)	Galactocentric distance (kpc)
Sculptor	−10.9	1.2±0.1	3×10^6	7×10^{-4}	78
Fornax	−13.6	3.1±0.3	2×10^7	10^{-4}	188
Leo I	−11.4	0.91±0.04	4×10^6	5×10^{-3}	220
Leo II	−9.8	0.65±0.10	10^6	2×10^{-1}	220
Draco	∼ −8	0.51±0.04	1.2×10^5	2×10^{-3}	80
Ursa Minor	∼ −8	1.2±0.5	10^5	8×10^{-1}	71
Carina	—	—	—	—	∼170
ωCen	−10.4	0.09	1.2×10^6	1.3×10^3	—
NGC 147	−14.6	2.2±1.0	6×10^7	—	—

galaxies. It is perhaps worthwhile, however, to differentiate between the few nearby elliptical galaxies that can be studied star by star and the multitude of distant unresolved ones. For this reason, the historically well-entrenched misnomer 'dwarf spheroidal' is used in this review for the Galaxy's companions.

Table 11.1 lists the seven known dwarf spheroidal (hereafter Dsph) companions of the Galaxy and some of their properties. For comparison, table 11.1 also gives the same data for the most massive globular cluster of the Galaxy, ω Cen, and the least massive elliptical galaxy of the Local Group, NGC 147. Very few data are available for the Carina galaxy because it was only recently discovered (Cannon, Hawarden & Tritton 1977). The other Dsph galaxies have been known for 20 years or more (see Hodge 1971).

The data in table 11.1 (primarily from Hodge 1971, 1976; Peterson & King 1975) are mostly crude estimates. Nonetheless, these data demonstrate that the Dsph galaxies are intermediate to globular clusters and objects that are universally called elliptical galaxies (e.g. NGC 147). The data in table 11.1 show that the Leo II, Draco and Ursa Minor systems overlap with globular clusters in terms of luminosity and mass. The radii in table 11.1, which are tidal radii in the sense of King's (1966) models, indicate, however, that all of the Dsph galaxies are much larger than globular clusters and have dimensions that are more similar to those of NGC 147. The Fornax system undoubtedly qualifies as a bona fide galaxy because it contains six globular clusters (Hodge 1971). A feeling for the low densities and surface brightnesses of the Dsph galaxies is perhaps best conveyed by the observation that an inexperienced or casual observer may have a difficult time locating these objects on the prints of the Palomar Sky Survey. The galactocentric distances of the Dsph galaxies are larger than those of either the LMC or the SMC (50 kpc and 60 kpc, respectively), but are considerably smaller than the distance to the next nearest galaxy, NGC 6822 ($\sim$ 600 kpc). One important quantity not given in table 11.1. is the relaxation time. The densities of the Dsph galaxies are so low that the relaxation times for even their cores are many times the age of the universe (Hodge 1966).

While it seems certain that at least most of the Dsph galaxies are satellites of the Galaxy (see Hartwick & Sargent 1978), the degree to which their evolution is linked to that of the Galaxy is not clear. Are the Dsph galaxies part of the Galaxy's halo, and hence probes of its formation and chemical evolution, as are the globular clusters; or are the Dsph galaxies examples of the evolution of isolated, primordial gas clouds? Part of the interest in the Dsph galaxies stems from this ambiguity about their relationship with the Galaxy.

The spatial distribution of the Dsph galaxies about the Galaxy may provide a clue to their origin. It was realized recently that the distribution is not random, since many of the Dsph galaxies and distant globular clusters and the Magellanic Clouds appear to define a plane which passes through the galactic centre but is inclined at a large angle to the galactic plane. Some of these objects lie in the directions to the Magellanic Stream and other H I gas clouds, which suggests that the plane occupied by the objects is the orbital plane of the Magellanic Clouds about the galactic centre and that the objects and gas clouds are the tidal debris of a past close encounter between the Magellanic Clouds and the Galaxy (Kunkel & Demers 1975; Lynden-Bell 1975, 1976). The criticism by Mathewson & Schwarz (1976) and the new radial velocity data for the objects (Hartwick & Sargent 1978) have essentially ruled out the early versions of this hypothesis, but the recent reformulation by W. E. Kunkel (preprint) seems more viable. Even if the Dsph galaxies and globular clusters are not tidal debris, it is still very interesting that they are not randomly distributed in space.

One would also like to know whether the Dsph galaxies are distributed throughout space or exist only as satellites of large galaxies. The discovery of three Dsph companions of M31 by van den Bergh (1972a, b) suggests that they are frequently satellites. The other galaxies of the Local Group are not known to have Dsph companions, but it is questionable whether they have been adequately surveyed. The intrinsic faintness and lack of central concentration of the Dsph galaxies make them exceedingly difficult to detect. It is perhaps significant that the new survey of the southern sky being carried out by the Science Research Council of United Kingdom, which goes substantially deeper than the Palomar Sky Survey, has added only one new object to the class, which moreover is not exceptionally distant from the Galaxy (Cannon *et al.* 1977). There is, therefore, some meager evidence that the Dsph galaxies are confined to relatively small volumes of space around large galaxies.

In this review I shall concentrate on the major properties of the Dsph galaxies that have been discussed in the recent literature and emphasize the differences and similarities between the Dsph galaxies and globular clusters. The excellent reviews by Hodge (1971) and van den Bergh (1975) contain references to older work, historical outlines and, of course, different perspectives of the subject.

11.2 Colour–magnitude diagrams

Colour–magnitude diagrams exist for the Sculptor (Hodge 1965b; Kunkel & Demers 1977), Leo II (H. H. Swope, private communication), Draco (Baade & Swope 1961; P. Stetson, private communication), Ursa Minor (van Agt 1967; Schommer, Olszewski & Kunkel 1977), and Fornax (S. Demers & W. E. Kunkel, private communication) systems. In *general* appearance, all of these diagrams resemble those of globular clusters. They have one other feature in common: the giant branches are steep, indicating that the systems have low metal abundances. The similarities stop there because in the Ursa Minor system most of the horizontal branch (hereafter HB) stars lie to the blue side of the RR Lyrae domain, whereas in the Sculptor, Leo II, and Draco systems they lie mostly to the red side. (The colour–magnitude diagram for the Fornax system does not reach the level of the HB.) The objects with red HBs are considered to be anomalous because they do not obey the relationship between metal abundance and HB type that is exhibited by most of the galactic globular clusters – namely that the fraction of the HB that lies to the red side of the RR Lyrae domain increases with the metallicity of the cluster. Stellar evolution calculations (see the reviews by Iben 1971 and Renzini 1977) predict the existence of this relationship if clusters have different Fe/H ratios but the same ages, helium abundances, and ratios of C, N, and O to Fe (i.e. Fe/H is the 'first parameter' determining HB type). The same calculations have shown that variations in one or more of these last three quantities, while holding the Fe/H ratio fixed, can produce the spectrum of HB types that is observed among clusters of similar metallicity. This has naturally led to speculation that one or more of these quantities is the 'second parameter' (Sandage & Wildey 1967; van den Bergh 1967; Kraft, p. 92, this volume) that is presumably responsible for the anomalously red HBs of Draco, Sculptor, etc., and globular clusters such as NGC 7006. The sensitivity of the HB to these quantities is in the sense that a metal-poor system with a red HB is thought to be either exceptionally young, helium-poor, or rich in carbon, nitrogen and oxygen.

There is a relationship between the presence of the second-parameter effect among the metal-poor globular clusters of the Galaxy and the distance of the clusters from the galactic centre. To see how the Dsph galaxies fit into this relationship, I have plotted, in fig. 11.1, HB type against galactocentric distance (r) for 40 objects for which colour–magnitude diagrams exist. Relatively metal-rich objects must be excluded from such a plot if it is to have any bearing on the second-parameter

Dwarf spheroidal galaxies

problem. I have therefore restricted the sample to those objects with $Z \leqslant 0.05\ Z\odot$ (taken from Kukarkin 1974; Butler 1975; Hesser, Hartwick & McClure 1977; Canterna & Schommer 1978; Searle & Zinn 1978; Zinn 1978). Following Mironov (1972), the HB classification is defined by the ratio $B/(B+R)$, where B and R are the numbers of blue and red HB stars respectively. Values of $B/(B+R)$ range, therefore, from near 1 for very blue HBs to near 0 for very red ones. The choice of this classification scheme over the many others that are available was solely a matter of personal preference, and does not affect the results.

Fig. 11.1 shows that the metal-poor clusters within 8 kpc of the galactic centre have only blue HBs, whereas those from 8 to 40 kpc from the galactic centre have a spectrum of HB types (see also Searle & Zinn 1978). It must also be noted that there is not a good correlation between Fe/H and HB type among the clusters in this second group. The objects beyond 40 kpc from the galactic centre (the Magellanic Clouds and their clusters are not plotted) also exhibit a range in HB characteristics, and it is interesting that this range is larger than the range found among the clusters between 8 and 40 kpc. Leo II, which has not been plotted in fig. 11.1 because of its large galactocentric distance, has an HB similar to Draco's. Thus, there are a total of five objects in the extreme halo (i.e. $r > 40$ kpc) that have redder HBs than any of the metal-poor clusters that lie within 40 kpc of the galactic centre. William E. Harris has presented a colour–

Fig. 11.1. Globular clusters and Dsph galaxies are plotted in a diagram of galactocentric distance (r) versus HB type ($B/(B+R)$). Only objects that have colour–magnitude diagrams can be plotted, and the sample has been restricted to objects that are more metal-deficient than one-twentieth of the solar abundance (see text). Note the differences in the ranges of HB type among the objects in the following three zones: $r < 8$ kpc, $8\ \text{kpc} < r < 40$ kpc, and $r > 40$ kpc.

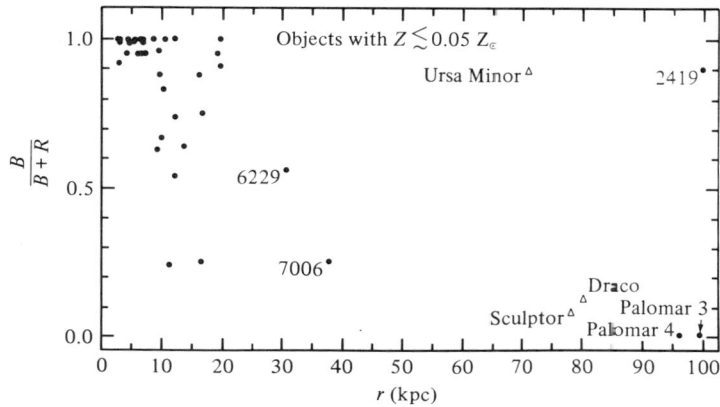

magnitude diagram for the globular cluster Palomar 12 ($r \sim 25$ kpc) which shows that it possesses a very red HB. The metal abundance of Palomar 12 appears to be low (Canterna & Schommer 1978; Cowley, Hartwick & Sargent 1978); hence there may be one metal-poor object with $r < 40$ kpc that is similar to Draco, etc., in terms of HB type. NGC 2419 and the Ursa Minor system are the only other distant objects for which the HB can be classified and they have very blue HBs. There is some doubt whether NGC 2419 formed in the extreme halo. The perigalacticon of its orbit is only 24 kpc (Racine & Harris 1975), and the gravitational perturbations of the Magellanic Clouds may have lengthened the semi-major axes of the orbits of a few galactic globular clusters by substantial amounts (Innanen & Valtonen 1977). The Ursa Minor system appears to be near its perigalacticon at the present time (Hodge & Michie 1969); thus a large range in HB type appears to be indigenous to the extreme halo.

It is also interesting to compare the objects in the extreme halo with the clusters of the Magellanic Clouds. The fact that the populous clusters of the Clouds, which all look globular in appearance, have a great dispersion in ages makes this a difficult proposition, however. It is possible, for example, to mistake a populous cluster of intermediate age (i.e. $t \sim 10^9$ yr) for a globular cluster with a red HB. The colour–magnitude diagrams for NGC 1466, 1841 and 2257 (Gascoigne 1966) and the Reticulum system (Demers & Kunkel 1976), which are probably members of the LMC, suggest that they are sufficiently old and metal-poor that they can be compared with the objects discussed above. NGC 1841 and NGC 2257 have very blue HBs, and NGC 1466 has an intermediate HB. For the Reticulum system $B/(B+R) = 0.38$, which signifies that it has a red HB, but one that is somewhat bluer than Draco's. This range in HB type is similar to that found in the Galaxy between 8 and 40 kpc, and it may be significant that these clusters lie in the outer environs of the LMC. Unfortunately, it is not possible at the present time to investigate whether there is a change in HB type in the LMC like the one at 8 kpc in the Galaxy. The only SMC cluster that is perhaps sufficiently old and metal-poor to be compared with the objects in the extreme halo, and for which a good quality colour–magnitude diagram exists, is NGC 121. Interestingly, its HB type resembles those of Draco, Sculptor, Leo II, and Palomar 3 and 4 (see Tifft 1963). NGC 121 lies in the outskirts of the SMC, but the significance of this fact is not clear.

For the galactic globular clusters, there is some observational evidence that the second parameter is the CNO to Fe ratio (Hartwick & McClure 1972; Cohen, Frogel & Persson 1978; Pilachowski 1978; Kraft, p. 92, this volume). It is not obvious, however, how the 8 kpc change in HB characteristics can be explained by CNO variations because the CNO abundances would have to be decoupled from that of Fe for clusters with $r > 8$ kpc, but tightly correlated with Fe, for some unknown reason, for clusters with $r < 8$ kpc. While this may have, in fact, occurred, it is perhaps worthwhile to consider an alternative explanation which is based on the hypothesis that age is the second parameter (see Searle & Zinn 1978). The observation that the metal-poor clusters inside 8 kpc have only blue HBs is explained by proposing that during the formation of the Galaxy the metal abundance of the gas in the central regions rose rapidly; hence, metal-poor clusters may have formed only over a relatively brief interval of time. Since the collapse of the outer parts of the Galaxy was at least roughly a free-fall collapse, there was probably not much mixing taking place between areas where metal enrichment had and had not occurred. Metal-poor clusters may have formed out of the uncontaminated gas over a considerable span of time, giving rise to the spectrum of HB types that is now observed in the outer halo. It is very difficult to test this hypothesis directly because the main-sequence dating procedure can, at best, give relative ages of globular clusters to a precision of $\sim 1 \times 10^9$ yr, which is only barely adequate. The data required to test whether the second parameter is CNO variations are now being gathered by several observers, and in the next few years I expect to see this issue resolved.

The relatively large distances to the Dsph galaxies make the observational problem of determining their CNO to Fe ratios very difficult; consequently, an explanation for their spectrum of HB types will not be quickly forthcoming. This is nevertheless an important problem to pursue, since the solution may tell us a great deal about nucleosynthesis within isolated systems or, because of the trends seen in fig. 11.1, it may tell us something more about the formation of the Galaxy.

11.3 Variable stars

The Sculptor, Draco, Ursa Minor, Leo I and Leo II galaxies have been surveyed for variable stars (see van Agt's 1973 review, and Hodge & Wright 1978). These surveys have discovered a great many RR Lyrae variables and some variables that are brighter than the HB. In some ways these stars are different from the variables found in globular clusters, and I shall concentrate on these differences here.

11.3.1 RR Lyrae variables

A number of authors have remarked that the periods of the RR Lyrae variables in Draco and Leo II are unusual because these systems cannot be placed in either one of the two Oosterhoff Groups that classify globular clusters. This is potentially a very significant observation because the periods of the variables depend on their masses, luminosities and effective temperatures, which presumably are functions of the compositions and ages of the systems.

The Oosterhoff effect is the observation that globular clusters can be divided into two groups according to the mean value of the periods of the type ab RR Lyrae variables (hereafter $\langle P_{ab} \rangle$). Clusters are said to belong to Oosterhoff Groups I and II if their values of $\langle P_{ab} \rangle$ are near $0\overset{d}{.}55$ and $0\overset{d}{.}64$, respectively. This dichotomy in $\langle P_{ab} \rangle$ is almost entirely the result of the clusters in the two groups having different transition periods (i.e. the period of the boundary line separating the type ab variables, the fundamental mode pulsators, from type c, the first harmonic pulsators). The transition period is typically $0\overset{d}{.}45$ and $0\overset{d}{.}53$ for Group I and II clusters, respectively, which means that Group I clusters contain type ab variables with periods between ~ 0.45 and 0.53 days while Group II clusters do not.

Table 11.2 lists some of the properties of the RR Lyrae variables in the four Dsph galaxies that have been surveyed for these stars and also the same data for the two Oosterhoff Groups. The transition period is customarily defined to be the period of the shortest period type ab variable (P_{ab}^{min} in table 11.2). The definition proposed by Cacciari & Renzini (1976) is perhaps better because it depends on the periods of more than one star.

Table 11.2. *Properties of the RR Lyrae variables*

	$\langle P_{ab} \rangle$	P_{ab}^{min}	P_{tr}
Sculptor	0.566	0.479	0.483
Leo II	0.593	0.487	0.522
Draco	0.617	0.537	0.545
Ursa Minor	0.636	0.490	0.560
Oosterhoff			
Group I clusters	0.551	0.454	0.453
Std. dev.	0.014	0.020	0.024
Oosterhoff			
Group II clusters	0.637	0.543	0.540
Std. dev.	0.008	0.023	0.024

Cacciari & Renzini first compute the periods the type c would have if they pulsated in the fundamental mode and then take as the transition period (P_{tr} in table 11.2) the period at which there are as many type c variables with fundamental periods greater than P_{tr} as there are type ab variables with periods less than P_{tr}.

The data in table 11.2 and inspections of the period–frequency histograms for the systems (see Cacciari & Renzini 1976; van Agt 1973) suggest that Sculptor and Ursa Minor belong to Oosterhoff Groups I and II respectively. As I indicated earlier, Draco and Leo II are not so easily placed in either group because their values of $\langle P_{ab} \rangle$ appear to be unusual. Their values of P_{ab}^{min} and P_{tr} suggest, however, that they belong to Group II.

This apparent ambiguity draws attention to the question of what should be used as the classification criterion. I do not believe that the traditional criterion, $\langle P_{ab} \rangle$, is always trustworthy. It is possible that the distribution of stars along a cluster's HB may be irregular as a result of uneven amounts of mass loss when the stars were red giants or as a result of some other process. The variation from cluster to cluster in the distribution of stars in the instability strip will produce differences in the period–frequency histograms for the type ab variable and hence a spread in the value of $\langle P_{ab} \rangle$. The transition period should not vary, however, as long as there are adequate numbers of variables in each cluster to define it. Until a few years ago the dispersion about the mean values of $\langle P_{ab} \rangle$ for the Oosterhoff Groups looked so small that there seemed to be little evidence for anything but more or less uniform distributions of stars across the instability strip. A number of clusters have been investigated for the first time only recently, and a few of these clusters have values of $\langle P_{ab} \rangle$ that are intermediate to the values for the Oosterhoff Groups (see Coutts, Dickens, Epps & Read 1975; Wehlau & Demers 1977). There is no ambiguity as to which group these clusters belong when they are classified according to the transition period, which suggests that it is perhaps a better criterion than $\langle P_{ab} \rangle$.

The colour–magnitude diagrams for Draco and Leo II show that in both systems there are large numbers of red HB stars, but only a very few stars to the blue side of the RR Lyrae domain. One can infer from this that the distribution of stars across the instability strip is likely to be skewed towards the red side. This should skew the period distribution of the type ab variables toward longer periods and make $\langle P_{ab} \rangle$ larger than it would be if the instability strip was evenly populated. One can then argue that since $\langle P_{ab} \rangle$ is $0\overset{d}{.}62$ and $0\overset{d}{.}59$ for Draco and Leo II respectively, these systems actually belong to Group I. I believe this hypothesis can be rejected on the basis of the values of the transition period. Draco and Leo II contain

6 and 8 type c variables respectively and a number of type ab variables with periods near $0^d.54$. There are therefore sufficient numbers of stars to define the transition period to as good or better accuracy than it is defined in most globular clusters. The fact that the transition periods of Draco and Leo II are similar to Group II clusters is evidence, in my opinion, that they belong to Group II. Castellani (1975) has reached the same conclusion with regard to Draco on the basis of its period–frequency histogram. The unusual values of $\langle P_{ab} \rangle$ are probably the result of uneven distributions of stars across the instability strip, but evidently the distributions do not have the form one imagines from looking at the colour–magnitude diagrams.

It is interesting to see how the Oosterhoff effect is related to the metallicities and HB types of the systems. This has been investigated before (e.g. Dickens 1972), but a number of recent improvements in the data make

Fig. 11.2. For globular clusters and Dsph galaxies, the transition period (P_{tr}) is plotted against [Fe/H] and HB type ($B/(B+R)$). The globular clusters in the two Oosterhoff Groups are depicted by different symbols, and the Dsph galaxies are depicted by open triangles. Note that in general the Oosterhoff Group I clusters are more metal-rich than the Group II clusters and that the Dsph galaxies obey this trend. Note also that there is not a good correlation between $B/(B+R)$ and P_{tr}.

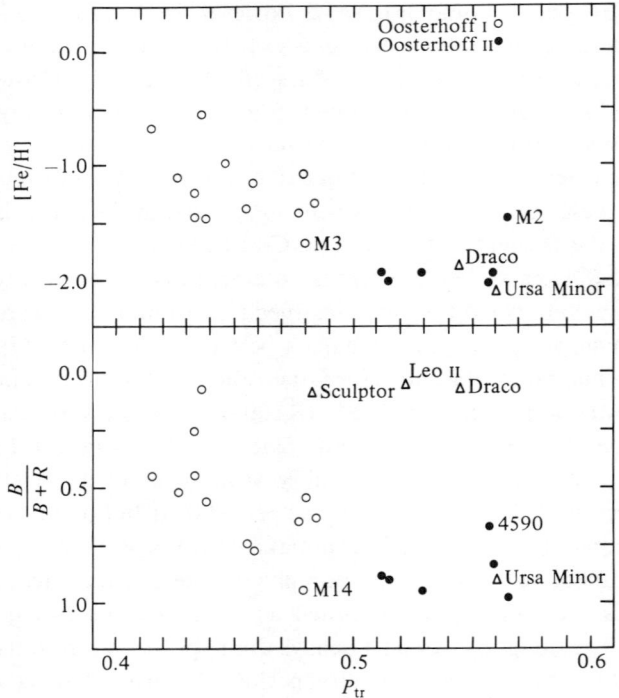

a repetition worthwhile. In fig. 11.2 I have plotted P_{tr} against [Fe/H] and $B/(B+R)$ for globular clusters and Dsph galaxies that contain ten or more type ab variables. The values of P_{tr} are therefore reasonably secure. These values were taken from Cacciari & Renzini's (1976) compilation, with the exception of the values for M72 and M2, which have been changed to reflect the fact that for each of these clusters Cacciari & Renzini mistook the longest-period type c variable for the shortest-period type ab variable. I have not included ω Cen in the sample because of the evidence that this peculiar cluster contains a mixture of Group I and Group II variables (Butler, Bell, Dickens & Epps 1977).

The upper half of fig. 11.2 shows that the Oosterhoff Group II clusters have, in general, lower metallicities than Group I clusters. This has been noted many times before, but I believe that until the new [Fe/H] data became available it was not realized that the division is quite sharp (see e.g. van Albada & Baker 1973). Fig. 11.2 shows that only M2 and M3 provide an overlap between the two groups, and it is possible that this is simply the result of the errors in the measurements of [Fe/H]. The observation that M3 red giants have unusually strong CO bands for their Fe/H ratios (Cohen *et al.* 1978) may be relevant here since the abundances of C, N, and O have a great influence on the position of the HB in the HR diagram. Only two of the Dsph galaxies, Draco and Ursa Minor, can be plotted in the upper half of fig. 11.2 because values of [Fe/H] are not available for the others (see section 11.4). Draco and Ursa Minor appear to obey the trend of Fe/H (or CNO/H) with Oosterhoff Group.

The bottom half of fig. 11.2 shows that there is not a good correlation between $B/(B+R)$ and Oosterhoff Group. The value of $B/(B+R)$ for NGC 4590 is not very firm because its colour–magnitude diagram (Harris 1975) contains only a few HB stars. In the case of M14, however, it is very unlikely that $B/(B+R)$ can be seriously in error (see Kogon, Wehlau & Demers 1974). Draco and Leo II also appear to deviate from any trend of HB type with Oosterhoff Group.

The Oosterhoff effect is widely throught to be caused by the so-called hysteresis phenomenon having to do with the direction that stars evolve across the instability strip and the effective temperatures at which the transitions from the fundamental to first harmonic mode and vice versa occur (see van Albada & Baker 1973; Stellingwerf 1975; Caputo, Castellani & Tornambè 1978). This theory predicts that Group I clusters should have redder HBs than Group II clusters, which was once consistent with the available observations (see van Albada & Baker 1973); but now, as fig. 11.2 illustrates, there are some contradictory data. The hypothesis that there is a difference in the abundance of He between the two groups (see Castellani, p. 83, this volume) can be formulated in a way that retains the hysteresis hypothesis and explains the lack of a trend in the bottom half of fig. 11.2. The details of this explanation are too lengthy to go into here, however.

Another important implication of fig. 11.2 is that since the Sculptor galaxy appears to belong to Group I, it may have a higher mean metal abundance than Draco, Ursa Minor, and Leo II. As I discussed earlier, the presence of a few type ab variables with periods between ~ 0.45 and 0.53 days identifies a system as Group I. It is conceivable that Sculptor contains variables of both groups, but it is classified as Group I on the basis of its variables that have periods near $0^d.48$. This hypothesis is consistent with the evidence that there is a large range in composition in Sculptor beginning near [Fe/H] ~ -2 (see section 11.4.2), assuming, of course, the trend of Fe/H with Oosterhoff Group is real.

11.3.2 Type II cepheids

The surveys of the Dsph galaxies have not discovered any variables that can be classified as Type II cepheids (also commonly called W Vir and BL Her stars). The Dsph galaxies do contain cepheid-like variables that are brighter than the HB, but at a given luminosity the periods of these stars are shorter than those of type II cepheids by substantial amounts. The properties of these unusual variables are discussed in the next section.

The absence of Type II cepheids in Sculptor, Leo II, and Draco may be related to the fact that these systems possess red HBs (see Norris & Zinn 1975). The short period Type II cepheids (i.e. the BL Her stars) are thought to have evolved from the blue HB (Strom, Strom, Rood & Iben 1970); hence they are expected to be absent in systems without blue HB stars. The post-HB stars that become the longer-period Type II cepheids (i.e. the W Vir stars) evolve rapidly through the instability strip, and the speed of their evolution increases with their masses (see e.g. Gingold 1976). The HB stars and post-HB stars in Sculptor, Leo II and Draco may be exceptionally massive since this would explain why these metal-poor systems have red HBs. If this is correct, then the post-HB stars may spend such short periods of time in the instability strip that there is only a small probability of observing one as a W Vir star.

The Ursa Minor system has a blue HB; consequently one might expect it to contain both BL Her and W Vir variables. It is not unusual, however, for a system with a blue HB to lack these stars, for several globular clusters (e.g. M92) share this property with Ursa Minor. This is probably a consequence of the short times scales of post-HB evolution and the fact that these low-mass systems contain relatively small populations of stars.

11.3.3 Anomalous cepheids

As I indicated above, the Dsph galaxies contain a group of cepheid variables that have some exceptional properties. The periods of these stars lie in the range of 0.3–3 days; consequently their periods overlap with those of the RR Lyrae and BL Her variables. They are brighter than the RR Lyrae variables by 0.5 to 2 magnitudes, and as a consequence of this and their periods, they do not obey the period–luminosity (P–L) relations of

Fig. 11.3. The anomalous cepheids in the Dsph galaxies (●), the SMC (×) and NGC 5466 (△) plotted in the period–luminosity diagram. The shaded areas depict the regions occupied by the well-known types of cepheid variables (the type II cepheids are called globular cluster cepheids). The data came from Hodge & Wright (1978) and from the references listed in Zinn & Searle (1976).

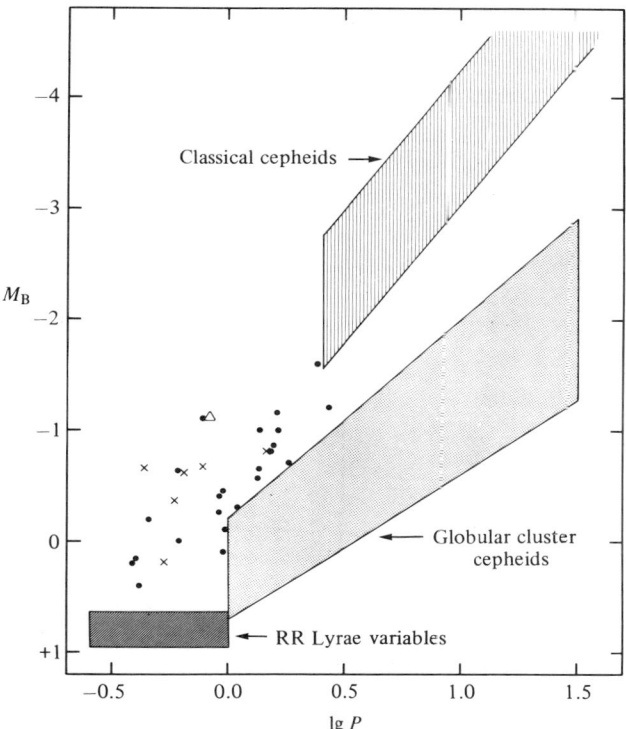

either the Type II cepheids or the classical cepheids. The $P-L$ relation that is defined by these stars is illustrated in fig. 11.3.

Every Dsph galaxy that has been searched for variable stars contains two or more of these anomalous cepheids. The Leo I galaxy appears to be unusually rich in these stars, for Hodge & Wright (1978) have classified 13 stars as anomalous cepheids and two other variables as probably anomalous cepheids. The SMC contains a number of anomalous cepheids (Tifft 1963; Graham 1975), but so far none have been found in the LMC despite the fact that searches have been made (Graham 1977). Anomalous cepheids appear to be very rare in globular clusters, for only NGC 5466 is known to contain one (Zinn & Dahn 1976).

Theoretical investigations (Christy 1970; Norris & Zinn 1975; Demarque & Hirshfeld 1975) have demonstrated that the periods, luminosities, and evolutionary histories of the anomalous cepheids can be understood if they are 2 to 3 times more massive than RR Lyrae variables, and observations

Fig. 11.4. The M_{bol} versus lg T_{eff} diagram for the ZAHB (i.e. initial core helium burning) models computed by Gross (1973). The three lines for $Z = 10^{-2}$, 10^{-3} and 10^{-4} connect models of different total mass (labelled) but identical core mass (0.45 $M_\odot$) and helium abundance ($Y = 0.35$). The approximate location of the instability strip is shown by the dashed lines. Note that stars of ~ 1.5 $M_\odot$ must be very metal-poor to populate the instability strip.

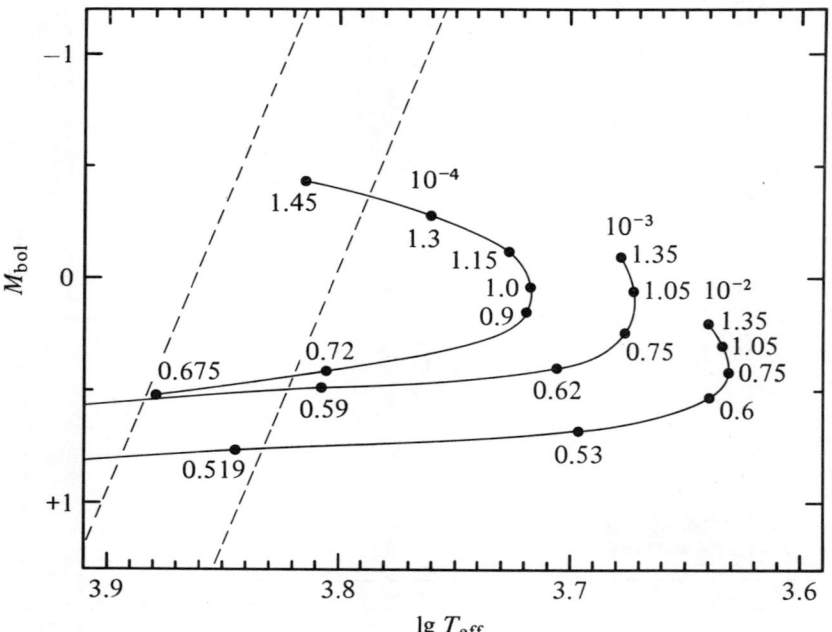

of the five anomalous cepheids and two RR Lyrae variables in the Draco galaxy have confirmed this mass ratio (Zinn & Searle 1976). Fig. 11.4, which shows the locations of the zero-age-horizontal-branch (ZAHB) models of Gross (1973) in the M_{bol} versus lg T_{eff} diagram, illustrates how stars of the predicted masses (i.e. ~ 1.5 M$_\odot$) can populate the instability strip if they are very metal-poor. The 1.45 M$_\odot$, $Z = 10^{-4}$ model that is plotted in fig. 11.4 has the correct luminosity and fundamental period to place it on the P–L relation of the anomalous cepheids (see Norris & Zinn 1975). The compositions of the anomalous cepheids have been investigated by Demarque & Hirshfeld (1975), who computed ZAHB models, and by Hirshfeld (1978), who considered in addition the evolution away from the ZAHB. Hirshfeld (1978) found that models with masses from 1.3 to 1.6 M$_\odot$, helium abundances near $Y = 0.24$, and metal abundances near $Z = 10^{-4}$ provide good fits to the observations of the shorter period ($P < 1.2$ days) anomalous cepheids. The longer period variables are not explained by these models and remain somewhat of an enigma.

It has been suggested that the large masses of the anomalous cepheids are the result of either relatively young ages ($t \sim 10^9$ years) or the transfer of mass in close binary systems (Norris & Zinn 1975; Demarque & Hirshfeld 1975; Zinn & Searle 1976). Against the young star hypothesis, one can argue that the Dsph galaxies cannot retain sufficient amounts of gas after the first generation of stars to have a second generation some 10^{10} years later (Zinn & Searle 1976). Also, radio observations indicate that these galaxies contain, at most, small amounts of H I gas at the present time (Knapp, Kerr & Bowers 1978). The binary star hypothesis has been given considerable support by the work of Renzini, Mengel & Sweigart (1977). These authors have provided a detailed explanation for how a binary system of two ~ 0.9 M$_\odot$ stars, separated by ~ 0.3–0.8 AU, can evolve into a system containing an anomalous cepheid of ~ 1.4 M$_\odot$. They have also given an explanation for why the anomalous cepheids are prevalent in the Dsph galaxies, but not in metal-poor globular clusters. They argue that in systems of high stellar density the orbits of any binaries that are present will shrink, due to close encounters with other stars, until the separations of the components are below the lower limit (~ 0.25 AU) for the production of anomalous cepheids. In the Dsph galaxies and in a few globular clusters, notably including NGC 5466, the stellar densities are sufficiently low that binaries of the proper separations to make anomalous cepheids can survive the required $\sim 12 \times 10^9$ yr. The only difficulty with this theory seems to be that while the presence of anomalous cepheids in the SMC is explained, their absence in the LMC is not. The other successes of this theory make it worthwhile, in my opinion, to extend the surveys of the LMC to see if some anomalous cepheids are indeed present. It is also important to note that this theory predicts the existence of anomalous cepheids in the Fornax system which has not yet been surveyed for variable stars.

11.4 Compositions

As we have seen, the abundances of helium, heavy elements, and carbon nitrogen and oxygen play important roles in determining the features of the stellar populations of the Dsph galaxies. The abundance of helium is difficult to measure in the Dsph galaxies for some of the same reasons that it is difficult to measure in globular clusters. The stars in which helium lines are present may have had their atmospheric compositions altered by gravitational settling or by convective mixing while on the giant branch. The method of estimating the helium abundance from the blue edge of the instability strip, which has been used successfully for some globular clusters, is thwarted in the case of the Dsph galaxies by the faintness of the HB stars (see Deupree & Hodson 1977). The only measurement of the helium abundance comes from the planetary nebula in the Fornax system, and its helium abundance is 'normal' (i.e. $He/H = 0.12$ by number, Danziger, Dopita, Hawarden & Webster 1978). The abundances of the elements heavier than helium can be estimated from observations of individual stars or inferred from the integrated light of the galaxies. In this section I review the data that have been gathered for each of the Dsph galaxies.

11.4.1 Fornax

Measurements of the integrated colours of the Fornax system (de Vaucouleurs & Ables 1968) and the integrated colours and spectra of its globular clusters (Hodge 1965a; van den Bergh 1969; de Vaucouleurs & Ables 1970; Danziger 1973) suggest that the majority of its stars are very metal-poor. However, one of the globular clusters appears to be substantially more metal-rich than the others (van den Bergh 1969; Danziger 1973), and the planetary nebula studied by Danziger *et al.* (1978) has ratios of N, O, and Ar to H that are only 2 to 3 times smaller than the corresponding values for the Orion nebula. It seems, then, that in Fornax there is a considerable range in metal abundance.

11.4.2 Sculptor

The giant branch in the colour–magnitude diagrams of Sculptor (Hodge 1965b; Kunkel & Demers 1977) is steep, a sign of low metal abundance, and it is also wider than the width that the observational errors alone could produce (J. Norris & M. S. Bessell, preprint), which is a sign of a range in metal abundance. The spectroscopic observations of four red giants (Cowley et al. 1978; J. Norris & M. S. Bessell, preprint) suggest that there is a wide range in metal abundance, which at the low-abundance end appears to be near the metallicities of M15 and M92 (i.e. [Fe/H] ~ -2). There is also some evidence for variations in the abundance of C or N among the stars, for one of the two stars observed by Norris & Bessell has an unusually strong $\lambda 3883$ CN band. The wide giant branch of Sculptor and the spectroscopic evidence for heavy element and C and N variations remind one of the peculiarities of ω Cen (see Freeman & Rodgers 1975; Dickens & Bell 1976; Bessell & Norris 1976).

11.4.3 Leo I and Leo II

The Leo I and Leo II galaxies have not been as thoroughly investigated as the other Dsph galaxies because they are somewhat more distant. Furthermore, the Leo I system lies so close to the bright star Regulus on the sky that detailed observations are very difficult to make, and for this reason nothing is known about its composition. For Leo II we have only the observation that it possesses the steep giant branch of a metal-poor system (H. H. Swope, private communication).

11.4.4 Ursa Minor

The colour–magnitude diagram of Ursa Minor is similar to that of M92 (Schommer et al. 1977), and the available data (Cowley et al. 1978; Canterna & Schommer 1978) indicate that Ursa Minor and M92 also have similar metallicities ([Fe/H] ~ -2.1).

One interesting result of Canterna & Schommer's broad-band photometry of four red giants was their discovery that one of them, star K, may have usually strong CH and CN bands. A comparison of Canterna & Schommer's photometry of star K with similar observations of the CN stars and CH stars in ω Cen reveals that star K has the colours of a CN star (J. Norris & M. S. Bessell, preprint). I have observed star K and several other red giants in Ursa Minor with the multichannel scanner of the 5-metre telescope. Star K does indeed have a peculiar spectrum, for the $\lambda 4216$ and $\lambda 3883$ CN bands and the G-band are unusually strong. There is no evidence of C_2 bands in the scans of star K (160 Å resolution); consequently Norris & Bessell are probably correct in identifying it as a CN star rather than a CH star. All of these observations suggest that the C and N variations in Ursa Minor may be as large as those in ω Cen.

11.4.5 Draco

The pioneering study of the Draco galaxy by Baade & Swope (1961) has greatly influenced the more recent investigations of the Dsph galaxies, and it is not simply a coincidence that more is known about the Draco system than any other. In recent years, the work of Hartwick & McClure (1974) has kindled a considerable amount of interest in Draco's metal abundance. From intermediate-band photometry of four red giants, Hartwick & McClure derived a value of $[Fe/H] = -2.8$ for Draco, which they cautioned was not very firm because it depended on an extrapolation of the calibration of the photometric indices with $[Fe/H]$. This value of $[Fe/H]$ was nonetheless very interesting because it suggested that Draco is more metal-poor than any globular cluster. The broad-band photometry by Canterna (1975) seemed to confirm this result, but the reanalysis of Hartwick & McClure's observations by Bell & Gustafsson (1975) suggested a much larger value of $[Fe/H]$. All of the more recent observations (Cowley et al. 1978; Canterna & Schommer 1978; Zinn 1978) have supported Bell & Gustafsson's conclusion that Draco is not exceptionally metal-poor.

In Baade & Swope's (1961) colour–magnitude diagram, the giant branch of Draco appears to be bifurcated. Normally the bluer of the two branches would be interpreted as the asymptotic branch, but it contains far too many stars in comparison with the redder branch for this to be the case. This unusual property of Draco has been investigated recently by Zinn (1978), who obtained scanner observations of 17 stars selected from the two branches. These observations suggest that there is a continuous range in colour among the stars rather than simply two giant branches and that the range in colour is caused by a range in metal abundance corresponding to the difference between M92 and M3, which is roughly a factor of three. A value of $[Fe/H] = -1.86$ was found for the mean metal abundance.

The low metal abundances of the Dsph galaxies are probably the result of gas being lost from the systems while star formation was taking place. Very crude models of the metal enrichment and mass loss in Draco have been constructed to match the distribution over metal abundance of the stars (Zinn 1978). These models provide a fit to the observed distribution if Draco has lost ~ 99 per cent of its initial mass. The mass-loss mechanism is not described in physical terms by these simple models. It is widely thought that the energy required to drive the gas out of Draco and other galaxies is supplied by supernova explosions.

So far, CN and CH variations have not been detected in Draco even thought 21 different red giants have been observed by the various observers listed above. This is a far larger number of stars than have been observed in either Sculptor or Ursa Minor, where CN and CH variations have been discovered. It is important to ask what size of variations would have been detected by the Draco observers. I believe it is safe to say that the sample of 21 stars does not contain any CH stars and CN stars similar to the extreme ones in ω Cen or any stars similar to star K in Ursa Minor. The abundances of C and N are therefore probably more uniform in Draco than in Ursa Minor and possibly in Sculptor as well. Whether this indicates that the primordial C and N to Fe ratios were more uniform in Draco or that the red giants in Draco have experienced less convective mixing is not clear. The CN and CH anomalies observed in the red giants in ω Cen also raise this issue (see e.g. Bessell & Norris 1976) and, until the more tractable problem posed by ω Cen is solved, the origin of these abundance variations will remain in doubt.

Spectroscopic surveys of 'normal' globular clusters such as M92, M13 and M5 have detected mild CH and CN variations that are probably caused by stellar evolution (see Norris & Zinn 1977; Zinn 1977; Bell, Dickens & Gustafsson 1978; and Kraft, p. 87, this volume). Nearly all of the Draco observations have insufficient resolution to detect variations of this kind, which are expected to be present.

11.4.6 Discussion

It is clearly important to extend the observations described above with the intention of better documenting the heavy element and C and N variations. The existing data are adequate, however, for one to argue that the atmospheric compositions of the stars in the Dsph galaxies are less uniform than the stars in most globular clusters. It is reasonable to believe that the relatively large masses of Fornax and Sculptor may have something to do with their ranges in composition. It is well known that mass and mean metal abundance, and presumably abundance range as well, are correlated among the more massive eliptical galaxies (see Faber's 1977 review). Fornax and Sculptor may represent the low mass and low metal abundance end of this relationship. It is not possible to extend this reasoning to the case of Draco, however. The mass of Draco is not larger than the masses of typical metal-poor clusters such as M92, which appear to have more uniform compositions. What then is the important difference between systems such as Draco and M92? The tidal hypothesis for the origin of the Dsph galaxies (see section 11.1) provides one way out of this dilemma, but not a particularly satisfying one since there is little other evidence to support it. The Dsph galaxies and the globular clusters have very different densities, which suggests the details of their formations from a gaseous state may be substantially different. The possibility that this may be related to the composition variations and the general problem of nucleosynthesis within systems of small mass need to be researched.

11.5 Summary

The most important properties of the Dsph galaxies can be summarized as follows:

(i) The masses and luminosities of the Dsph galaxies range from those of massive globular clusters to low-mass elliptical galaxies. The Dsph galaxies have much larger dimensions than globular clusters and hence much smaller stellar densities.

(ii) The Dsph galaxies appear to be satellites of the Galaxy, and their distribution in space does not appear to be random.

(iii) The Dsph galaxies are pure Population II systems, for their colour–magnitude diagrams closely resemble those of globular clusters. With one exception, the Dsph galaxies have usually red HBs for metal-poor systems; consequently they are said to suffer from the second-parameter anomaly.

(iv) The variable stars in the Dsph galaxies are different from the variables found in globular clusters in some important respects. Although the RR Lyrae variables in the Dsph galaxies can be classified by Oosterhoff Group, the Dsph galaxies do not obey the expected correlation between Oosterhoff Group and HB type. The Dsph galaxies do not contain any Type II cepheids, but they do contain a class of unusual cepheid variable which is very rare in globular clusters.

(v) The Dsph galaxies have low mean metal abundances, but the latest measurements do not indicate that they are more metal-poor than the most metal-deficient globular clusters. The heavy element, carbon and nitrogen abundances in the stars of the Dsph galaxies appear to be less uniform than in the stars in most globular clusters.

Further study of these properties is likely to increase our knowledge of the evolution of the Galaxy and of low-mass systems.

References

Baade, W. & Swope, H. H. (1961). *Astron. J.* **66**, 300.
Bell, R. A., Dickens, R. J. & Gustafsson, B. (1978). Presented at symposium on Important Advances in 20th Century Astronomy, Copenhagen.
Bell, R. A. & Gustafsson, B. (1975). *Roy. obs. Bull.* **182**, 109.
Bessell, M. S. & Norris, J. (1976). *Astrophys. J.* **208**, 369 (erratum in **210**, 618).
Butler, D. (1975). *Astrophys. J.* **200**, 68.
Butler, D., Bell, R. A., Dickens, R. J. & Epps, E. (1978) In *The HR Diagram*, IAU Symposium no. 80., ed. A. G. Davis Philip & D. S. Hayes, p. 183. Dordrecht: D. Reidel.
Cacciari, C. & Renzini, A. (1976). *Astron. Astrophys. Suppl.* **25**, 303.
Cannon, R. D., Hawarden, T. G. & Tritton, S. B. (1977). *Mon. Not. Roy. astron. Soc.* **180**, 81P.
Canterna, R. (1975). *Astrophys. J. Lett.* **200**, L63.
Canterna, R. & Schommer, R. A. (1978). *Astrophys. J. Lett.* **219**, L119.
Caputo, F., Castellani, V. & Tornambè, A. (1978). *Astron. Astrophys.* **67**, 107.
Castellani, V. (1975). *Mon. Not. Roy. astron. Soc.* **172**, 59P.
Christy, R. F. (1970). *J. Roy. astron. Soc. Canada*, **64**, 8.
Cohen, J. G., Frogel, J. A. & Persson, S. E. (1978). *Astrophys. J.* **222**, 165.
Coutts, C., Dickens, R. J., Epps, E. & Read, M. (1975) *Astrophys. J. Lett.* **197**, L45.
Cowley, A. P., Hartwick, F. D. A. & Sargent, W. L. W. (1978). *Astrophys. J.* **220**, 453.
Danziger, I. J. (1973). *Astrophys. J.* **181**, 641.
Danziger, I. J., Dopita, M. A., Hawarden, T. G. & Webster, B. L. (1978). *Astrophys. J.* **220**, 458.
Demarque, P. & Hirshfeld, A. W. (1975). *Astrophys. J.* **202**, 346.
Demers, S. & Kunkel, W. E. (1976). *Astrophys. J.* **208**, 932.
Deupree, R. G. & Hodson, S. W. (1977). *Astrophys. J.* **218**, 654.
de Vaucouleurs, G. & Ables, H. D. (1968). *Astrophys. J.* **151**, 105.
de Vaucouleurs, G. & Ables, H. D. (1970). *Astrophys. J.* **159**, 425.
Dickens, R. J. (1972). *Mon. Not. Roy. astron. Soc.* **157**, 281.
Dickens, R. J. & Bell, R. A. (1976). *Astrophys. J.* **207**, 506.
Faber, S. M. (1977). In *The Evolution of Galaxies and Stellar Populations*, ed. B. M. Tinsley & R. B. Larson, p. 157. New Haven: Yale University Observatory.
Freeman, K. C. & Rodgers, A. W. (1975). *Astrophys. J. Lett.* **201**, L71.
Gascoigne, S. C. B. (1966). *Mon. Not. Roy. astron. Soc.* **134**, 59.
Gingold, R. A. (1976). *Astrophys. J.* **204**, 116.
Graham, J. A. (1975). *Publ. astron. Soc. Pacific*, **87**, 641.
Graham, J. A. (1977). *Publ. astron. Soc. Pacific*, **89**, 425.
Gross, P. G. (1973). *Mon. Not. Roy. astron. Soc.* **164**, 55.
Harris, W. E. (1975). *Astrophys. J. Suppl.* **29**, 397.
Hartwick, F. D. A. & McClure, R. D. (1972). *Astrophys. J. Lett.* **176**, L57.
Hartwick, F. D. A. & McClure, R. D. (1974). *Astrophys. J.* **193**, 321.
Hartwick, F. D. A. & Sargent, W. L. W. (1978). *Astrophys. J.* **221**, 512.
Hesser, J. E., Hartwick, F. D. A. & McClure, R. D. (1977). *Astrophys. J. Suppl.* **33**, 471.
Hirshfeld, A. (1978). Ph.D. thesis, Yale University.
Hodge, P. W. (1965a). *Astrophys. J.* **141**, 308.
Hodge, P. W. (1965b). *Astrophys. J.* **142**, 1390.
Hodge, P. W. (1966). *Astrophys. J.* **144**, 869.
Hodge, P. W. (1971). *Ann. Rev. Astron. Astrophys.* **9**, 35.
Hodge, P. W. (1976). *Astron. J.* **81**, 25.

Hodge, P. W. & Michie, R. W. (1969). *Astron. J.* **74**, 587.
Hodge, P. W. & Wright, F. W. (1978). *Astron. J.* **83**, 228.
Iben, I., Jr (1971). *Publ. astron. Soc. Pacific*, **83**, 697.
Innanen, K. A. & Valtonen, M. J. (1977). *Astrophys. J.* **214**, 692.
King, I. R. (1966). *Astron. J.* **71**, 64.
Knapp, G. R., Kerr, F. J. & Bowers, P. F. (1978). *Astron. J.* **83**, 360.
Kogon, C. S., Wehlau, A. & Demers, S. (1974). *Astron. J.* **79**, 387.
Kukarkin, B. V. (1974). *The Globular Star Clusters*. Moscow: Nauka.
Kunkel, W. E. & Demers, S. (1975). *Roy. Obs. Bull.* no. 182, 241.
Kunkel, W. E. & Demers, S. (1977). *Astrophys. J.* **214**, 21.
Lynden-Bell, D. (1975). *Roy. Obs. Bull.* No. 182, 235.
Lynden-Bell, D. (1976). *Mon. Not. Roy. astron. Soc.* **174**, 695.
Mathewson, D. S. & Schwarz, M. P. (1976). *Mon. Not. Roy. astron. Soc.* **176**, 47P.
Mironov, A. V. (1972). *Soviet Astron.* **16**, 105.
Norris, J. & Zinn, R. (1975). *Astrophys. J.* **202**, 335.
Norris, J. & Zinn, R. (1977). *Astrophys. J.* **215**, 74.
Peterson, C. J. & King, I. R. (1975). *Astron. J.* **80**, 427.
Pilachowski, C. A. (1978). *Astrophys. J.* **224**, 412.
Racine, R. & Harris, W. E. (1975). *Astrophys. J.* **196**, 413.
Renzini, A. (1977). In *Advanced Stages in Stellar Evolution*, 7th Course of the Swiss Society of Astronomy and Astrophysics, Saas-Fee, ed. P. Bouvier & A. Maeder, p. 149. Geneva Observatory Publ.
Renzini, A., Mengel, J. G. & Sweigart, A. V. (1977). *Astron. Astrophys.* **56**, 369.
Sandage, A. R. & Wildey, R. (1967). *Astrophys. J.* **150**, 469.
Schommer, R. A., Olszewski, E. W. & Kunkel, W. E. (1977). In *The HR Diagram*, IAU Symposium no. 80, ed. A. G. Davis Philip & D. S. Hayes, p. 269. Dordrecht: D. Reidel.
Searle, L. & Zinn, R. (1978). *Astrophys. J.* **225**, 357.
Stellingwerf, R. F. (1975). *Astrophys. J.* **195**, 441.
Strom, S. E., Strom, K. M., Rood, R. T. & Iben, I., Jr (1970). *Astron. Astrophys.* **8**, 243.
Tifft, W. G. (1963). *Mon. Not. Roy. astron. Soc.* **125**, 199.
van Agt, S. L. T. J. (1967). *Bull. astron. Inst. Netherlands*, **19**, 275.
van Agt, S. L. T. J. (1973). In *Variable Stars in Globular Clusters and in Related Systems*, IAU Colloquium no. 21, ed. J. D. Fernie, p. 35. Dordrecht: D. Reidel.
van Albada, T. S. & Baker, N. (1973). *Astrophys. J.* **185**, 477.
van den Bergh, S. (1967). *Astron. J.* **72**, 70.
van den Bergh, S. (1969). *Astrophys. J. Suppl.* **19**, 145.
van den Bergh, S. (1972a). *Astrophys. J. Lett.* **171**, L31.
van den Bergh, S. (1972b). *Astrophys. J. Lett.* **178**, L99.
van den Bergh, S. (1975). *Ann. Rev. Astron. Astrophys.* **13**, 217.
Wehlau, A. & Demers, S. (1977). *Astron. Astrophys.* **57**, 251.
Zinn, R. (1977). *Astrophys. J.* **218**, 96.
Zinn, R. (1978). *Astrophys. J.* **225**, 790.
Zinn, R. & Dahn, C. C. (1976). *Astron. J.* **81**, 527.
Zinn, R. & Searle, L. (1976). *Astrophys. J.* **209**, 734.

12
Globular clusters as extragalactic distance indicators

DAVID A. HANES

12.1 Introduction

The establishment of the far-field extragalactic distance scale necessitates the measurement of the distances of galaxies whose peculiar velocities are small compared to the cosmic expansion rate at that distance. In practice this means studying galaxies at least as remote as those in the Virgo cluster, where a precise estimate of the mean recession velocity (~ 1100 km s^{-1}) can be obtained by averaging over a large number of cluster members. Ideally, the procedure is simply to compare luminous Local Group objects with their apparently fainter counterparts in Virgo galaxies and thereby directly to obtain the luminosity distance. However, most distance determinations to date have relied on indicators insufficiently luminous to allow this single step determination; or, more correctly, the bright indicators used in the remote mapping – such as the Sc I galaxies themselves – are not found in the Local Group, and their absolute calibration follows in steps from a series of distance determinations for nearby groups using the well-known indicators: cepheids, diameters of H II regions, brightest stars within galaxies, and the like. Sandage & Tammann (1976, and earlier papers referenced therein) have presented such a comprehensive study in an impressive series of closely-argued presentations.

It has long been recognized that globular clusters should permit single-step evaluations of the Virgo cluster distance modulus: they are seen in great numbers around many Virgo galaxies. However, the large intrinsic spread in globular cluster luminosities (from $M_V \sim -10$ to $M_V \sim -3$ in our own galaxy) does not permit straight-forward comparisons without a clearer understanding of what constitutes a reliable fiducial mark: the mean magnitude? (Kron & Mayall 1960); the absolute magnitude of the brightest single globular cluster associated with a galaxy? (Racine 1968b; Sandage 1968); the absolute magnitude of the brightest cluster, empirically corrected to account for the fact that brighter galaxies possess larger numbers of globular clusters? (de Vaucouleurs 1977a); the full luminosity function? (Hanes 1977a); or some other?

In this chapter I will present an historical perspective of research in this area, culminating with what I now believe to be a reliable measurement of the Virgo cluster distance modulus, based on a single stride to this cosmologically important distance.

12.2 Advantages of using globular clusters

There are three advantages. The first of these I have already mentioned, and it remains the most appealing of all: namely, that globular clusters will permit a single-step determination of the distance of the Virgo cluster. Fig. 12.1 underscores this point, showing (a) a direct photograph of the galactic globular cluster M13 and (b) a direct photograph of M87, the well-known giant elliptical galaxy in the Virgo cluster, surrounded by its retinue of globulars. The alternative treatment, that of establishing a chain of indicators to successively greater distances (Tammann 1973), is less appealing in that one broken or weak link may invalidate the whole chain (de Vaucouleurs 1977b); and of course the errors propagate multiplicatively link-by-link through the chain of reasoning (see e.g. de Vaucouleurs 1979, and the previous papers in that series for an exhaustive re-analysis of the Sandage and Tammann chain).

The second advantage is that the globular cluster distance modulus will rely finally only upon pure Population II indicators: the clusters themselves, and RR Lyrae stars to set the calibration in the sample of globulars in our own galaxy. The distance scale eventually arrived at will thus be essentially completely independent of the usual methods, based as they are on Population I indicators such as cepheids, H II regions, brightest supergiants, and the Sc I galaxies whose Population I spiral structure is used as an indicator of luminosity class (van den Bergh 1960). In principle then the use of globular clusters will afford an independent check on the more well-known techniques.

Finally, the globular cluster distance scale is in principle independent of the effects of interstellar obscuration. This is because the difference between the apparent magnitude V_{HB} of a horizontal branch star and the integrated magnitude V_{cl} of the globular cluster to which it belongs is independent of any intervening absorption (provided simply that the obscuration is not patchy on scales smaller than the cluster, or that such patchiness can be dealt with). Thus from two measurable quantities V_{HB} and V_{cl} and the known absolute magnitude $M_{V,HB}$ of RR Lyrae stars,

Globular clusters as extragalactic distance indicators 215

Fig. 12.1. (a) NGC 6205 (M13), a globular cluster in our own galaxy. This figure is a reproduction of a Hale Observatories 5-metre prime focus photograph. (b) NGC 4486 (M87), a giant elliptical galaxy in the Virgo cluster, surrounded by its retinue of globular clusters. This plate is a reproduction of a Hale Observatories 5-metre prime focus photograph.

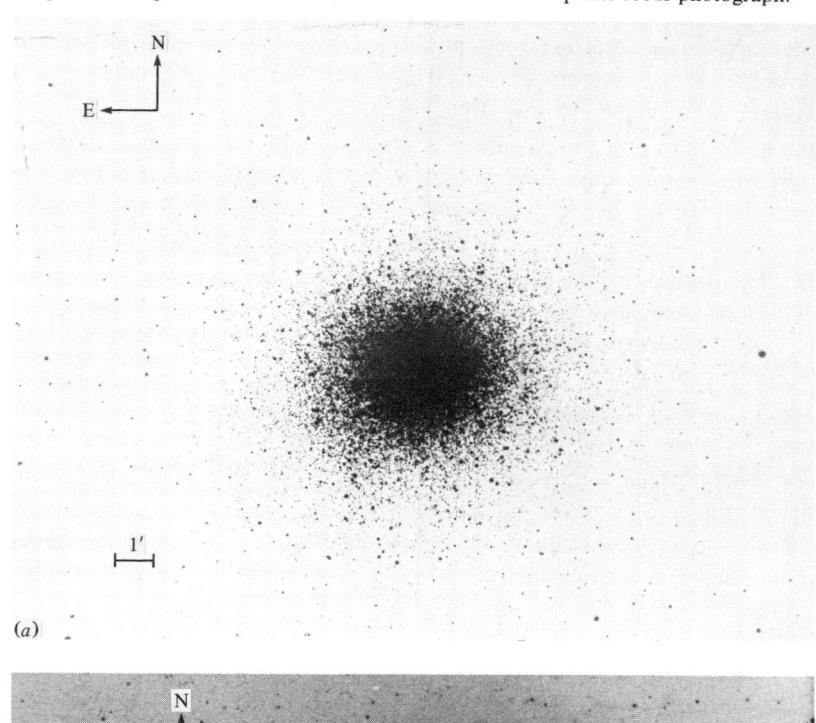

(a)

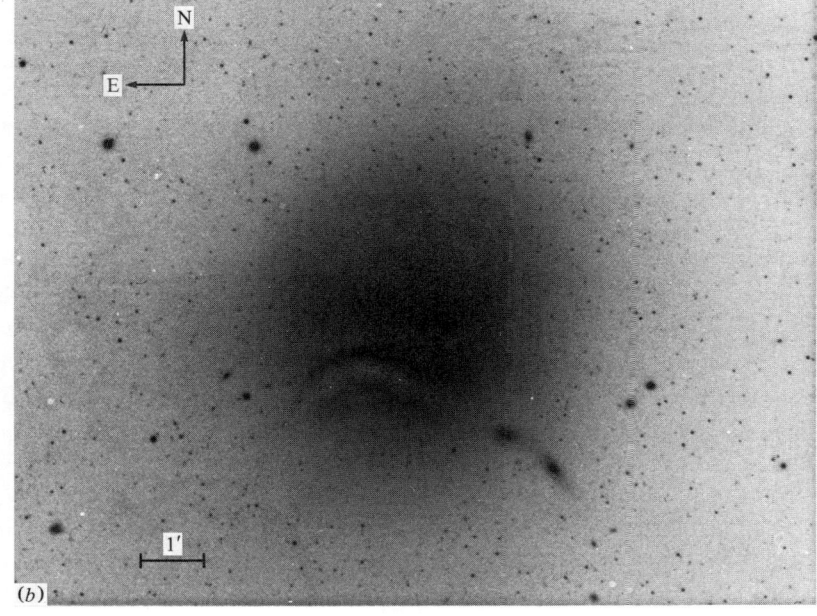

(b)

calibrated locally (van Herk 1965; Woolley & Savage 1971; Heck 1973; Hemenway 1975), the absolute magnitude of globular clusters can be derived regardless of the amount of obscuration present.

Interestingly this last advantage applies too in a later stage of the calculation: once the absolute magnitude is known for some representative globular cluster associated with (say) NGC 4472, the brightest Virgo cluster elliptical galaxy, then the absorption-independent difference between the apparent magnitude of such a globular and that of the galaxy itself will yield the absolute magnitude of NGC 4472, thereby setting the zero-point in the Hubble diagram and establishing the cosmic distance scale. As it happens, the Virgo cluster lies at high galactic latitude where the obscuration may be negligible (Sandage 1973; but see de Vaucouleurs 1978a). Nevertheless, the argument is perfectly valid and may be important in future studies elsewhere.

12.3 Disadvantages encountered

The obvious disadvantage is that globular clusters span such a wide range of absolute magnitudes. In the next section I will relate how various attempts to use globulars as distance indicators foundered on the rock of an imperfect understanding of the luminosity functions of globular cluster families. I believe that we now have such an understanding, and that the problem is tractable.

A second difficulty is that of the identification of the globular clusters associated with the Virgo cluster galaxies. A complication is that they are expected to be unresolved at the distance of the Virgo cluster: the mean diameter of the isophote containing 90 per cent of the integrated light of globular clusters in our own galaxy is 23 pc (Woltjer 1975), and at even the rather small Virgo cluster distance of 10 Mpc (corresponding to a Hubble constant near 100 km/s/Mpc) the angular subtense of such a cluster is only about 0.5 arcsec. Thus the discrimination between globulars and field stars is a problem. The Virgo cluster is actually rather well placed in that it is at high galactic latitude and the foreground contamination is minimized; nevertheless our own galaxy, if at the distance of the Virgo cluster, would show only marginal evidence for the existence of its known globular cluster population, in the purely statistical sense of a significant number of unresolved objects being left after field object subtraction (Hanes 1975).

Improvements on this are possible, of course. Multicolour photometry aids in discriminating between clusters and stars, as restricting the study

Globular clusters as extragalactic distance indicators

to the relevant colour range effectively increases the contrast. Unfortunately, such photometry is not yet in hand in the fields of the spiral galaxies in the Virgo cluster. Globular clusters are centrally concentrated in the galaxies to which they belong, and an excess of dubious or marginal statistical significance may be more convincing if it displays this behaviour. Still, close examination of the numbers involved reveals that at present the Virgo cluster spirals are tantalizingly just beyond the feasible establishment of a reliable luminosity function for their associated globular clusters.

The obvious solution is to study the clusters associated with the giant elliptical galaxies, such as M87 (fig. 12.1 (b)). Here the globular cluster populations are so large that in fact moderately large uncertainties in the field star correction are essentially unimportant, which is all to the good. However, a different question now arises: are the globular clusters associated with giant elliptical galaxies very like those found in spiral galaxies? – for, after all, it is within the spiral galaxies of the Local Group (and in our own galaxy in particular) that the calibrating luminosity function is established. There are theoretical reasons for expecting the luminosity functions to be different. For example, globulars are subject to disruptive events (disc passages) not experienced by their counterparts in elliptical galaxies (Fall & Rees 1977; and Fall, p. 309, this volume). The brighter, more massive globular clusters may thus be preferentially destroyed in spirals, leading to a carving away of the primordial luminosity function into one rather dissimilar to the function for globular clusters in elliptical galaxies – even assuming that these were the same primordially, which is a different (and unanswered) question yet again. However, on empirical grounds I believe that the differences between these various luminosity functions are small, as I will explain in section 12.5.2. It is certainly true, however, that this uncertainty remains the biggest problem in the whole approach of using globular clusters as extragalactic distance indicators.

12.4 An historical review

12.4.1 The earliest attempts

The earliest work that I am aware of in this area is that of Baum (1955), who made two important observations. First, he commented that the objects around M87 – which he identified without elaboration as globular clusters – were distributed roughly as the distribution of the underlying surface brightness, clearly indicative of a real physical relationship.

Secondly, he compared the brightness of the clusters around M87 to those around M31 to deduce a Virgo cluster apparent distance modulus of 30.2 mag and a Hubble constant near 150 km/s/Mpc. He gave no details of the photometric techniques used or clusters studied, but Sandage (1968) tells us that Baum intercompared photoelectrically the brightest single clusters in each galaxy. I do not know how Baum selected the appropriate cluster candidate in M87.

Kron & Mayall (1960) presented photometry for globular clusters in several Local Group galaxies and were the first to test critically the similarity of the globular clusters in different systems. They recognized strong observational selection effects but stated, when comparing the cluster luminosity functions in M31 and in our own galaxy, that 'about

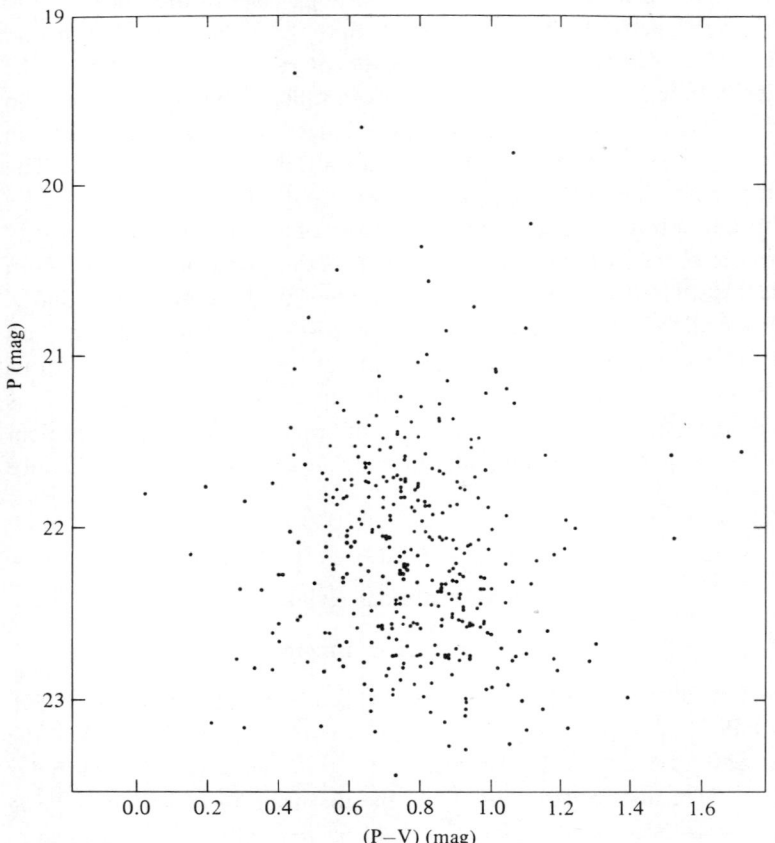

Fig. 12.2. The P versus P−V colour-magnitude diagram for 360 unresolved objects in the M87 field (from Racine 1968a).

all we are willing to conclude is that the present data do not indicate any substantial difference... between M31 and the Galaxy'. Kron & Mayall assumed that the observed maxima in the frequency–magnitude distributions for different galaxies corresponded to the same absolute magnitude, and thereby derived imprecise distance moduli for M31, the Magellanic Clouds, and the centre of our own galaxy; but of course the uncertainties were large because of the (admitted) selection effects.

A year later, Sandage (1961) retested the usefulness of globular clusters as distance indicators by again comparing Kron & Mayall's (1960) photometry for globulars in our own galaxy and in M31 to derive the distance of the latter. Sandage got discrepant values (relative to estimates from the more usual methods), whether intercomparing single brightest clusters or entire luminosity functions. The globular cluster distance moduli were too small, which is qualitatively as expected if the single brightest cluster in M31 is brighter than that in our own galaxy (say, by a sample-size effect) and if the M31 function was incompletely sampled as a function of magnitude, as Kron & Mayall (1960) themselves noted. Both of these are in fact the case, as we shall see.

Two extremely important contributions came next from Racine (1968a, b). He carried out photographic (P,V) photometry on several hundred objects associated with M87 and showed (Racine 1968a) that the mean colour of globular clusters in that galaxy was not significantly different from that for other cluster families, given the observational uncertainties. This encouraging demonstration of the gross similarity of the clusters associated with M87 and with the Local Group galaxies was an important starting point for future applications. Fig. 12.2 reproduces Racine's (1968a) colour–magnitude diagram for 360 unresolved objects in the M87 field. In this photometric system a typical Local Group globular cluster colour ($B-V \sim 0.75$ mag) corresponds to $P-V \sim 0.65$ mag, close to the ridge line in the 'Christmas tree' distribution of points (which may be somewhat reddened) in the figure. The limited numbers studied and the uncertain contamination by field stars precluded the accurate establishment of the cluster luminosity function, so Racine (1968a) did not derive a Virgo cluster distance modulus.

Thus in a subsequent analysis Racine (1968b) enlarged the scope of the study, deriving magnitudes in the P bandpass for more than 1000 objects in a somewhat larger field around M87. The resulting luminosity function is reproduced in fig. 12.3(a), and the striking feature is the abrupt rise at the $P \sim 21.2$ mag level. This 'step' is in fact somewhat overemphasized by Racine's numerical treatment: he subtracted an outer annulus from an

Fig. 12.3. (a) The luminosity function in B for the globular clusters associated with M87. N_T represents Racine's estimated total cluster population, rescaled to an observed total of 1150. The data are from Racine (1968b). (b) The same function with statistical uncertainties (vertical bars) indicated. The horizontal bars indicate the bin sizes. ΔN represents the difference observed after the annular subtraction explained in the text.

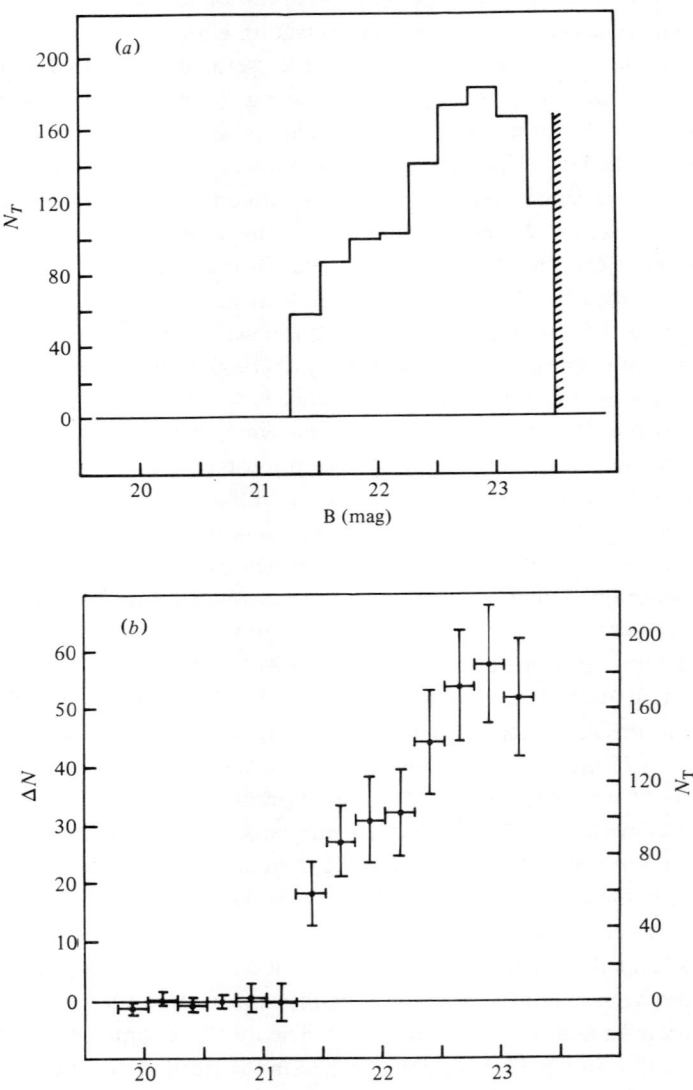

(appropriately normalized) inner annulus to leave a pure globular cluster luminosity function, which he then rescaled to a total population of 1150 (his estimate of the total cluster count). Thus while the first non-zero bin in fig. 12.3(a) contains an estimated 58 globular clusters, the appropriate uncertainty in the count is nearer ± 18 than the simple-minded $\pm \sqrt{58}$ suggested by the plot. I have shown the effect of including the true statistical uncertainties in fig. 12.3(b). The abrupt rise in the luminosity function, while less dramatic, still seems to be present, and on the basis of fig. 12.3(a) Racine was prompted to suggest that the bright cutoff for globular clusters is not a sensitive function of the total number of clusters or of the mass of the parent galaxy. His direct intercomparisons of the brightest single clusters in our own galaxy and in M31 with those in M87 then yielded Virgo cluster apparent distance moduli near 30.8 mag.

Sandage (1968) reapplied Racine's method under the same assumptions to derive a distance modulus of 31.1 mag and a Hubble constant near $H = 75$ km/s/Mpc. Sandage went on to say that the latter estimate must be an upper limit to the true value of H because if the working assumption (the equality of first-ranked globular clusters) was wrong, then M87, with a larger cluster population, should have intrinsically brighter clusters. However, this statement was not justified because of the uncertain identification of the single brightest cluster in M31: the likely candidate, cluster M$_{\rm II}$, was excluded on the grounds of photometry contaminated by nearby bright stars. (Eventually van den Bergh (1969) was able to confirm the correctness of the photometry of M$_{\rm II}$ and redo Sandage's calculations). Thus amendments of H in either direction should not have been ruled out at the time.

Finally van den Bergh, writing in 1969 a review which appeared in final form in the long-awaited Volume 9 (van den Bergh 1970, 1975) compared nine different methods of determining the value of the Hubble constant. That relying on globular clusters gave the smallest value of all, and van den Bergh rightly pointed out just how critically the method used then depended upon the strict equality of absolute luminosity for the first-ranked clusters in the giant spiral and giant elliptical galaxies.

12.4.2 The recognition of sample-size effects

After the warnings that had been sounded so many times, it was de Vaucouleurs (1970) who first demonstrated within the Local Group the existence of a loose correlation between the absolute magnitude $M_{\rm B}(1)$ of the first-ranked cluster in some family and the total absolute magnitude

M_B (G) of the galaxy to which that cluster family belongs. This empirical relationship is shown in fig. 12.4, and the upper envelope is given by de Vaucouleurs as

$$M_B(1) = -10.0 + 0.2[21.0 + M_B(G)] \qquad (1)$$

(where, following Hodge (1974), I have corrected some misprints) as calibrated in the Local Group galaxies.

Following a suggestion which he attributed to Kron & Mayall (1960), de Vaucouleurs (1970) assumed that the universal globular cluster population was normally distributed with a mean $\langle M_B \text{(cluster)} \rangle = -7.2$ mag and an intrinsic dispersion $\sigma = 0.8$ mag. He then interpreted eq. (1) as reflecting a sample-size effect, with samples of size N drawn from a universal population according to the prescription:

$$\lg N \approx -0.3[10.0 + M_B(G)]. \qquad (2)$$

De Vaucouleurs tested the hypothesis in the light of the data then available and concluded that the agreement was good (though this is not entirely surprising since eq. (2) was generated empirically to match the data; its functional form and the coefficients entering were not restricted

Fig. 12.4. The correlation between the absolute magnitude $M_B(1)$ of the brightest cluster in a galaxy and the absolute magnitude $M_B(G)$ of that galaxy, as observed in the Local Group. The dots represent clusters with $(B-V) > 0.5$, while the crosses represent those with $(B-V) < 0.5$. The solid line is the empirical upper limit (eq. (1), text) while the dotted line shows the expected relationship following the population scaling (eq. (2)) explained in the text (from de Vaucouleurs 1970).

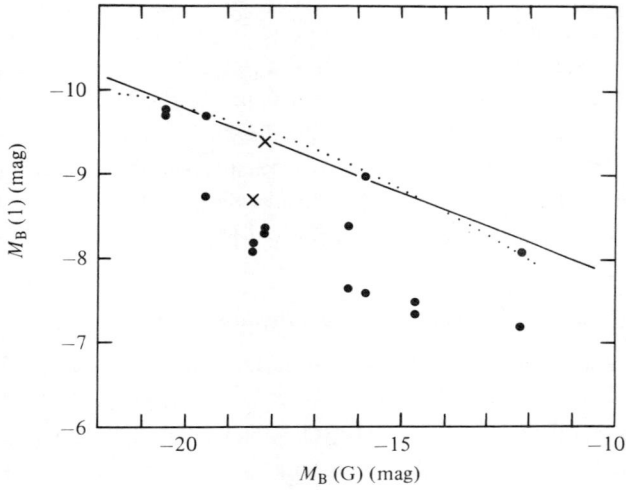

by any theoretical constraints). Finally, he applied eq. (1) to Racine's (1968b) photometry for the globulars around M87 to deduce a Virgo cluster apparent distance modulus of 31.5 mag and Hubble constant near 50 km/s/Mpc – in those days a surprisingly small value.

However, the analysis had its problems. First, de Vaucouleurs combined clusters of various types so that (for example) a total population of ∼ 1000 clusters was assigned to our own galaxy (Alter, Ruprecht & Vanysek 1958) while M31 possessed only ∼ 300 (Veteŝnik 1962), in constrast to the true situation for the globular clusters alone: M31 has at least twice (Harris 1974) and probably several times (Sargent, Kowal, Hartwick & van den Bergh 1977) the cluster population of our own galaxy. If it is argued that combining clusters of all types is the statistically soundest approach, then the analysis is open to the further criticism that the mean magnitude and intrinsic dispersion adopted may be reasonable parameters for globular clusters (and indeed were so chosen) but are not for open clusters. Moreover, we know now (Harris 1974; de Vaucouleurs 1977a) that the mean magnitude and intrinsic dispersion used by de Vaucouleurs (1970) are *not* good estimates for those statistics within our own galaxy. But, as I will show later (section 12.4.4), the worst failing in this kind of modelling is that there is *no* simple scaling of total cluster population with galaxian luminosity (that is, eq. (2) does not hold for all galaxies), and the behaviour of the first-ranked cluster is thus unpredictable (that is, eq. (1) in turn does not hold for all galaxies).

One puzzling question left unaddressed by de Vaucouleurs (1970) was the explanation of the very sharp cutoff seen by Racine (1968b) in the luminosity function of globular clusters associated with M87. If the parent population truly was normally distributed, then the observed field-corrected function in this rich sample should be expected to fall asymptotically towards zero at the bright end.

As it happens, some evidence for just such a smooth falloff was found by Hanes (1971), who carried out UBV photographic photometry on more than 800 objects in the M87 field. I will discuss the multicolour results in the next section, but here I simply present Hanes' B luminosity function in fig. 12.5. It differs from the original (Hanes 1971) in that it has been field-corrected in the way Racine (1968b) treated his data, with an outer annulus subtracted from an inner, though the annuli used here do not correspond to those used by Racine. The Hanes study covered a wider area so the statistics are somewhat better, but the crucial difference is that the Hanes luminosity function is restricted to objects in the colour range $0.2 \leqslant B-V \leqslant 1.2$. This additional restriction sharpens the contrast between the globular clusters and the field stars. The plot suggests that the luminosity function is falling off more smoothly than thought by Racine (1968b) and may in fact be starting to climb out of the background at a level fully as bright as $B \approx 20.5$ mag, though that point of view could hardly be argued with great conviction. The data are clearly much affected by the relatively large uncertainties unavoidable in small-number statistical samples. The point I wish to emphasize is that the apparently very sharp cutoff found by Racine (1968b) was much less convincing in this later work. However, it is clear that what was needed was an independent estimate of the field contamination so that the low-accuracy annular field correction methods could be bypassed: more Virgo cluster fields needed studying, but that advance still lay some way off (Hanes 1977b).

Meanwhile, Hodge (1974) reconsidered the photometry of globular clusters in the Local Group galaxies in the M31 group (NGCs 147, 185, 205, 224; and the Fornax and Sculptor dwarf spheroidals). In this restricted sample of galaxies, the globular cluster searches are probably complete to comparable magnitude levels. Hodge was able to show that the general correlations established by de Vaucouleurs (1970) persisted, and he recast the dependences of first-ranked cluster brightness and total

Fig. 12.5. The luminosity function in B for the globular clusters in M87. ΔN represents the difference after annular subtraction, as in fig. 12.3(b), but here the data (from Hanes 1971) were restricted to objects with $0.2 \leqslant B-V \leqslant 1.2$. The vertical bars represent the statistical uncertainties and the horizontal bars show the bin sizes.

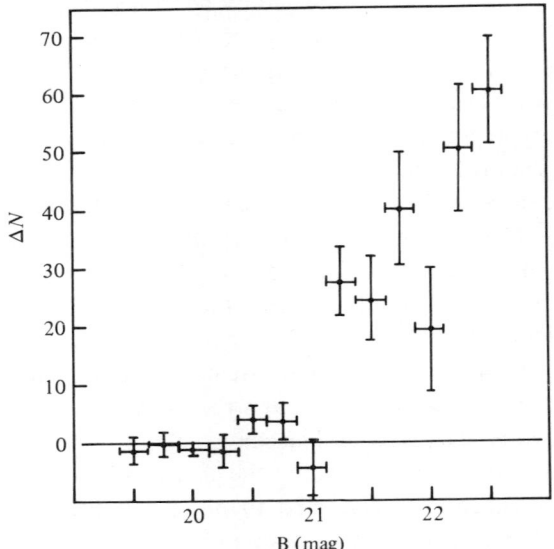

cluster population upon galaxian magnitudes in the different forms suggested by these new data. At least within the Local Group, it was now clear that the globular cluster luminosity function did not truncate sharply at the bright end, but the immediate application of Hodge's new-found relationships to the limited Virgo cluster data then available was not obvious.

12.4.3 The colours of globular clusters associated with M87

Hanes (1971) presented B,V photometry for more than 800 objects in the M87 field. Fig. 12.6(a) shows a colour–magnitude diagram in the B, (B−V) system for 478 objects having at least three independent measures in each of B and V. The familiar 'Christmas tree' distribution of points reappears, as in Racine (1968a) and fig. 12.2, with a ridge line near $B-V \approx 0.7$ mag. For the subset of points in the colour range $0.2 \leqslant B-V \leqslant 1.2$, the mean colour is plotted in fig. 12.6(b) in 0.5 mag bins from $B = 20.5$ to $B = 23.0$. Also shown is the overall mean for the M87 data and the mean for 62 well-studied globular clusters in our own galaxy (Racine 1973). These mean colours are closely similar, though at first viewing it is puzzling that the fainter clusters in M87 seem somewhat bluer than their bright brethren (a trend which can be discerned in fig. 12.6(a)). No strong correlation of the kind has been found in other cluster systems, though Frogel, Persson & Cohen (p. 170, this volume) now suggest that there is a weak tendency for $(V-K)_0$ to be numerically larger for the brighter clusters in M31 and in our own galaxy. However, there are two other (perhaps more likely) explanations for the trend in the M87 photometry. First, the photographic measurements were made by Hanes with an iris photometer of limited sensitivity (a problem overcome in later studies) and scale errors of ± 0.05 mag cannot be ruled out near $B \approx 23$ mag. Secondly, it must be remembered that the points plotted contain an admixture of field stars. The numbers of field stars do not rise as rapidly as the numbers of globulars beyond $B = 20.5$ mag (Hanes 1977b), and at the latitude of M87 the field stars are redder in the mean than the clusters (Weistrop 1972). Thus at the bright end of the luminosity function the field star contamination will lead to an erroneously large value of $\langle B-V \rangle$ while at fainter levels a more representative estimate will result. On balance, it is clear that at least in the B−V colour the globulars associated with M87 are very like those in our own galaxy.

For 189 objects, Hanes (1971) presented U photometry as well, but noted that it was of poorer quality than the B and V data. Moreover, the B, V

Fig. 12.6. (a) The B versus B−V colour–magnitude diagram for 478 objects in the M87 field with at least three measures in each of B and V (Hanes 1971). (b) The dependence of mean cluster colour $\langle B-V \rangle$ upon B for the objects in fig. 12.6(a). The mean colours have been calculated for bins of width 0.5 mag in B and for objects with $0.2 \leqslant B-V \leqslant 1.2$; the number of objects entering each determination is shown at right. The error bars are the standard errors of the means. Also shown is the overall mean for 426 objects in the M87 field and the mean for 62 well-studied clusters in our own galaxy (Racine 1973).

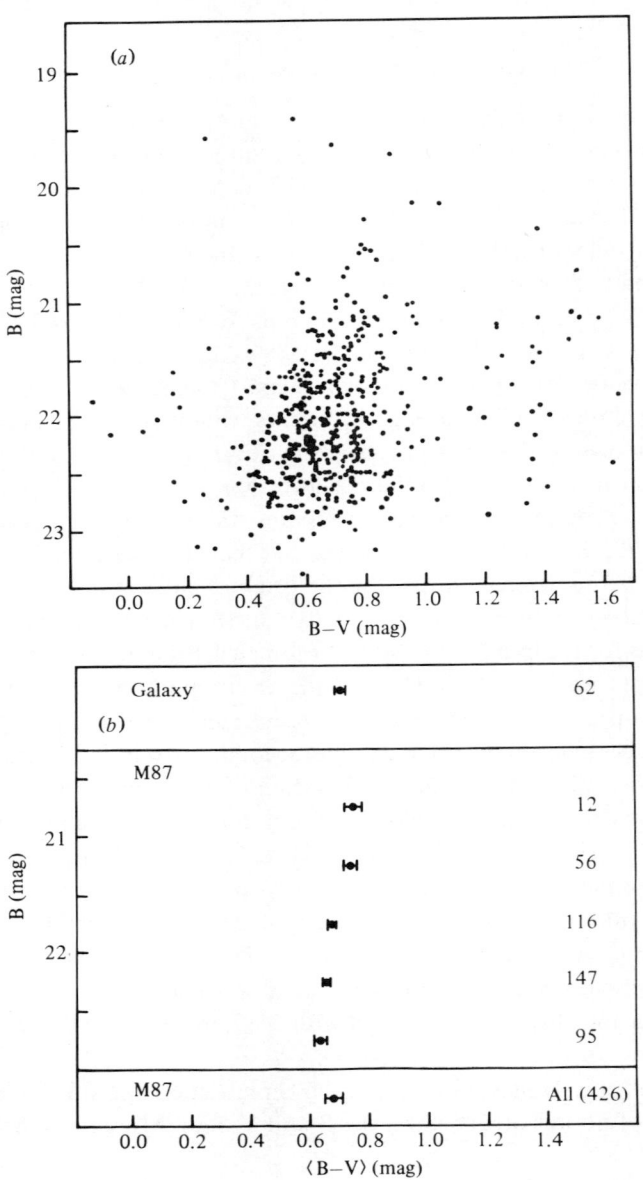

data could be compared to Racine's (1968a) values, which were calibrated in a completely independent way. No such independent check could be made for the U photometry, and Hanes (1971) could not rule out zero-point errors near ±0.2 mag.

With these reservations in mind, let us examine in fig. 12.7 the distribution of 62 points in the (U−B, B−V) colour–colour plane. Here the data have been further restricted to objects with at least three measures in B and V and at least two in U; and only the objects fainter than B = 20.5 mag have been plotted. Only the moderately bright part of the cluster luminosity function is sampled, of course, and the admixture of field stars will be of relatively more importance here than in fig. 12.6(a). The main sequence colour–colour relation (Johnson 1963) and the subdwarf envelope

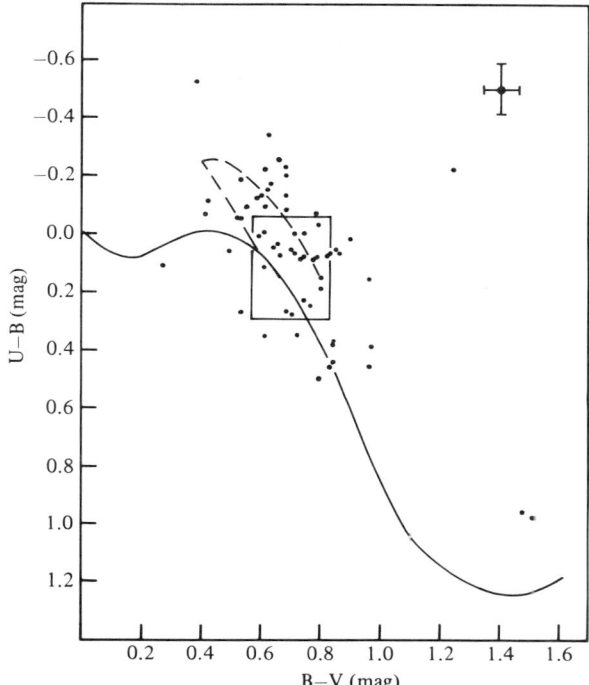

Fig. 12.7. The U−B versus B−V colour–colour diagram for 62 objects in the M87 field. The points plotted have two measures of U and at least three in each of B and V, and are fainter than B = 20.5 mag. The main-sequence colour–colour relation (Johnson 1963) is shown as a solid line, and the subdwarf envelope (Dixon 1965) is represented by a broken line. The box represents the area within which 90 per cent of Racine's (1973) well-studied galactic globular clusters lie, after reddening corrections. The typical error in the determination of U−B and B−V for a single object is indicated at top right. The data are from Hanes (1971).

(Dixon 1965) are indicated. The box represents the area within which 90 per cent of Racine's (1973) well-studied reddening-corrected galactic globular clusters are found. Finally, the probable errors in the determination of $U-B$ and $B-V$ for an individual object are shown in the upper right of the figure. Given the uncertainties, it seems entirely plausible that the points plotted represent a mixture of galactic halo stars and globular clusters in M87, the latter being very like the clusters in our own galaxy.

Ables, Newell & O'Neil (1974) restudied the B, V broadband colours of the M87 globulars in a programme of electronographic photometry. In general, this photometry (limited to 86 objects) confirmed the conclusions of Hanes: the globulars in M87 were similar to those in our own galaxy. There were some differences, however. Ables *et al.* (1974) found essentially perfect agreement in the V photometry but a 0.2 mag zero-point difference in B with respect to Hanes (1971). This led them to conclude that the clusters in M87 were on average redder than found by Hanes, a conclusion which they interpreted as reflecting a higher mean cluster metallicity. However, their study relied on transfers for calibration while Hanes' plates were internally calibrated through the use of the Racine wedge. Hanes (1975) reconfirmed his earlier results, and Racine, Oke & Searle (1978) have also confirmed Hanes' (1971) colours in a spectrophotometric study of clusters in M87. Thus there is no strong evidence that the mean cluster colours are different.

A second difference was that Ables *et al.* (1974) found no dependence of mean cluster colour upon radial distance in the M87 field, but pointed out that such trend was present in the Hanes (1971) data. The programme objects are seen projected on M87 itself and the photometry is sensitive to errors in the subtraction of the underlying luminosity, which of course has a strong radial gradient. Ables *et al.* (1974) have argued that their electronography is more successful in this than the conventional iris photographic photometry of Hanes (1971); and indeed Hanes (1975), though reconfirming his own background correction calibration, could not completely rule out an effect of the size suggested by Ables *et al.* (1974).

These are differences which I hope to resolve at some future time; and Bill Harris and his co-workers are also exploring the question of the colours of globular clusters in Virgo galaxies. But in any event the demonstration of the close similarity of $B-V$ for clusters in M87 and in the Local Group galaxies was an important and encouraging finding. These studies reinforced our belief that the objects around M87 are not grossly unlike the globular clusters in our own galaxy.

12.4.4 A survey of globular clusters in twenty Virgo galaxies

Hanes (1975, 1977b) presented photographic photometry in the G(= 103aJ+GG13) system for objects in the fields of twenty Virgo galaxies. Here I will briefly summarize those findings.

(i) Fig. 12.8 shows the luminosity functions of unresolved objects in the fields of a few of the galaxies studied (see Hanes 1977b for the complete sample). The field contamination is now well understood. It can be derived by studying plates in essentially blank fields or plates centred on faint

Fig. 12.8. The luminosity functions (number of objects N as a function of G magnitude) of unresolved objects in the fields of five Virgo galaxies. The dotted lines show the functions after correction for field stars. The galaxies are identified in each panel, and the galaxy types are also shown. The data are from Hanes (1977b).

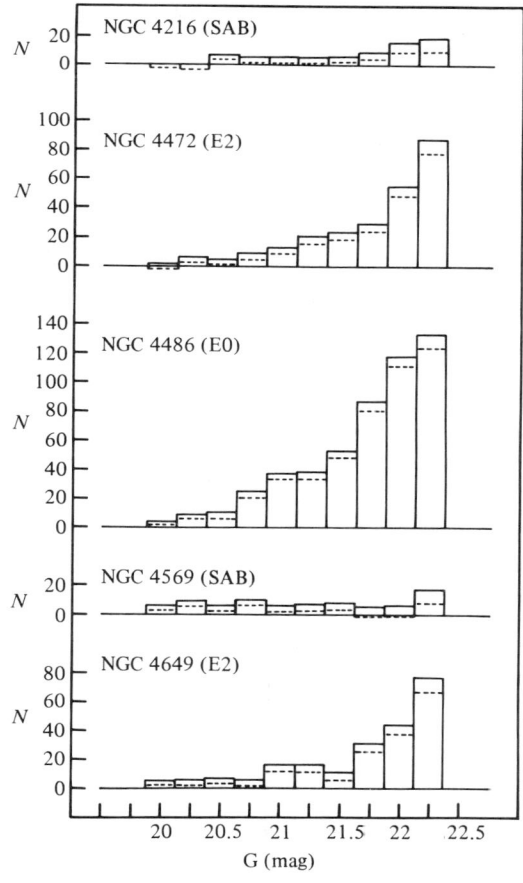

galaxies with no conspicuous central concentration of objects; alternatively, it can be estimated by differencing annuli in the many fields now available and summing over all fields to improve the statistics. The results are identical (Hanes 1975), and in the figure corrections have been made for the field contamination, as indicated.

The bright elliptical galaxies (NGCs 4472, 4486, 4649) have large numbers of excess objects after field correction but the spirals (NGCs 4216, 4569) do not, as I anticipated in section 12.3. Notice that in the bin of width 0.25 mag centred on $G = 22.25$ mag, M87 possesses as many globular clusters as have yet been identified in our whole galaxy!

Fig. 12.9. The surface density σ of objects (number per arcmin^2), counted in annuli in five Virgo galaxy fields, as a function of radius r. Objects from $18.00 \leq G \leq 22.25$ are included; the field contribution in this range amounts to 0.5 stars/arcmin^2, as indicated by a broken line in each panel. The data are from Hanes (1977b).

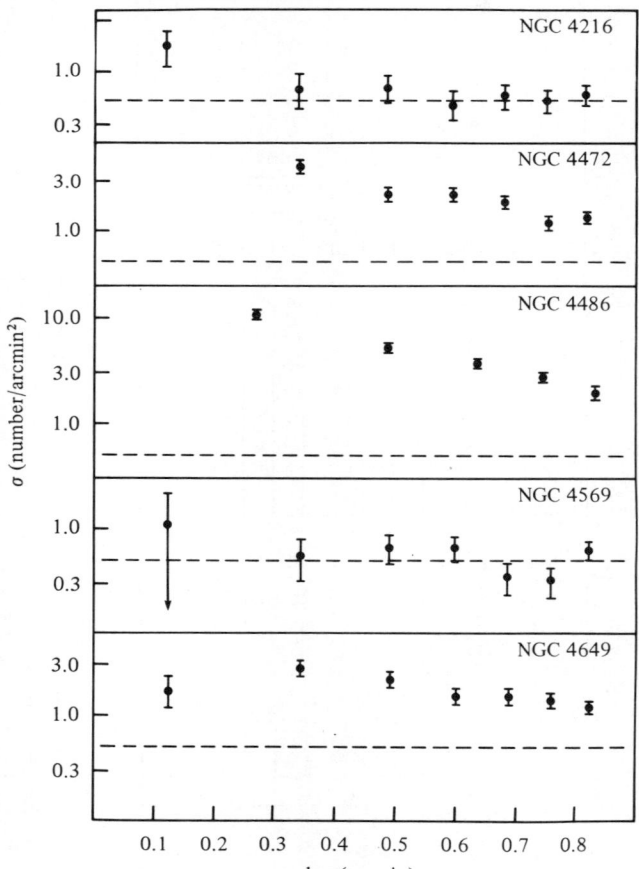

(ii) Fig. 12.9 presents the surface density of objects as a function of radius for the same five galaxies. The objects are clearly concentrated to the galaxies themselves, at least for the ellipticals; and the marginal central excess in NGC 4216 may make the numerical excess near $G = 22.0$ mag (fig. 12.8) slightly more believable. It is clear though that the clusters in the spiral galaxies will not be useful distance indicators given the present data.

In the elliptical galaxies, the surface densities of the globular cluster populations fall off like the distributions of underlying luminosity (Baum 1955; Hanes 1975; Harris & Smith 1976; Harris & Petrie 1978). I am at present exploring the significance, if any, of deviations from this close correlation.

Fig. 12.10. The total observed cluster population, N, integrated over $20.00 \leq G \leq 22.25$ and corrected for the field, as a function of galaxy $B(0)$ magnitude (de Vaucouleurs & de Vaucouleurs 1964). E and S0 galaxies are plotted as open symbols and spirals as filled symbols. The error bars represent $\pm N^{\frac{1}{2}}$. The fitted lines and the point in parentheses are explained in the text.

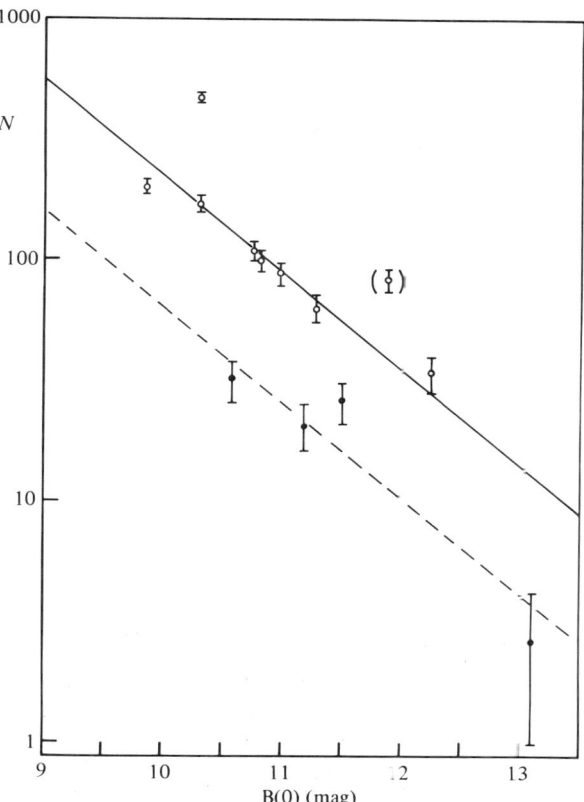

(iii) Fig. 12.10 shows how the total observed cluster population depends on the luminosity of the parent galaxy. The solid line plotted has the functional form $\lg N = \text{const} - 0.4B(0)$, which is the behaviour expected if the total cluster population is directly proportional to the mass of the parent galaxy and if the various galaxies have identical mass-to-luminosity ratios. The constant has been chosen such that the line is a good fit to seven of the nine E/S0 data points plotted. An eighth point, enclosed in parentheses, corresponds to the S0 galaxy NGC 4596. Here the total population is uncertain: though an excess above the field seems to be present, no central concentration is seen and the identification of a globular cluster population is problematic. However, no such doubts can be expressed for the ninth point plotted at the top of the figure. This is M87, and it shows that, for whatever reason (van den Bergh 1977), this well-studied galaxy is overabundant by a factor of about three in globular clusters compared with other Virgo galaxies of the same luminosity. Distance determination techniques which rely on population scaling arguments (de Vaucouleurs 1970, 1977a) will be vitiated by this effect.

The broken line in fig. 12.10 shows the same function displaced by 1.35

Fig. 12.11. The observed differential luminosity function for the globular clusters associated with NGC 4486 (filled symbols, with $\pm N^{\frac{1}{2}}$ error bars), compared with the function summed over the four bright elliptical galaxies NGCs 4374, 4406, 4472, 4649 (open symbols). No vertical scaling has been applied (from Hanes 1977b).

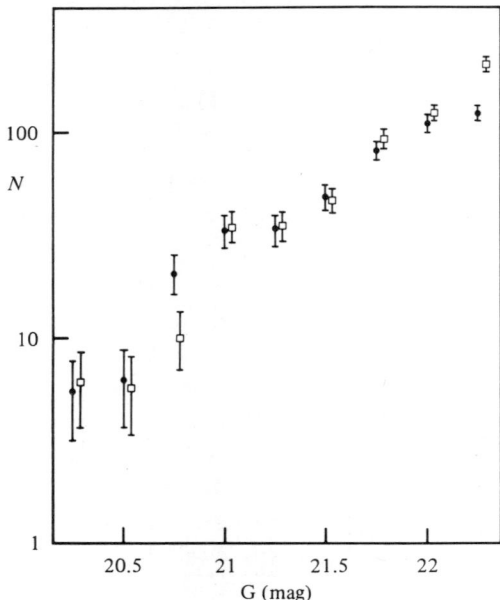

mag in B(0) to fit the points which correspond to the spiral galaxies (but note that the deduced populations are small and the uncertainties large). The fits again looks good, though the constraints are minimal, and the displacement of 1.35 mag suggests that the same scaling of cluster population with total galaxy mass holds for the spirals as well as for the ellipticals (Wakamatsu 1977), provided that the spiral mass/luminosity ratio is smaller by a factor of ~ 3.5 (de Vaucouleurs 1974).

(iv) The cluster population in M87 is fortuitously close to that summed over four other bright elliptical galaxies (NGCs 4374, 4406, 4472, 4649). The unscaled luminosity functions are compared in fig. 12.11. The functions plotted are *differential*, so that all data points are independent, and it is striking that the luminosity functions are so similar over the full range studied (at $G \sim 22.25$ mag, incompleteness may start to become important, especially in the dark M87 field). At least for the Virgo cluster elliptical galaxies, the globular cluster luminosity functions seem to be identical, with only a scaling by the total population distinguishing them.

If an assumed universal globular cluster luminosity function has sufficient structure to provide some reliable fiducial mark (say, some point of inflection, or a characteristic curvature) then the fitting of the observed Virgo globular cluster luminosity function against that for the globular clusters in the Local Group galaxies should provide a useful estimate of the Virgo cluster distance modulus. Such an application presupposes the similarity of the properties of globular clusters in the different galaxies studied. What evidence is there for this point of view?

12.4.5 The uniformity of globular cluster populations

Hanes (1977c) summarized photometric evidence for the uniformity of globular cluster populations. Since then, Racine, Oke & Searle (1978) have shown further that the globular clusters in M87 are spectroscopically similar to those in M31, though perhaps somewhat metal-richer in the mean than those in our own galaxy. Frogel *et al.* (p. 163, this volume) have shown that the cluster systems in M31 and in our own galaxy are indistinguishable in their dependence of infrared colour upon metallicity and in the observed ranges of these parameters. However, van den Bergh (p. 183, this volume) has contested Hanes' (1977c) conclusions that the clusters in our own galaxy and in M31 are photometrically similar, pointing out that the frequency distributions in the reddening-free metallicity-indicative photometric parameter $\mathscr{R}$ (Racine 1973) are not the same according to various Kolmogorov–Smirnov statistical tests. The difference is that van den Bergh has compared *observed* distributions without considering the effects of random errors of measurement. As these are larger in M31 than in our own galaxy (and larger still in M87) the smearing of the intrinsic $\mathscr{R}$ frequency distribution must be considered. When this is done it becomes clear that the distributions are not significantly different (Hanes 1977c). Van den Bergh (1969, and p. 183, this volume) suggested that M31 possesses relatively more metal-rich globulars than does our own galaxy, partly on the photometric evidence and partly because of the observed distributions of line-strengths of clusters. The second of these is subject to various systematic uncertainties, as van den Bergh (1969) himself admits, but if his contention is correct then the Hanes (1977c) results simply tell us anew that UBV photometric indices are not very useful metallicity indicators – or at least, not given the present quality of the photometry.

The evidence put forward so far suggests that globular clusters within galaxies are not greatly dissimilar in metallicity, however it is evidenced. It is of more importance here to enquire about the similarity of the cluster luminosity functions. Hanes (1977a) considered this question; and a recent summary is to be found in Harris & Racine (1979), who find that 'the hypothesis of a universal [luminosity function] $\phi(M)$ for globular clusters is entirely reasonable for a remarkably wide range of parent galaxy types'. Van den Bergh's (1969) suggestion that the mean luminosity for globular clusters in M31 was about twice that for the clusters in our own galaxy was based upon a very incomplete sample, as van den Bergh himself and co-workers (Sargent *et al.* 1977) later showed. Racine & Shara (1979) have now demonstrated the virtual equality of the cluster luminosity functions in M31 and in our own galaxy.

In the previous section I showed how the functions are closely similar within various Virgo elliptical galaxies. Unfortunately the globular cluster luminosity functions in Virgo galaxies are not sampled deeply enough (Hanes 1977a) or in sufficient photometric detail (Harris & Smith 1976; Harris & Petrie 1978) to test the similarity to the Local Group functions over the full range of cluster absolute magnitude. The working assumption must be made of such an equivalence; I will present some consistency arguments to justify this procedure in section 12.5.2.

12.5 Distance determinations

12.5.1 Methodology and results

The new data presented by Hanes (1975, 1977b) permitted renewed attempts to derive Virgo cluster distance moduli. De Vaucouleurs (1977a) used the Hanes data to deduce a Virgo cluster true distance modulus of 30.45 ± 0.15 mag, implying a Hubble constant of $H = 86 \pm 9$ km/s/Mpc; but his approach was open to some of the same criticisms I have levelled against his earlier treatment (de Vaucouleurs 1970). A side issue is the fact that de Vaucouleurs again adopted a simple relationship between the total cluster population and the absolute magnitude of a parent galaxy, despite Hanes' (1975, 1977b) demonstration that no such simple scaling exists. Still, the precise formulation of such a relationship represents something of a luxury since (if it is true) the effect will always manifest itself in an empirical correlation of $M_B(1)$, the absolute magnitude of the brightest cluster, with $M_B(G)$, the absolute magnitude of the galaxy to which it belongs. Given that, observations of apparent magnitudes B(1) and B(G) should lead iteratively to estimates of the distance modulus in the manner described by de Vaucouleurs. But since no such scaling exists, the method fails. For example, M87 is no brighter than NGC 4649 but has thrice the cluster population and a much more luminous first-ranked cluster. In fact, de Vaucouleurs considers two subsets of the Virgo data. The first combines the data for six galaxies *including* M87 and implies an apparent distance modulus of 30.0 mag. The second considers five fainter galaxies (*excluding* M87) and finds a modulus of 30.4 mag. Though M87 alone has almost half the cluster population of all the galaxies studied, with consequently good statistics (and the six brightest have $\gtrsim 90$ per cent of the total), de Vaucouleurs rejects the first of these in favour of the second.

Hanes (1977a) has taken the view that the shape of the luminosity function may be universal, with scalings in total population that may be arbitrary. Granted this working assumption, how does one then derive a Virgo cluster distance modulus? The obvious response is to intercompare the observed functions in our own galaxy and in Virgo, with adjustments in the magnitude scale to account for the different distances and adjustments in the observed numbers to allow for different total populations. Of course, we would like to retain *all* the advantages pointed out in section 12.2: namely to derive a distance modulus which depends purely upon Population II indicators, which is obscuration independent, and which comes in a single stride from our own galaxy to the Virgo cluster. The implication is that we would like to use the galactic globular cluster family as our *sole*

236   D. A. Hanes

calibrating sample. Including the clusters in M31, for example, reintroduces the Population I distance scale at an early state and reraises the question of the appropriate reddening corrections for clusters within M31 (Harris 1974; van den Bergh 1969; Martin & Shawl 1979).

This presents a problem. The clusters in our own galaxy are limited in

Fig. 12.12. (*a*) A comparison of the integrated luminosity function for the globular clusters summed over five Virgo galaxies (solid symbols with $\pm N^{\frac{1}{2}}$ error bars) with that for the globular clusters in our own galaxy (open symbols), as the latter would appear at an apparent distance modulus of 30.0 mag. The ordinate scale refers to the galaxy sample, and the curves are vertically scaled to minimize the deviations in the region of overlap. The apparent G

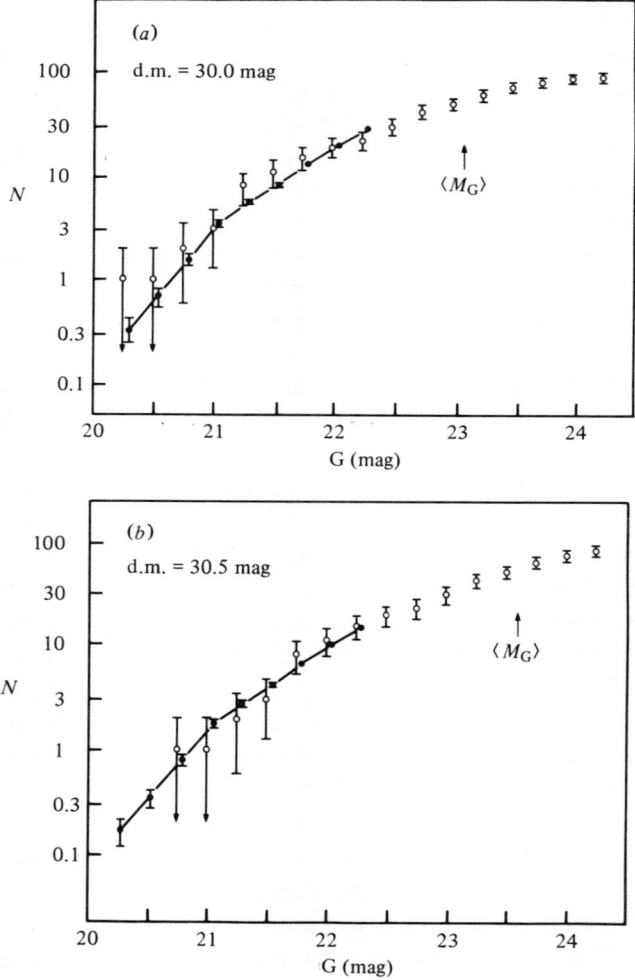

Globular clusters as extragalactic distance indicators

number, and at quite reasonable Virgo cluster distance moduli all but a few of them would be fainter than the completeness limit in the Hanes (1977b) Virgo data. Thus the overlap is limited, and the goodness of various estimates of the Virgo cluster distance modulus becomes difficult to assess. Still, the test can be done. Examine fig. 12.12, for instance, where

magnitude of a cluster of luminosity $\langle M_G \rangle$, the assumed universal mean absolute magnitude, is indicated. The data are from Hanes (1977a). (b) As in fig. 12.12(a), with a distance modulus of 30.5 mag. (c) As in fig. 12.12(a), with a distance modulus of 31.0 mag. (d) As in fig. 12.12(a), with a distance modulus of 31.5 mag. The average cluster is now fainter than $G = 24.5$ mag and lies off the diagram to the right.

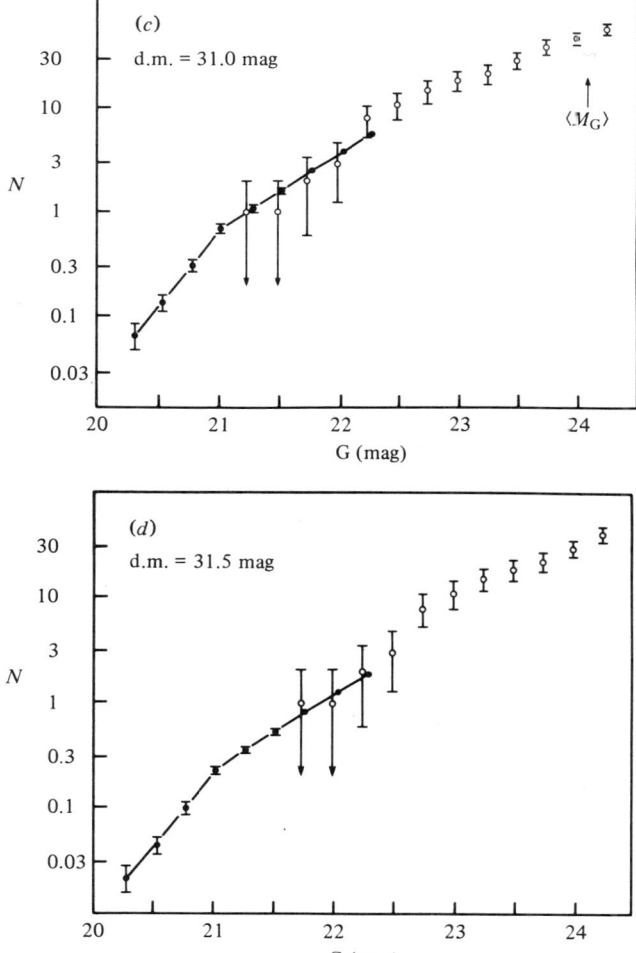

the Virgo cluster data (summed over five galaxies to provide a statistically secure sample) are compared with the globular cluster luminosity function in our galaxy as it would appear at apparent distance moduli of 30.0, 30.5, 31.0 and 31.5 mag, as indicated (integrated luminosity functions are used because the differential function in our own galaxy is very noisy). Inspection of the plots suggests that a distance modulus of ~ 30.5 mag is to be preferred, with reasonable uncertainties of ± 0.5 mag. However, if all but the region of overlap is covered in panel (d), the fit (to three points, representing two globular clusters in our galaxy!) looks quite good. The unsatisfactory feature in the overall match of the two functions is the implication that the composite luminosity function is kinked. Moreover, the kink, if real, occurs fortuitously just fainter than the Hanes completeness limits, which is an improbable piece of bad luck – though such things do happen. On continuity arguments, at least, it is psychologically more satisfactory to conclude that a Virgo cluster distance modulus near 30.5 mag is correct. Unfortunately it is hard to quantify such arguments.

Is there any way of extending the region of overlap? Surely; there are two. Either we can attempt to add more bright clusters to our calibrating sample or we can try to sample the Virgo globular clusters to fainter levels. I will defer the second of these approaches to section 12.5.2; let us here consider the first.

Hanes (1977a) has taken two different views:

(i) In an effort to retain all the advantages peculiar to the galactic globular cluster sample alone, I assumed that some idealized analytic representation of the cluster luminosity function could be extrapolated beyond the observed domain. Fig. 12.13 demonstrates that an excellent

Fig. 12.13. The luminosity function of globular clusters in our own galaxy, and its representation as a standard normal curve with mean $\langle M_G \rangle = -6.91$ mag and intrinsic dispersion $\sigma = 1.10$ mag (from Hanes 1977a).

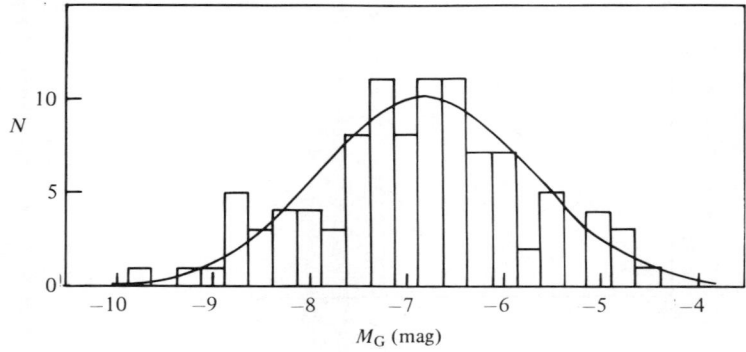

representation of the observed globular cluster luminosity function in our own galaxy is a normal distribution of mean $\langle M_G \rangle = 6.91 \pm 0.10$ mag and intrinsic dispersion $\sigma = 1.10 \pm 0.10$ mag. The adoption of such an analytic representation nicely smoothes the calibrating sample so that in fig. 12.14 it can be used in differential form and fitted to the Virgo cluster data (again summed over five galaxies). As the figure shows, uncertainties of ± 0.1 mag in the adopted dispersion imply concomitant uncertainties of ± 0.4 mag in the apparent mean magnitude of Virgo globular clusters, given that the population scaling is arbitrary (the implied total globular cluster population changes by only 30–40 per cent and cannot be ruled out on any astrophysical arguments). Thus the formal uncertainties in our knowledge of the intrinsic dispersion and of the mean magnitude suggest that the

Fig. 12.14. The differential luminosity function for globular clusters, summed over five Virgo cluster galaxies, and its representation by standard normal curves of intrinsic dispersion (a) 1.00 mag (b) 1.10 mag and (c) 1.20 mag, as shown. The mean magnitude is indicated for each curve.

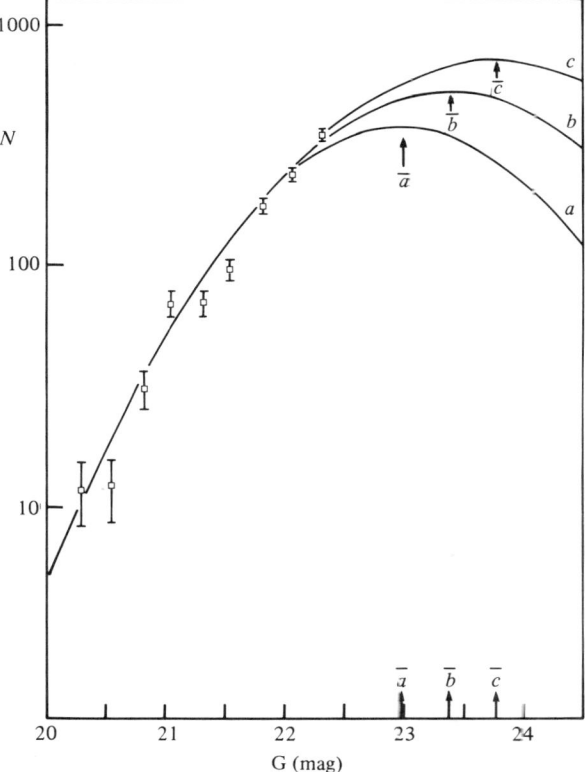

ultimate accuracy attainable in this approach (with the limited data used here) is ±0.4 mag. Only larger calibrating samples can reduce the formal uncertainties in $\langle M_G \rangle$ and σ, which are *purely statistical* limitations.

Of course it must be remembered that there may be two *systematic* errors. The first of these is the uncertain validity of extrapolating the adopted representation brightward. Hanes (1975) tested this by fitting normal curves to the bright end of the cluster luminosity function in M31 and found that such an extrapolation seemed valid, as Racine & Shara (1979) have since confirmed. The second possible systematic error is that the calibrating cluster sample may be incomplete at the faint end; such a sample would then yield too bright an estimate of $\langle M_G \rangle$ and too small a value of σ. These errors work in opposite senses: underestimating σ yields an underestimate of the apparent mean magnitude of the Virgo globular clusters (as in fig. 12.14) but the erroneously bright mean absolute magnitude redresses the balance to an extent in the derivation of the Virgo cluster distance modulus, although the strong dependence on σ ultimately dominates and the derived distance modulus will still be too small. Hanes (1977a) has presented arguments to suggest that such an effect may be marginally present in his calibrating sample, with consequent small adjustments needed of about 0.2 mag in distance modulus.

Notice incidentally that these potential systematic errors are in no way peculiar to this analytic approach, but will also bias the simple comparison of Virgo and galactic data shown in fig. 12.12. Magnitude-dependent incompleteness will obviously alter the shape of the calibrating luminosity function, and the smooth brightward extrapolation of the galactic sample enters in some subtle way in our subjective assessment of the goodness-of-fit in the various panels of fig. 12.12. The adoption of the analytic representation thus introduces no new uncertainty, except that it restricts the range of allowed extrapolations for the globular cluster luminosity function.

So far I have compared the calibrating sample with the data summed over several Virgo galaxies. Fig. 12.15 now shows the results of fitting the adopted analytic function to the separate data for several globular-cluster-rich Virgo galaxies. The apparent distance moduli derived are plotted versus the (non-physical) NGC numbers of the objects, and we see that the galaxy-to-galaxy scatter is small. The error bars are the purely formal uncertainties in the fitting procedure, but when we recognize that the centre-to-edge depth of the Virgo cloud may amount to 0.2 mag in distance modulus it is clear that the internal agreement is excellent: the galaxies are at the same distance.

(ii) An alternative method of augmenting the bright end of the calibrating luminosity function is to add the bright part of the luminosity function for globular clusters in M31. This is an approach which I use with reluctance, since it introduces many additional uncertainties: the appropriate Population I distance modulus of M31 (de Vaucouleurs 1977c, 1978b); the individual extinction corrections internal to M31 (Harris 1974), which are subject in turn to the further uncertainty of the appropriate ratio of total-to-selective absorption (Kron & Mayall 1960; van den Bergh 1969; Martin & Shawl 1979); and the extent and onset of magnitude-dependent incompleteness in the M31 sample (Sargent et al. 1977; Harris & Racine 1979). Some of these problems are more easily handled than others – we can restrict the sample to little-reddened clusters, for example – and Hanes (1977a) did include the M31 clusters in the establishment of an extended calibrating function. The recent study of Racine & Shara (1979) confirms the correctness of this treatment. Hanes (1977a) presented figures very like the various panels of fig. 12.12, here, and again deduced Virgo cluster apparent distance moduli near 30.4 ± 0.5 mag, in close agreement with the results of the analytic approach.

Fig. 12.15. The apparent mean magnitude, $\bar{G}$, of globular clusters associated with seven cluster-rich Virgo galaxies. The means are those implied by fitting with standard normal curves of intrinsic dispersion 1.10 mag. The corresponding apparent distance moduli $(m-M)_G$ are shown at the right, under the assumption that $\langle M_C \rangle = -6.91$ mag and with the small systematic corrections described in the text. The rightmost point corresponds to a fit to the data summed over the five richest cluster samples (NGCs 4374, 4406, 4472, 4486, and 4649), and does not differ from a weighted mean of the individual determinations.

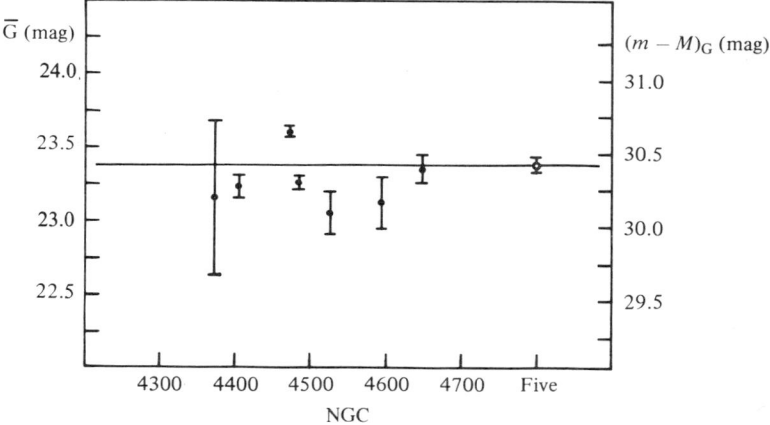

Fig. 12.16. (a) Equivalent to fig. 12.12(a), with some small differences. Here the solid points represent data summed over various Local Group galaxies, including M31, as explained in Hanes (1977a). The Virgo data (open boxes) have been scaled to cross the Local Group data at G = 22.0 mag. A new faint Virgo data point has been added at G ≈ 23.8 mag. The vertical extent of each symbol represents the $\pm N^{\frac{1}{2}}$ uncertainty, while the horizontal extent is of no

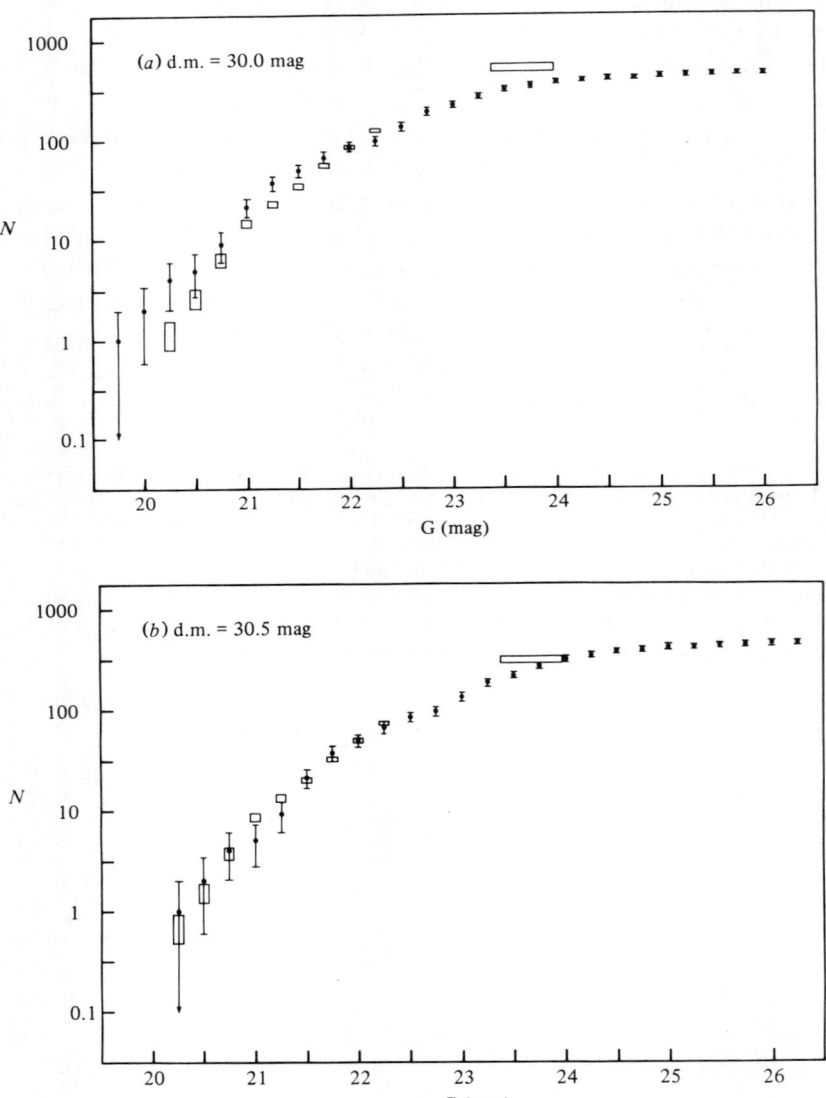

meaning except for the photometrically imprecise rightmost Virgo data point. The Local Group function is plotted as it would appear at an apparent distance modulus of 30.0 mag. (*b*) As in fig. 12.16(*a*), with a distance modulus of 30.5 mag. (*c*) As in fig. 12.16(*a*), with a distance modulus of 31.0 mag. (*d*) As in fig. 12.16(*a*), with a distance modulus of 31.5 mag.

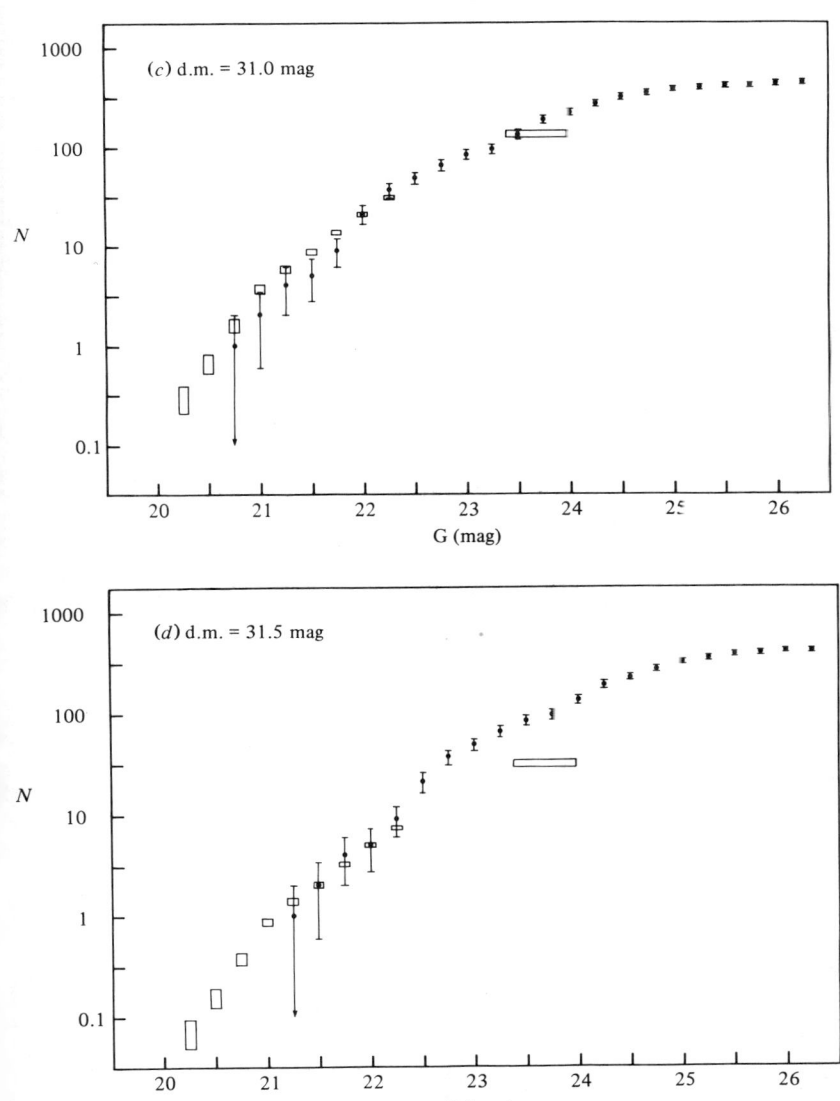

12.5.2 Recent data: a new constraint

In the preceding section I noted the desirability of extending the photometry of globular clusters around Virgo galaxies to considerably fainter levels. This important advance came in studies of the M87 field by Harris & Smith (1976) and of the NGC 4472 (M49) field by Harris & Petrie (1978). These authors demonstrated, among other things, that the clouds of globular clusters around these two giant elliptical galaxies extended to very large radii ($\gtrsim 20$ arcmin) and that the total cluster populations were consequently extremely large indeed.

The immediate combination of these new data with the results of Hanes (1977a) was not straightforward – there were some differences. First, the new studies did not incorporate detailed photometry object-by-object; rather, the authors simply carried out counts of unresolved objects to their plate limits (at $B \sim 24$ mag) within annuli centred on the programme galaxies. No differential luminosity functions are available, therefore. The authors also had plate/filter combinations different from those used by Hanes (1977b) and studied larger angular regions around each galaxy, using prime focus plates from the Cerro Tololo 4-metre reflector.

However, with care all these differences can be reconciled, as I have described in detail in Hanes (1979). What can be derived is an estimate of the *integrated* Virgo cluster population to some faint level – that is, in the panels of fig. 12.12 we will be able to add a Virgo cluster data point somewhere to the upper right. This will clearly provide an important discrimination between the possible distance moduli. Before doing so, let us recognize that fig. 12.12 and the assumption of the universality of the globular cluster luminosity function together implicitly *predict* what total cluster population is expected in the Virgo galaxies for various distance moduli. In Hanes (1977a) I suggested that a distance modulus near 30.5 mag was to be preferred. Is this borne out by the new observations of Harris & Smith (1976) and Harris & Petrie (1978)?

The answer is yes. Fig. 12.16 shows plots equivalent to those of fig. 12.12, but with the new datum added. The four panels correspond to Virgo cluster apparent distance moduli of 30.0, 30.5, 31.0 and 31.5 mag, as indicated. In Hanes (1979) I concluded on the basis of these plots that a distance modulus of 30.7 ± 0.3 mag was to be preferred, confirming the earlier results (Hanes 1977a).

I think that it is important to stress the fact that these results came *after* the studies which led to fig. 12.12 but that the new counts completely confirm the predictions implicit in that figure. It seems to me that this is strong evidence that the globular cluster luminosity function may truly be universal. As long as the overlap was limited (as in fig. 12.12) there was a real concern that different functions were being compared; perhaps, as Fall & Rees (1977) stated, this explained the 'too small' Virgo cluster distance modulus found by Hanes (1977a). But now we have a proven similarity in shape over almost four magnitudes; and moreover the distance modulus implied is within the usual realm of discourse (Bottinelli & Gougenheim 1976; Tully & Fisher 1977; de Vaucouleurs 1979). If the luminosity functions are in fact sensibly different, there is no reason at all to expect that a quite reasonable Virgo cluster distance modulus should be found to yield the good agreement shown in fig. 12.16.

12.5.3 The Hubble constant

This is not the place to consider all the historical (and on-going) arguments about the extragalactic distance mapping and the relevance of the Virgo cluster distance modulus to the global value of the Hubble constant; see for instance Tammann (1973, 1977), van den Bergh (1975), de Vaucouleurs (1977b, 1979), Peebles (1978), Hanes (1979), and the many other authors mentioned in those works whose research makes this such a colourful area.

Following Hanes (1979), let me simply note that, granted the precepts of Sandage & Tammann (1976), the evidence from the globular cluster studies is that the present value of the Hubble constant is 80 ± 11 km/s/Mpc. Under the same assumptions Sandage & Tammann have themselves found a Hubble constant of 50.3 ± 4.2 km/s/Mpc. The difference is significant, and for the various reasons put forward in this chapter (in section 12.2) – and also considering the recent difficulties brought to light in the conventional chain of reasoning (de Vaucouleurs 1979) – I am firmly confident that the former value is the more nearly correct.

12.6 Conclusions

My principal conclusions are as follows:

(i) Globular clusters within galaxies seem to be photometrically similar on average, and similar in at least the *range* of metallicities observed (Hanes 1977c; Frogel *et al.*, p. 163, this volume; Racine *et al.* 1978; Harris & Racine 1979).

(ii) The total globular cluster population associated with a given galaxy is to first approximation directly proportional to the mass of that galaxy (Hanes 1977b; Wakamatsu 1977; Harris & Racine 1979). However, there are galaxies – such as M87 – which are outstanding exceptions. The reasons are as yet unclear.

(iii) There is strong evidence that the luminosity functions of globular clusters are closely similar in the galaxies of the Local Group (Hanes 1977a; Harris & Racine 1979; Racine & Shara 1979), and that these functions are likewise closely similar within the elliptical galaxies of the Virgo cluster (Hanes 1977b). There is now strong circumstantial evidence that the luminosity functions of globular clusters in spiral galaxies and elliptical galaxies are closely similar (Hanes 1979).

(iv) The apparent distance modulus of the Virgo cluster of galaxies is 30.7 ± 0.3 mag (Hanes 1979). Insofar as this value can be used to determine the Hubble constant, the present value of that parameter is found to be 80 ± 11 km/s/Mpc.

Globular clusters seem at last to be realizing their full potential as extragalactic distance indicators. Further studies in remote galaxies known to be rich in globular clusters (Smith & Weedman 1976; Dawe & Dickens 1976; Smith 1977; de Vaucouleurs 1977a) are full of promise.

I have benefited greatly from conversations with and suggestions from too many people to mention them all individually, but as this is at least partly an historical review I would like especially to acknowledge the part played in this field by René Racine, who first enthusiastically introduced me ten years ago to the study of extragalactic globular clusters.

References

Ables, H. D., Newell, E. B. & O'Neil, E. J. (1974). *Publ. astron. Soc. Pacific*, **86**, 311.
Alter, G., Ruprecht, J. & Vanysek, V. (1958). *Catalogue of Star Clusters and Associations*, 1st edn. Prague: Verlag Akad. Wiss.
Baum, W. A. (1955). *Publ. astron. Soc. Pacific*, **67**, 328.
Bottinelli, L. & Gougenheim, L. (1976). *Astron. Astrophys.* **51**, 275.
Dawe, J. A. & Dickens, R. J. (1976). *Nature*, **263**, 395.
de Vaucouleurs, G. (1970). *Astrophys. J.* **159**, 435.
de Vaucouleurs, G. (1974). In *The Formation and Dynamics of Galaxies*, IAU Symposium no. 58, ed. J. R. Shakeshaft, p. 1. Dordrecht: D. Reidel.
de Vaucouleurs, G. (1977a). *Nature*, **266**, 126.
de Vaucouleurs, G. (1977b). In *Décalages vers le Rouge et Expansion de l'Univers*, IAU Colloquium no. 37, ed. C. Balkowski & B. E. Westerlund, p. 301. Paris: CNRS.
de Vaucouleurs, G. (1977c). In *Décalages vers le rouge et Expansion de l'Univers*, IAU Colloquium no. 37, ed. C. Balkowski & B. E. Westerlund, p. 39. Paris:CNRS.
de Vaucouleurs, G. (1978a). *Astrophys. J.* **223**, 351.
de Vaucouleurs, G. (1978b). *Astrophys. J.* **223**, 730.
de Vaucouleurs, G. (1979). *Astrophys. J.* **227**, 380.
de Vaucouleurs, G. & de Vaucouleurs, A. (1964). *Reference Catalogue of Bright Galaxies*. University of Texas Press.
Dixon, M. E. (1965). *Mon. Not. Roy. astron. Soc.* **129**, 51.

Fall, S. M. & Rees, M. J. (1977). *Mon. Not. Roy. astron. Soc.* **181**, 37F.
Hanes, D. A. (1971). M.Sc. thesis, University of Toronto.
Hanes, D. A. (1975). Ph.D. thesis, University of Toronto.
Hanes, D. A. (1977a). *Mon. Not. Roy. astron. Soc.* **180**, 309.
Hanes, D. A. (1977b). *Mem. Roy. astron. Soc.* **84**, 45.
Hanes, D. A. (1977c). *Mon. Not. Roy. astron. Soc.* **179**, 331.
Hanes, D. A. (1979). *Mon. Not. Roy. astron. Soc.* **188**, 901.
Harris, W. E. (1974). Ph.D. thesis, University of Toronto.
Harris, W. E. & Petrie, P. L. (1978). *Astrophys. J.* **223**, 88.
Harris, W. E. & Racine, R. (1979). *Ann. Rev. Astron. Astrophys.* **17**, 241.
Harris, W. E. & Smith, M. G. (1976). *Astrophys. J.* **207**, 1036.
Heck, A. (1973). *Astron. Astrophys.* **24**, 313.
Hemenway, M. K. (1975). *Astron. J.* **80**, 199.
Hodge, P. W. (1974). *Publ. astron. Soc. Pacific*, **86**, 289.
Johnson, H. L. (1963). *Basic Astronomical Data*, ed. K. Strand, p. 204. University of Chicago Press.
Kron, G. E. & Mayall, N. U. (1960). *Astron. J.* **65**, 581.
Martin, P. G. & Shawl, S. J. (1979). *Astrophys. J. Lett.* **231**, L57.
Peebles, P. J. E. (1978). *Comm. Astrophys.* **7**, No. 6, 197.
Racine, R. (1968a). *Publ. astron. Soc. Pacific*, **80**, 326.
Racine, R. (1968b). *J. Roy. astron. Soc. Canada*, **62**, 367.
Racine, R. (1973). *Astron. J.* **78**, 180.
Racine, R., Oke, J. B. & Searle, L. (1978). *Astrophys. J.* **223**, 82.
Racine, R. & Shara, M. (1979). *Astron. J.* **84**, 1694.
Sandage, A. R. (1961). In *Problems of Extragalactic Research*, IAU Symposium no. 15, ed. G. C. McVittie, p. 359. New York: MacMillan.
Sandage, A. R. (1968). *Astrophys. J. Lett.* **152**, L149.
Sandage, A. R. (1973). *Astrophys. J.* **183**, 711.
Sandage, A. R. & Tammann, G. A. (1976). *Astrophys. J.* **210**, 7.
Sargent, W. L. W., Kowal, C. T., Hartwick, F. D. A. & van den Bergh, S. (1977). *Astron. J.* **82**, 947.
Smith, M. G. (1977). In *Décalages vers le Rouge et Expansion de l'Univers*, IAU Colloquium no. 37, ed. C. Balkowski & B. E. Westerlund, p. 75. Paris: CNRS.
Smith, M. G. & Weedman, D. W. (1976). *Astrophys. J.* **205**, 709.
Tammann, G. A. (1973). In *Confrontation of Cosmological Theories with Observational Data*, IAU Symposium No. 63, ed. M. S. Longair, p. 47. Dordrecht: D. Reidel.
Tammann, G. A. (1977). In *Décalages vers le Rouge et Expansion de l'Univers*, IAU Colloquium No. 37, ed. C. Balkowski & B. E. Westerlund, p. 43. Paris: CNRS.
Tully, R. B. & Fisher, J. R. (1977). *Astron. Astrophys.* **54**, 661.
van den Bergh, S. (1960). *Publ. David Dunlap Obs.* **2**, 159.
van den Bergh, S. (1969). *Astrophys. J. Suppl.* **19**, 145.
van den Bergh, S. (1970). *Nature*, **225**, 503.
van den Bergh, S. (1975). In *Galaxies and the Universe* (vol. 9 in Stars and Stellar Systems), ed. A. R. Sandage, M. Sandage & J. Kristian, p. 509. University of Chicago Press.
van den Bergh, S. (1977). In *The Evolution of Galaxies and Stellar Populations*, ed. B. M. Tinsley & R. B. Larson, p. 19. New Haven: Yale University Observatory.
van Herk, G. (1965). *Bull. astron. Inst. Netherlands*, **18**, 71.
Vetešnik, M. (1962). *Bull. astron. Inst. Czechoslovakia*, **13**, 180 and 218.
Wakamatsu, K. I. (1977). *Publ. astron. Soc. Pacific*, **89**, 504.
Weistrop, D. (1972). *Astron. J.* **77**, 366.
Woltjer, L. (1975). *Astron. Astrophys.* **42**, 109.
Woolley, R. v. d. R. & Savage, A. (1971). *Roy. Obs. Bull.* **19**, No. 170, 363.

13
Luminosity distributions and density profiles

IVAN R. KING

13.1 Introduction

In this chapter I will discuss the luminosity distributions and density profiles of globular clusters, with special reference to the data that bear upon their structure. The quality of these data will affect the depth of our understanding of the dynamics of clusters and of their luminosity functions. While this will be a rather general outline, I would like to concentrate on the accuracy obtainable and the practical limitations – not only what we can do, but also what we *cannot* do. With telescope time as scarce as it is, it is important to try to optimize our observations.

For a more general review of the subject, see King (1975). In this chapter I will be considering some particular aspects in more detail than in that earlier discussion.

13.2 Basic observations

There are two ways of determining density distributions in star clusters. We can simply count the stars; alternatively, we can do surface photometry and use the luminosity distribution as an index of the star distribution.

Before we consider the details of these techniques, let us examine a very standard globular cluster, namely M13, shown in fig. 13.1. A few things are immediately obvious. First, there are crowding problems. I will consider these problems quantitatively later, but it is quite clear that as you near the centre you will run into trouble: either you must use shorter-exposure plates and count to a less faint level, or else you must rely on surface photometry in this region. Later I will consider the relative statistical accuracy of surface photometry and star counts down to different limiting magnitudes.

Secondly, the cluster extends quite far – it more than fills this picture. In fact, I would never do star counts on a plate like this, because there is no way of determining the background level that must be subtracted.

For long exposures with a small-field reflector, you have to use plates that are centred somewhere out near the edge of the cluster. Relative to this picture, the density of cluster stars becomes equal to background at almost twice the distance from centre to edge of the picture. Valid counts go a little further than that, and the extrapolated limiting radius is about 50 per cent further still.

Finally, you should be aware that in the photograph we have lost latitude. On the original 5-metre plate, you can see the buildup of light to the centre, while the figure here shows only a burnt-out region. From the core radius shown in fig. 13.1 you can see that we are losing something in the print.

13.3 Star counts

In making star counts one has to deal with both systematic problems and statistical problems. Let me consider the systematic problems first.

I have already alluded to crowding. Another problem that has never

Fig. 13.1. The globular cluster M13 (NGC 6205), reproduced from a 5-metre prime-focus plate. The core radius, r_c, is indicated (Peterson & King 1975). Photograph courtesy of Hale Observatories.

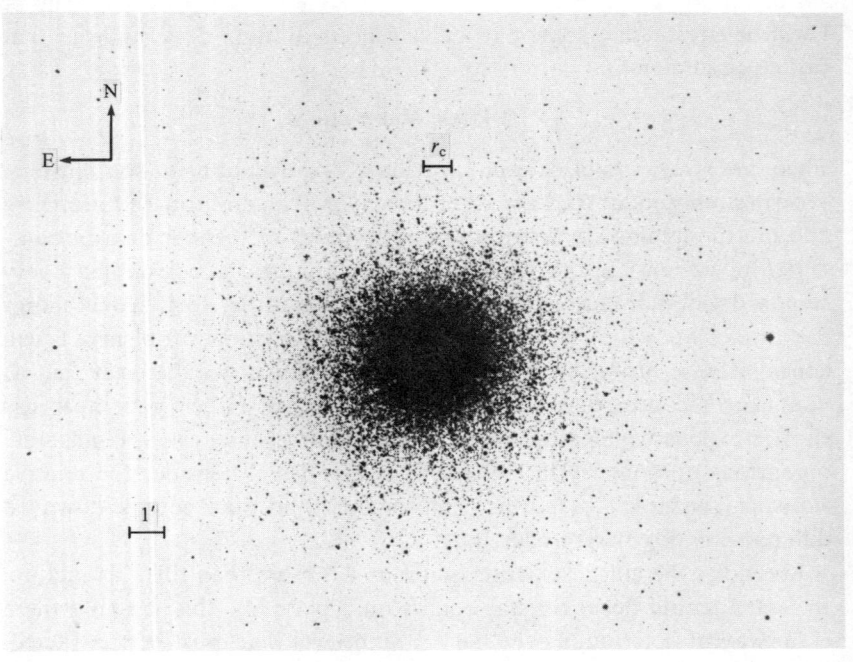

been rigorously studied (and which in practice is probably lumped under the empirical crowding corrections) is that in the central region of the cluster you are counting bright stars superimposed on a background of faint light due to unresolved fainter stars. On a photographic plate, this certainly has the effect of pushing stars up above the threshold.

Then there are the systematic biases in the observer. You cannot avoid fatigue, so you must arrange the observations in such a way that fatigue cancels out. For example, when counting by rectangular reseaus I have always counted every fifth line, filling in the next set of lines later, and so on. Similarly, in circular counts I examine corresponding sectors in each quadrant, then do another sector, etc., so that if fatigue induces imaginary counts, or decreases my efficiency, no false ellipticities can result. More subtly, there is a tendency in the outer parts of the cluster, where stars are less frequent, not to look quite so carefully. Ray White has examined this problem in detail, counting in blank fields to see if he could get constant density everywhere. It is important to count in the same way in the large outer sectors as in the tiny inner sectors.

Then there are the statistical problems. In a region containing N stars, suppose N_c of these belong to the cluster while N_b belong to the background. Your net count is

$$N - N_b = N_c \pm (N_c + N_b)^{\frac{1}{2}}.$$

The effect of statistical fluctuations squeezes you from both sides. In the centre of the cluster, where you are restricted to counting bright stars, the numbers are necessarily small and you have poor statistical results; and in the outer parts the background tends to swamp the cluster, making the uncertainty very large. The background determination itself is not very difficult, however, because you can just count a large region around the cluster.

Irregular backgrounds can occur in two different ways and may give rise to serious problems. One of these is exemplified by the clusters 47 Tuc and NGC 362, which are on the edge of the SMC: you get a very strong background gradient across these fields. It can be dealt with, but – carefully! The other problem is that many clusters, some of them extremely interesting (as X-ray sources, for example; see Lewin, p. 315, this volume), are down low in the galactic plane where you have both a rich background and irregular absorption. There are some clusters where that may be an insuperable problem – the background is different on all four sides.

13.4 Surface photometry

This can be done either photoelectrically or photographically. Most good work has been done photoelectrically, but I am going to end up saying here that we need more photographic surface photometry for specific reasons.

But first, how does one do photoelectric photometry of globular clusters? I believe that there is no reason why we should not do it in the easiest possible way: use a photoelectric photometer with several aperture sizes, centre each on the cluster and measure it, and then difference the apertures to get annuli. Differencing observational data is normally a deadly sin, but here the statistical fluctuations of the data themselves, due to the limited number of stars, are larger than the observational errors of the photometry, so in this almost unique case it is perfectly all right to difference. The alternative, to run a smaller spot through the cluster from centre to edge, is subject to serious statistical errors and gives poorer results.

Some especially interesting clusters, such as those containing X-ray sources, have extremely small centres. This makes photoelectric photometry difficult for two reasons. First, it is very hard to centre the spot in exactly the same way every time; in some published work the biggest scatter in the photometry is due to different centring of successive measurements. Secondly, we are talking about sizes of a few seconds of arc, and you simply cannot do successful photoelectric photometry with diaphragms as small as that. The seeing is time-variable, so different amounts of light spill in and out of the aperture; and, to make matters worse, in clusters there may be rather bright stars sitting at the edge of the diaphragm.

So there are certainly good reasons to supplement the photoelectric work with photographic surface photometry, especially now that so many people know how to do it correctly. In particular we should use photographic methods for the high-resolution work near the centres of globular clusters. The tradeoff of course is that the reduction becomes very expensive, what with scanning the plates and doing careful intensity calibrations.

I should mention in passing two other things you can try to do on direct photographs. The first, which is risky, is to dissect the photograph into individual star images by fitting some standard image profile. Then even in the crowded central regions where the images overlap you can measure stars one by one. If you are interested in specific stars near the centre of a cluster, this is the only way you can get at them. The second, which may be more dangerous still, is to fit all the stars you can and subtract out their intensity, leaving (you hope) the profile due to the underlying fainter stars. This has been tried once or twice, notably by a group at the Pic-du-Midi Observatory working on M15 (Leroy, Aurière & Laques 1976). They removed about 100 stars and were left with a smooth-looking background which agrees with the profile given by the total light. However, the worry is that one might be identifying stars better in less-crowded regions than in the crowded regions, with obvious systematic effects.

13.5 Statistical limitations in photometry

Let us return to the photoelectric work for a moment to consider the sampling statistics. The problem is rather more complicated than for the simple star counts; nevertheless it is the same square-root-of-n phenomenon, because the light that you get in a surface photometry measurement is contributed by a finite number of stars, which have an associated sampling error.

Suppose your measurement includes n_i stars of type i, each contributing a luminosity L_i. The total light is

$$l = \Sigma L_i n_i.$$

The variance is simply the sum of the variances, since the quantities are independent. (Note that this summation-of-variances rule is a general theorem of statistics and has nothing to do with the distributions being Gaussian – which these certainly are not.) The variance of n_i is n_i, whence

$$\sigma_{n_i} = n_i^{\frac{1}{2}},$$

$$\sigma_{L_i n_i} = L_i n_i^{\frac{1}{2}},$$

$$\mathrm{Var}(L_i n_i) = L_i^2 n_i$$

and $$\mathrm{Var}\,(l) = \Sigma L_i^2 n_i,$$

whereas $$l = \Sigma L_i n_i.$$

The different appearance of L_i in the two expressions makes it clear that the bright stars cause the problems. In globular clusters most of the integrated light comes from the brighter stars, so these variances are quite serious.

It is easy to quantify these effects. I have used the cluster luminosity function of M3, for which $M_V \approx -9$, to answer the following question: 'Given an infinite stellar population representative of M3, by how much

254   *I. R. King*

Fig. 13.2. (*a*) The V surface-brightness distribution for M15. The solid line shows the average seeing profile, determined from the images of isolated stars. This figure is reproduced from Newell & O'Neil (1978); see their paper for a detailed explanation of the various symbols. (*b*) As in fig. 13.2(*a*), except that the profile is in B (from Newell & O'Neil 1978).

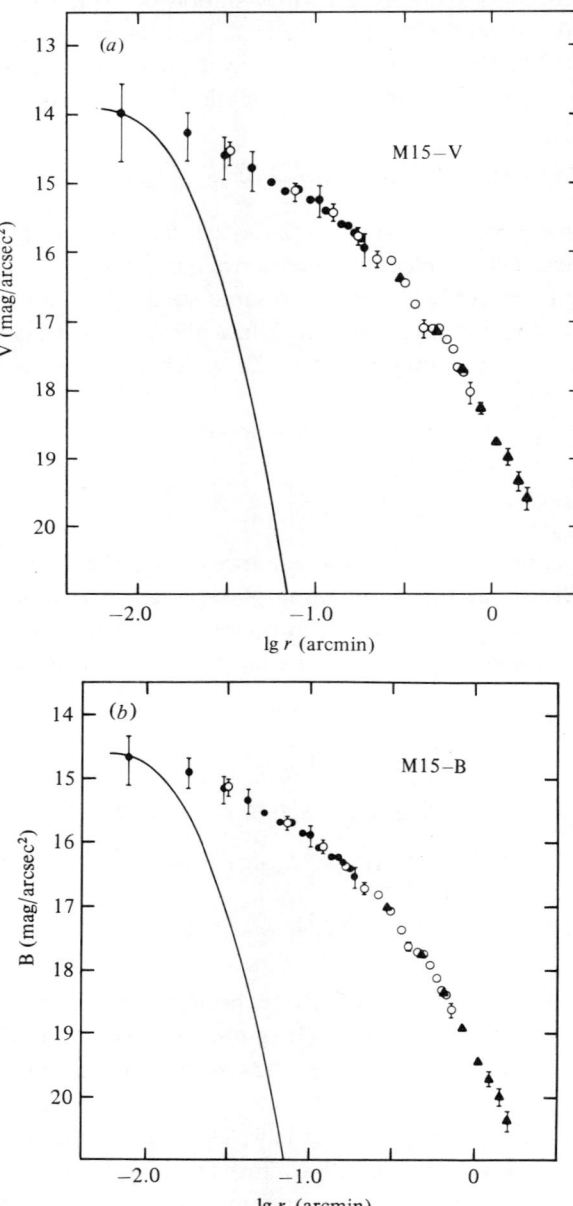

would random stellar samples, each the size of M3, differ in magnitude and colour one from another?' The answer is that in blue light, $\sigma_B = 0.022$ mag, while in the visual, which is dominated more by the individual red giants, you find $\sigma_V = 0.028$ mag. The calculation for $B-V$ is a little more complicated (because in each case the same stars contribute both the B and the V light) but reveals that $\sigma_{(B-V)} = 0.011$ mag (King 1966b). And this is for a rather bright globular cluster. When you get to lower-luminosity clusters, there are statistical uncertainties of about 0.05 mag in $B-V$ in the integrated light of the entire cluster. Clearly one should be very careful in the interpretation of photometric population measures.

An interesting implication is that you will get considerably more weight by doing your surface photometry in the blue rather than in the yellow. In fig. 13.2 I have presented two graphs from the paper by Newell & O'Neil (1978), showing surface-brightness distributions in B and in V at quite high resolving power for the central parts of M15. You can see that the curve is much less well delineated in the V data than in the B data. Newell

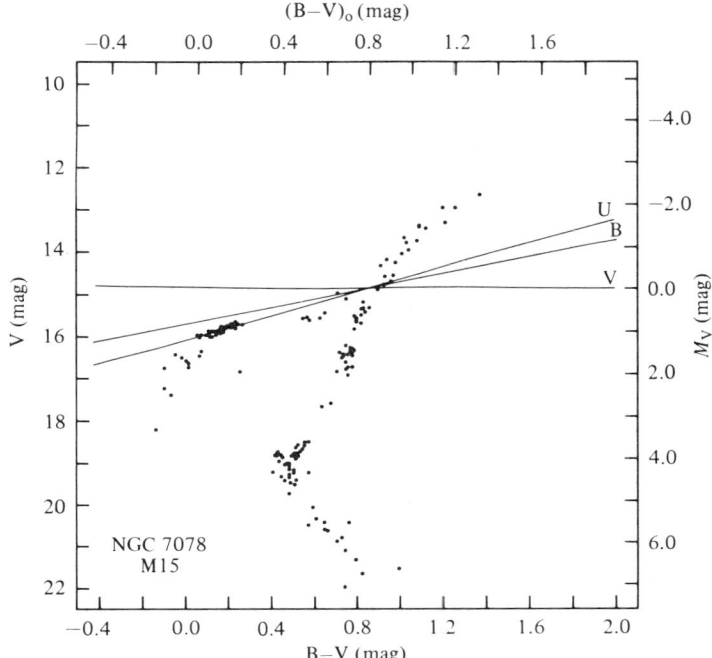

Fig. 13.3. The colour–magnitude diagram for M15 (NGC 7078) as determined by Sandage (1970). The lines superimposed are loci of constant V, B, and U, as indicated. Note that in B and U photometry the HB stars contribute more than in V photometry, thereby improving the effective sample (from Alcaino 1973).

& O'Neil state that their ultraviolet variances are no better than those in B, which is hard to understand. The effect is shown graphically in fig. 13.3, where the colour–magnitude diagram for M15 (Sandage 1970) is plotted with lines superimposed representing loci of roughly constant V, B and U (as indicated). You can see that as you go from V through B to U you effectively flatten the giant branch and also get a considerable contribution from the horizontal branch (HB) stars: more stars contribute, the effective N is larger, and the statistics should be somewhat better. By contrast, the light in the V distribution is contributed predominantly by the rather few red giants, giving large variances.

Reducing the noise is particularly important for the high-concentration globulars, including those that have X-ray sources. The problem there is that you are interested in a small central region, in which the *total* information is small, and it would be sinful not to use it in the best possible way.

Finally, one can ask how the statistical accuracies of surface photometry and star counts compare. There are two questions one can ask here. First, how many stars would you have to have counted to get the same statistical errors as you would get from surface photometry? The answer is that the effective star in surface photometry is around the absolute magnitude of the HB. This means that the accuracy of surface photometry is as if you had counted N stars, where N is the ratio of the luminosity of the cluster to the luminosity of an HB star. The next question is, how faint do you actually have to go to find that many stars? The tradeoff comes in fact about 2.5 mag below the HB. Unless you can count to at least this level, you are going to get better accuracy by doing surface photometry.

13.6 Luminosity functions

I would like to make a few cautionary remarks on the subject of the determination of luminosity functions. First, when doing star counts it is moderately straightforward, if sufficient care is taken, to get a uniform limiting magnitude in a given count; but it is difficult to state with any assurance exactly what that limiting magnitude is. Errors in the adopted magnitude will lead to spurious bumps and features in the final luminosity function, because successive magnitude bins will be of different widths. Chris Wilson and I have handled this problem in the following way: instead of counting very many stars over large areas, with uncertain magnitude limits, we have restricted the area examined to include a statistically significant number of stars at each magnitude level but a small enough sample to permit quick iris photometry of each star – not precision photometry, but sufficiently accurate to tie down the limiting magnitude in a sufficient way.

That is an observational problem; the second remark relates to dynamics. When you determine a luminosity function you necessarily measure the bright stars in the centre of the cluster and the faint stars in the outer parts, because the faint stars are too crowded at the centre and the bright stars are too sparse at the edge. Thus your luminosity function is sensitive to the relative distribution of bright and faint stars. Fig. 13.4 illustrates this point. It shows results from a dynamical model which I made to fit Sandage's (1954) counts for M3. (The actual counts are given by Oort & van Herk (1959).) The solid lines show the actual observations, and the inward extrapolations, shown by dashed lines, are predicted by the model. The extrapolation of the bright stars poses little uncertainty, but the extrapolation of the faint stars is totally dependent on having a correct model, to the extent of making a difference of a factor of 2 in the relative numbers of giants and main sequence stars that one deduces from these particular counts. Any interpretation of the observed luminosity functions in terms of the time scales of various evolutionary stages will contain this same factor, so that it is certainly important.

The principal uncertainty in the application of these models is the degree of anisotropy in the velocity distribution. In fig. 13.4, I have used an

Fig. 13.4. Star counts as a function of radius for M3 (NGC 5272). The solid lines represent the actual observations (Sandage 1954; Oort & van Herk 1959) while the dashed portions are the model-dependent inward extrapolations (King 1975). For the faint stars the extrapolation is entirely dependent on having the correct model, which here was taken to be an isotropic one (from King 1975).

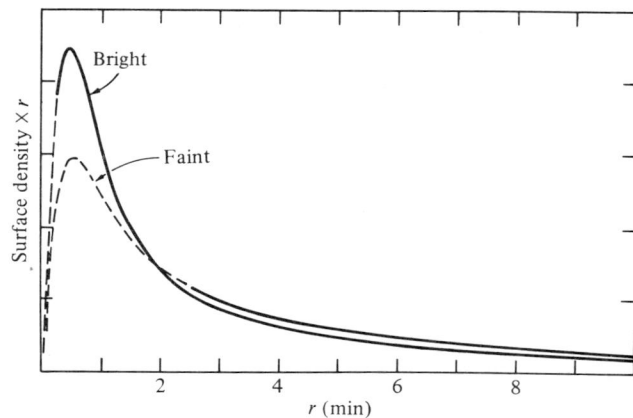

isotropic velocity distribution, but we know from the work of Jim Gunn (p. 271, this volume) and Kyle Cudworth (1978) that, at least in M3, there is an appreciable anisotropy. In fact Chris Wilson and I found that the luminosity function had uncertainties of the order of 30–40 per cent due to anisotropies in the range that we had to admit was possible. Moreover, the problem is not easily solved, because the anisotropies deduced so far apply to giant stars. Since we do not really understand what determines the degree of anistropy of the velocity distribution, I do not see how we are going to deduce how anisotropic the velocity distribution is for the dwarf stars.

Finally I would like to mention an interesting technique that Allan Sandage has been using just lately. He has been making photoelectric measurements between the stars in a globular cluster, selecting regions that contain no stars visible down to the plate limit. The measurements should give us interesting insights into the fainter stellar population, though the interpretation may be difficult.

13.7 The effects of crowding on star counts

Briefly, I can state that the problem of crowding is worse than it looks. The effects are easily gauged – you simply study some poor plate material (a plate taken in bad seeing or at a small scale) and ask how well the results reproduce what you get from good plate material. The effect goes in the direction of losing counts. My own investigations suggest that when about a fiftieth of the total area in your region is covered by star images you begin to suffer from appreciable crowding effects. Intuitively it seems shocking that the images have to be so well separated (but I note that the result is similar to the experience of radio astronomers, even though all the physical parameters are different). The effects have been quantified in our big star-count paper (King et al. 1968), but I would like to warn against extrapolating our formula unquestioningly. Anyone who has crowding problems should make his own determination of them.

13.8 Cluster ellipticities

Surface distributions in star clusters can also yield measures of cluster ellipticities. We do know that some of the clusters in the Magellanic Clouds have greater ellipticities than those in the Milky Way, but investigations in this area are quite limited. The eye-estimates of ellipticities by Shapley and Sawyer-Hogg (Shapley 1930) for galactic globular clusters are surprisingly good: the eye can see a pretty good isophote in the burned-out part of the image. Surface photometry could be used to provide a quantitative measure, at least for the inner regions of the cluster.

It is important to draw a distinction between the ellipticity in the inner parts and that in the outer parts, however, because they come from two different physical causes. Any ellipticity in the inner parts must be due to the cluster's rotation. Although in elliptical galaxies the ellipticity seems to be due mainly to anisotropic velocity dispersions and little to rotation, in globular clusters the residual velocities must be pretty close to isotropic, at least in the centres, where the clusters are completely relaxed. In the outer parts, by contrast, the ellipticity should be due to the tidal effects of the Milky Way.

There should in fact be a rather sharp demarcation between the areas where these different physical effects dominate. The internal forces obey (roughly) an inverse-square law, while the tidal force goes linearly across the cluster; so the relative size of these forces goes as r^3. Near the edge of the cluster the dominance should go over from the internal effects connected with the rotation to the external tidal force.

Disappointingly, such behaviour is not conclusively present in my own star counts, although the first cluster in which I looked for outer ellipticity behaved beautifully. This was M13, a cluster with a rather modest ellipticity in its central parts ($r \lesssim 5$ arcmin); see fig. 13.1. In the outer parts ($r \sim 15$ arcmin), the ellipticity turns right around and points towards the galactic centre, with a statistical significance level of more than 3σ. However, no other cluster has so far displayed this behaviour. In ω Cen there is an ellipticity that points towards the galactic centre, but in this case it is presumably a coincidence: the central ellipticity is also aligned the same way, and I cannot believe that the galactic tidal force reaches into the centre of ω Cen.

I think that this would be a fruitful area of study, especially on 4-metre plates, which will go much fainter than my Palomar Schmidt counts.

13.9 Observed properties of globular clusters

Before I discuss the results that have been acquired so far, I would like to re-emphasize the fact that there are considerable differences between globular clusters. This is illustrated in fig. 13.5, which shows M53 and NGC 5053 in the same Schmidt field. The two clusters have rather similar tidal radii and distances, whereas their core radii are not quite so similar (that of NGC 5053 being four times that of M53) and their surface brightnesses are obviously very different.

Now, one can fit some kind of curves to the star counts or to the luminosity distributions in globular clusters, and the interesting thing I found (King 1962, 1966a) is that you can fit all the clusters quite well simply by specifying a core radius and a limiting radius, plus of course some sort of central surface brightness to normalize the curve. Fig. 13.6 shows a whole family of such curves, and in fig. 13.7 a suitably chosen curve has been applied to the data for NGC 5053. You can consider the curves themselves as defining the quantities; but, as it happens, the core radius is rather close to the half-brightness point, the point where the surface brightness is down to half its central value, at least for the case where (r_c/r_t) is rather small. Of course the limiting radius has the physical significance of being connected with the galactic tidal field.

Given this kind of treatment, where do we stand for the clusters in the Galaxy? To begin with, it is arguable how many clusters there are: the catalogues probably have a total of about 130 at the present time. We know core radii for 105 clusters, a number which could be slightly improved; we know limiting radii to moderate accuracy for 60 clusters; and we know

Fig. 13.5. The globular clusters M53 (NGC 5024, upper right) and NGC 5053 (lower left). M53 has a core radius of 2.7 pc and a tidal radius of 127 pc; for NGC 5053 the corresponding numbers are 11 pc and 63 pc (Peterson & King 1975). The clusters lie at distances of 17.2 kpc and 15.1 kpc from the sun for M53 and NGC 5053 respectively (Harris 1976). This figure is a reproduction from a blue Palomar Sky Survey print.

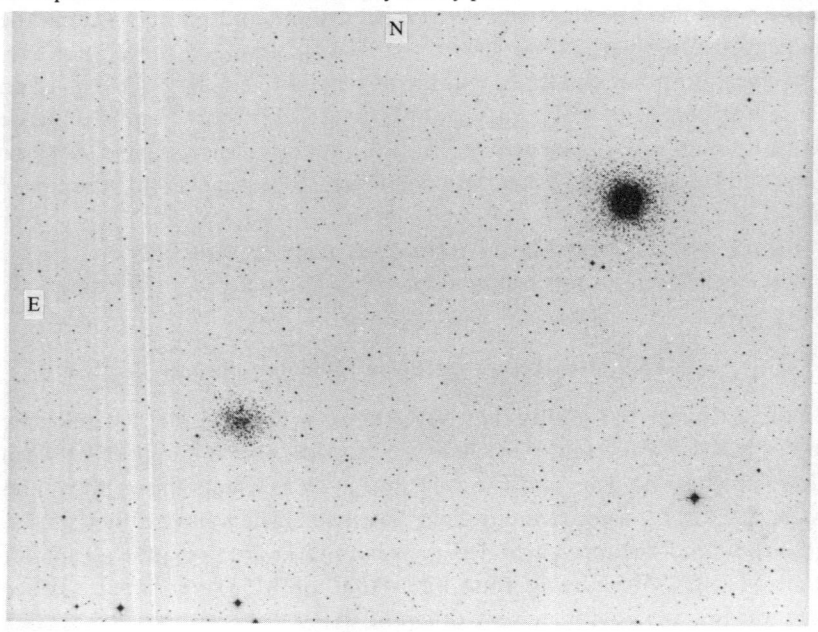

Luminosity distributions 261

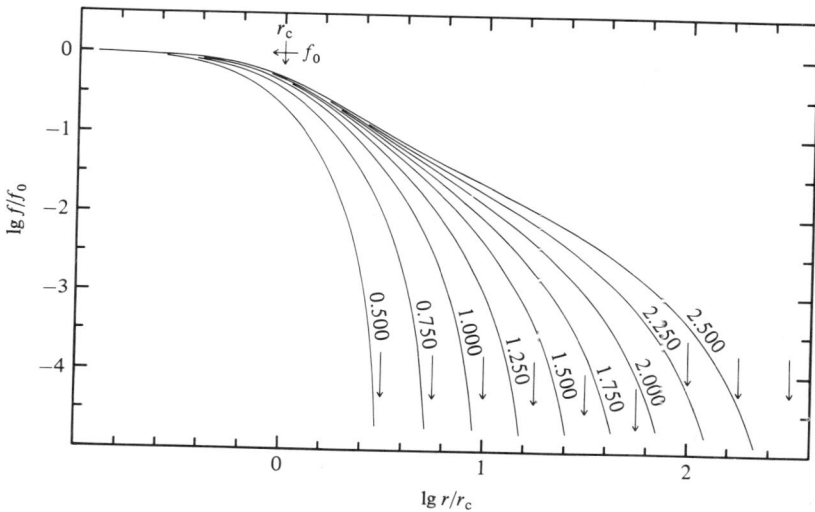

Fig. 13.6. A family of standard curves representing self-consistent dynamical models of star clusters. Here r_c and f_0 represent the core radius and the central surface brightness, and the tidal radius for each curve is represented by a short vertical arrow. This figure has been reproduced from King (1966a), with a small modification (an incorrectly drawn line, marked '∞' in the original, has been removed).

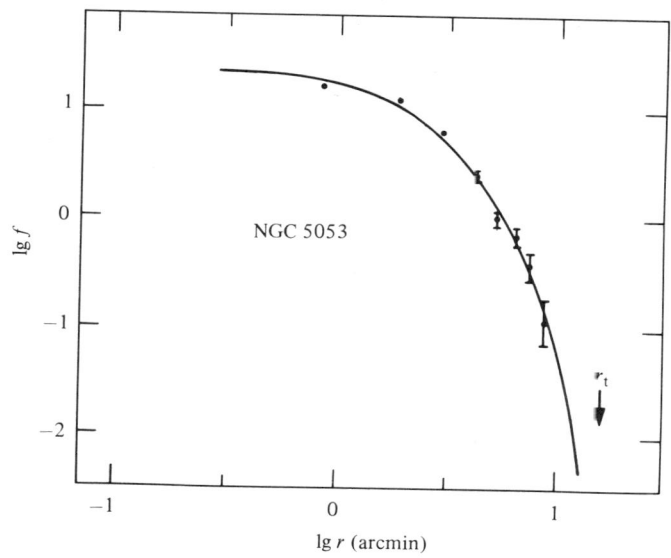

Fig. 13.7. The star counts for NGC 5053, fitted with a suitably chosen standard curve. This figure has been reproduced from King (1975).

core radii, tidal radii and central surface brightnesses for 51 clusters (Peterson & King 1975; Peterson 1976). The tidal radius is usually the limiting quantity, but there are actually 9 clusters that have very low central surface brightnesses so that we lack that value. For statistical studies then we have observational data on a rather good sampling of the globular clusters in the Milky Way. The quality of the data is much harder to assess, but it is my feeling that 15 to 20 clusters have data good enough that we are never going to make them much better: these of course are the nearby clusters, especially those at high galactic latitudes where field stars present little interference. For the others, the quality of the data will certainly improve.

13.10 The X-ray clusters

The X-ray clusters (NGC 1851, NGC 6440, NGC 6441, NGC 6624, NGC 6712, M15, Liller 1, and possibly Palomar 2 and NGC 5824) are of special interest (see Jernigan, p. 351, and Lewin, p. 315, this volume). For a number of these clusters we have fairly good observations – but for other clusters they are not nearly good enough.

M15 is of course very well observed: we have the Newell & O'Neil (1978) study in several colours on electronographic plates of the central regions, and there are fairly good star counts reaching far out. Some of the other clusters are in very badly crowded fields, and the limiting radii which have been determined for a couple of them are shaky because of the rich backgrounds. NGC 1851 has been studied in the central regions by Bahcall, Lasker & Wamsteker (1977), who claim that there is an excess of light over the central curves that fit other clusters; such an excess is seen in M15 as well (Newell & O'Neil 1978). However, the star counts in NGC 1851 differ from the surface photometry, and purely on statistical grounds without alleging any systematic effects I would prefer to trust the surface photometry. Thus I regard a central excess in NGC 1851 as still dubious.

There has been a good deal of work done on NGC 6624 without as yet providing us with a really reliable photometric profile (Bahcall 1976; Harvel & Martins 1977; Faÿ *et al.* 1977; Canizares *et al.* 1978). All of these clusters deserve further study, especially within radii of about 5 arcsec. I think the work has to be done photographically; I would *insist* that it at least be done in the blue and recommend that it be done in the ultraviolet if possible. I have some ultraviolet plates myself for a few of these clusters, plus some others of similarly high concentration which have not yet been identified as X-ray sources. I hope to analyse these plates and would encourage other workers to do the same with their plate material. The identification and study of (for example) luminosity excesses like that seen in M15 may help us to understand what makes a cluster an X-ray source.

Still, we may not get an answer because of statistical limitations or because too many bright stars disturb our photometry. It may well be that we shall have to wait for the Space Telescope to study the centres of these interesting clusters at much higher resolution.

13.11 Orbits of globular clusters

The determination of the tidal limits of globular clusters permits us to determine something about the clusters' orbits in the Milky Way. The reasoning is straightforward. A cluster orbits the Milky Way faster than it can relax and fill out its outer parts, so it is sheared off at the time when it comes closest to the galactic centre. Thus the size of the tidal limit indicates what its perigalactic distance is. This sounds simple enough, but I would like to point out some difficulties in practice.

First of all, there is some doubt about the identification of the actual quantitative value of the tidal radius of the cluster. There are two reasons for this. One of them is that the curves which I have used (section 13.9) in the plot of logarithm of surface density against logarithm of radius were

Fig. 13.8. The effect of anisotropic velocity distributions on the projected surface density of a globular star cluster. This figure has been reproduced from King (1975).

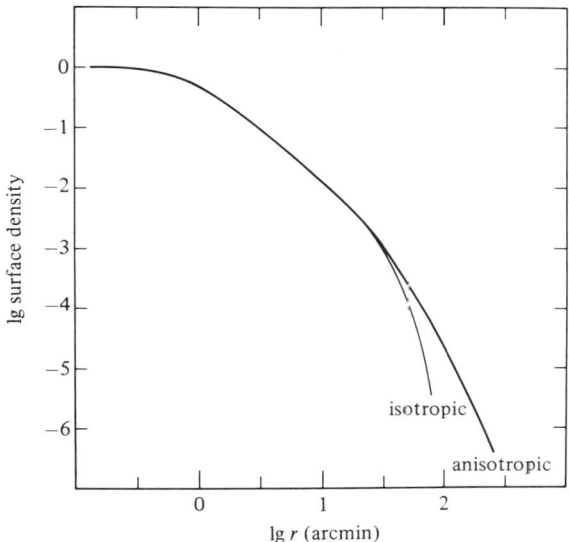

calculated from dynamical models of star clusters with isotropic velocity distributions. I have in fact calculated models with anisotropic velocity distributions, and the effect, shown in fig. 13.8, is to increase the real tidal limit; worse than that, the difference can happen beyond the last observed point. The second source of uncertainty works in the opposite direction, and is due to the fact that stars which have enough energy to get out beyond the tidal limits of the cluster can still be bound by a pseudo-integral. Hénon (1969) and Jefferys (1976) have investigated this numerically, and Dan Keenan (Keenan, Innanen & House 1973) has studied it in clusters as they actually go around the galactic centre. A surprising number of stars that should escape get held in the cluster, giving rise to an envelope of stars beyond what is properly the tidal limit. Thus the tidal limit *as observed* is very uncertain; it should perhaps be larger because of anisotropy; it should perhaps be smaller because of artificially bound stars.

Finally, let us remember that in determining perigalactic points we are assuming something about the force field of the Milky Way, still a very controversial subject in itself.

Fig. 13.9. The B surface brightness profile for M15 as adopted by Newell & O'Neil (1978); see their paper for a complete description of the various symbols used. The solid lines represent the computed surface-brightness distributions for two single-mass near-isothermal models. Each curve is labelled with the parameter $c = \lg(r_t/r_c)$ (from Newell & O'Neil 1978).

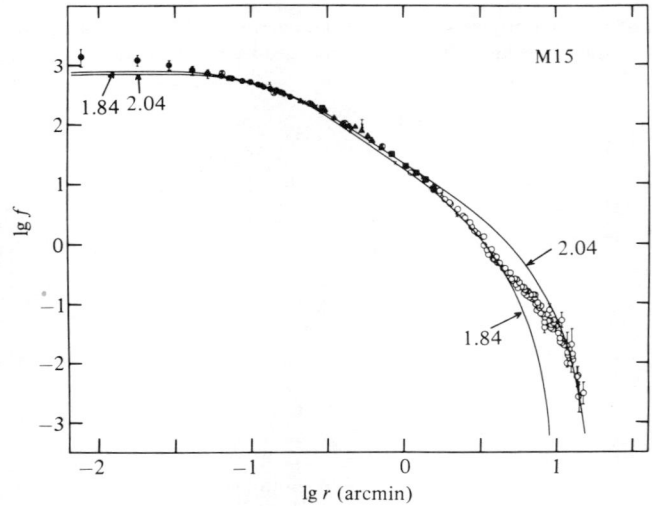

13.12 Comparisons with theory

In this last section I would like to discuss the comparisons of observed density distributions with theory: how does one deduce masses, mass/luminosity ratios, and so forth? Needless to say, there are problems of interpretation.

Let us consider first the central brightness excess seen in M15 by Newell & O'Neil (1978). Fig. 13.9 shows the data and a couple of standard star cluster curves, and you simply cannot fit the brightness excess unless you do something to the star cluster model. Newell, Da Costa & Norris (1976) adjusted the model by including a massive ($\sim 800\,M_\odot$) black hole at the cluster centre, and did in fact achieve a more reasonable fit (shown in fig. 13.10).

Fig. 13.10. The M15 surface-brightness profile compared with several multiple-mass near-isothermal models (Newell, Da Costa & Norris 1976). Each curve is labelled with the mass (in solar masses) of the central object included in the computation. The computed curves have not been convolved with the seeing profile, which is compared with the profile of the nucleus of M15 in the inset. See Newell, Da Costa & Norris (1976), from which this figure is reproduced, for a fuller description.

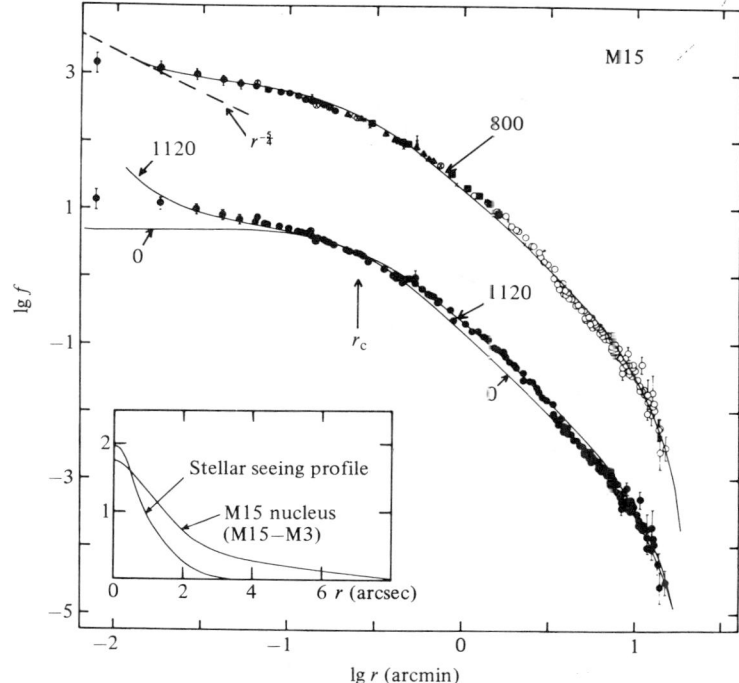

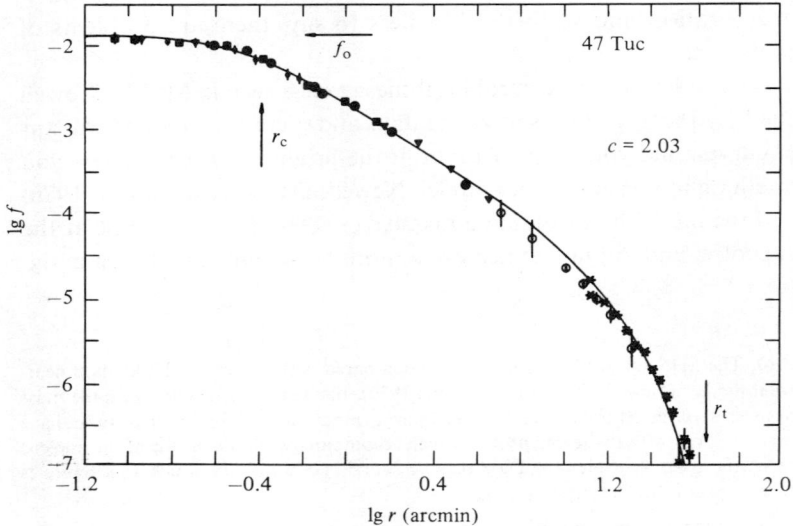

Fig. 13.11. The observed surface-brightness/star-count distribution for 47 Tuc, with the best fitting curve, defined by $c = \lg(r_t/r_c)$, from King's (1966a) family of theoretical surface density distributions. This figure is reproduced from Illingworth & Illingworth (1976), where the various sources of data are identified.

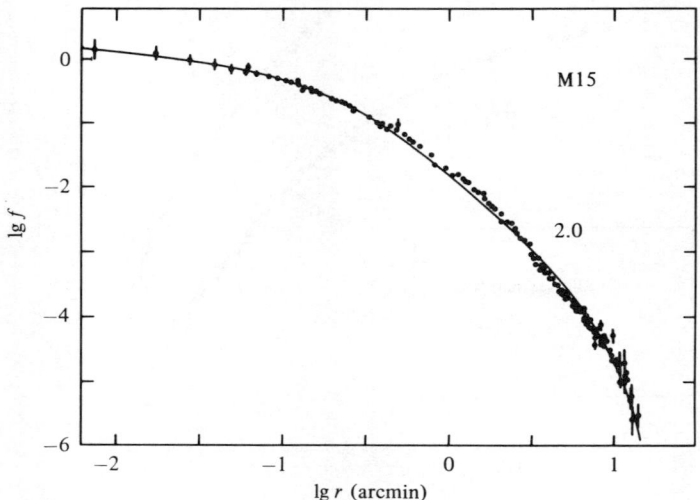

Fig. 13.12. The surface brightness and star distribution in M15 fitted from a model in which the neutron star mass is 2.0 $M_\odot$. The data points are from Newell, Da Costa & Norris (1976) and the figure is reproduced from Illingworth & King (1977).

But before I consider this further, let us go back to examine the simple level of fitting exemplified in fig. 13.11. Here we see the surface brightness distribution for 47 Tuc fitted by a model calculated from a single stellar mass (Illingworth & Illingworth 1976). While there are small systematic deviations you can see that by and large it is not a bad fit, especially considering that it is over a range of 10^5 in surface brightness. However, one would like to do better than this, because real clusters have mass functions, unlike the model used in fig. 13.11, and the use of a mass function changes the distribution derived and the mass/luminosity ratio implied. For instance, a mass-to-luminosity ratio of about 1.5 solar units with a single-mass model turns into ~ 5 solar units when fitted to a reasonable mixed model. The difference is appreciable.

With this in mind Illingworth and I have fitted M15 in a fairly natural way (Illingworth & King 1977) as shown in fig. 13.12. We were constrained both by the velocity dispersion and by the luminosity distribution. For instance, when we tried to add mass in the form of red dwarfs, to reproduce the observed velocity dispersion, then we could not reproduce the luminosity distribution correctly: the red dwarfs go out to the envelope, making it too red, they make the red giants heap up too much to the centre, and so on. Instead we had to put the mass in white dwarfs – in fact almost 50 per cent of the mass of the cluster. This requires only a modest turnup of the mass function at the brighter end, and the interesting implication was that when we extrapolated the mass function beyond the white dwarf region we found that there ought to be some neutron stars as well. Because of their high mass, these stars prefer to be at the cluster centre and provide enough extra mass to deform the standard curve so that it fits the observed profile nicely. While this may not be the complete or final truth about M15, it does illustrate just how important it is to use models that have a proper mass function: it can make a big difference.

It would be extremely interesting to use the observations to resolve some of the uncertainties in the theory. A good example is the segregation of stars of different kinds within globular clusters. I have already mentioned (in section 13.6) problems which beset studies of the relative distributions of bright and faint stars: you tend to study bright stars at the centre and faint stars at the edge. I should say that some (as yet unpublished) work by Gary Da Costa shows differences in the expected sense; hopefully these interesting results will appear soon. Another interesting segregation problem is in the distribution of the HB and other evolved stars: do they evidence their mass loss by having different distributions? A number of people have looked into this, myself included. In my case there were obviously systematic differences between the photometric and the star-count data, so I had to give up (King 1972). This question remains largely unanswered.

Finally, it would be interesting to deduce something about anisotropies in velocity distributions from the surface distributions. As we saw in fig. 13.7, there are real differences, but these tend to be far out where you do not get really good star counts. Since for a given central concentration there is a whole family of models (as in fig. 13.6), you can find different tradeoffs of central concentration and anisotropy to fit the same data. Perhaps the best we can hope for is to set a reasonable limit to the anisotropy, because if you put in too much you will find a curve which extrapolates outward to some absurd limiting radius. This at least provides a limit, but I believe strongly that we will learn much more about the anisotropy from the velocity distributions which we can actually observe, though we will have to resort to the star counts for the fainter stars.

13.13 Conclusions

I hope that these remarks have given a flavour of the subject, in particular its complicated and in some ways grubby nature. But note that it is out of the grubbiness of the data and the careful attention to all the details I have discussed that one gets the facts on which everything else depends.

I would like to express my thanks to Dave Hanes and Barry Madore, without whose efforts this chapter would not have appeared. From an audio tape they prepared a first draft on which I needed to do only a little editing. I take full responsibility for the content of the chapter, but the credit for writing it goes to them.

References

Alcaino, G. (1973). *Atlas of Galactic Globular Clusters with Colour–Magnitude Diagrams.* Universidad Católica de Chile, Santiago.
Bahcall, N. A. (1976). *Astrophys. J. Lett.* **204**, L83.
Bahcall, N. A., Lasker, B. M. & Wamsteker, W. (1977). *Astrophys. J. Lett.* **213**, L105.
Canizares, C. R., Grindlay, J. E., Hiltner, W. A., Liller, W. & McClintock, J. E. (1978). *Astrophys. J.* **224**, 39.
Cudworth, K. (1978). Results reported at NATO Advanced Study Institute on Globular Clusters, Cambridge.
Faÿ, T. D., Mufson, S. L., Duncan, B. J., Hoover, R. B., Sanford, P. W., Charles, P. A., White, N. E., Wisniewski, W. & Wamsteker, W. (1977). *Astrophys. J.* **211**, 152.
Harris, W. E. (1976). *Astron. J.* **81**, 1095.
Harvel, C. A. & Martins, D. H. (1977). *Astrophys. J. Lett.* **213**, L49.
Hénon, M. (1969). *Astron. Astrophys.* **1**, 223.

Illingworth, G. & Illingworth, W. (1976). *Astrophys. J. Suppl.* **30**, 227
Illingworth, G. & King, I. R. (1977). *Astrophys. J. Lett.* **218**, L109.
Jefferys, W. H. (1976). *Astron. J.* **81**, 983.
Keenan, D. W., Innanen, K. A. & House, F. C. (1973). *Astron. J.* **78**, 173.
King, I. R. (1962). *Astron. J.* **67**, 471.
King, I. R. (1966a). *Astron. J.* **71**, 64.
King, I. R. (1966b). *Astron. J.* **71**, 276.
King, I. R. (1972). In *Evolution of Population II Stars*, ed. A. G. Davis Philip, p. 31. Dudley Observatory, Albany, New York.
King, I. R. (1975). In *Dynamics of Stellar Systems*, IAU Symposium no. 69, ed. A. Hayli, p. 99. Dordrecht: D. Reidel.
King, I. R., Hedemann, Jr, E., Hodge, S. M. & White, R. E. (1968). *Astron. J.* **73**, 456.
Leroy, J. L., Aurière, M. & Laques, P. (1976). *Astron. Astrophys.* **53**, 227.
Newell, B., Da Costa, G. S. & Norris, J. (1976). *Astrophys. J. Lett.* **208**, L55.
Newell, B. & O'Neil, Jr, E. J. (1978). *Astrophys. J. Suppl.* **37**, 27.
Oort, J. H. & van Herk, G. (1959). *Bull. astron. Inst. Netherlands*, **14**, 299.
Peterson, C. J. (1976). *Astron. J.* **81**, 617.
Peterson, C. J. & King, I. R. (1975). *Astron. J.* **80**, 427.
Sandage, A. R. (1954). *Astron. J.* **59**, 162.
Sandage, A. R. (1970). *Astrophys. J.* **162**, 841.
Shapley, H. (1930). *Star Clusters*, Harvard Observatory Monograph no. 2. New York: McGraw-Hill.

14
On the dynamics of globular clusters†

JAMES E. GUNN

14.1 Introduction

The topic of the dynamics of globular clusters has been the subject of active research and speculation for many years, but until recently the only striking feature of relevant dynamical data was its all-too-apparent absence. It was to remedy this situation that Dr R. F. Griffin and I undertook in late 1971 to construct a photoelectric radial-velocity spectrometer at Palomar (Griffin & Gunn 1974) using the techniques pioneered by Griffin at Cambridge for the accurate determination of the radial velocities of stars too faint for ordinary photographic techniques. This instrument has now been in operation for seven years, and we have obtained about seven hundred globular cluster radial velocities, to a characteristic accuracy of about 1 km s^{-1}. We will mostly discuss about a quarter of these data here, 166 observations of 111 stars in the cluster M3.

We will discuss the data briefly, and concentrate on attempts to make *ad hoc* dynamical models which describe the present state of the cluster. Models which accurately mimic the observed properties of the cluster are in fact not difficult to come by, and one acquires enough feeling for the various parameters at one's disposal to have some confidence in the final results. There is one glaring problem left, the very high-energy stars, which we have not faced squarely yet – probably our methods are not really suited to this task.

Our models are not evolutionary ones, but the subject of globular cluster evolution has recently been the centre of much activity. We hope that the existence of high-quality dynamical data will motivate the construction of realistic evolutionary models which include the effects of stellar evolution.

† Supported in part by the National Science Foundation (AST76-80801 A01).

14.2 The observations of M3

The radial velocities of 111 giants were obtained over several seasons from 1972 to 1976 with the Palomar spectrometer. The velocity reference star chosen for the cluster was HD 126778, a nearby K giant with a similar radial velocity (-137.3 ± 0.25 km s^{-1} as compared to our derived cluster mean of -146.9 km s^{-1}). The spectrometer scans over a total range of about 80 km s^{-1}. Observing times on M3 stars ranged between 2 and 10 min, the usual criterion for the length of an observation being that the formal standard error be of order 1 km s^{-1}, as judged in a straightforward way (see Griffin & Gunn (1974) for a discussion of the statistical errors in this instrument) from the real-time cross-correlation display.

Full details of the observations appear in Gunn & Griffin (1979) so we will only summarize the salient points here. The observations represent a reasonably magnitude-limited sample, and extend to the very centre of the cluster and out to a radius of about 17 arcmin, about 36 core radii. The only unambiguous velocity variations occur in the two stars (von Zeipel 318 and 803), which are known to be a long-period variable and a W Vir star, respectively. We are sensitive to orbital variations for stars of about one solar mass over the time scale involved for semi-major axes smaller than about 10 AU. In the field 25–30 per cent of red giants display velocity variations of a type which would have appeared in our M3 data had they been present. In the thirty stars for which we have two or more measurements, there are none. In the relevant range (0.5–10 AU) the binaries are 'hard' enough (see Heggie, p. 282, this volume) that they would be little influenced by their environment (Hills 1975). The inference is that star formation has proceeded in a fashion which is very different from that in the solar neighbourhood, in that (at least) binaries are strongly suppressed. We shall see that the mass function is also rather different from the one in the solar neighbourhood, a result anticipated by the work of Da Costa (1977) on four southern clusters. It is also noteworthy that the agreement among multiple measurements of the same stars, while showing no clearcut variation, is rather worse than one would predict on the basis of the formal errors. We think that we understand the errors in the machine well enough, based on much more accurate work on the Hyades and other galactic clusters, that we can at least strongly suggest that the additional noise is not in the observations but in the stars. Small-amplitude (~ 0.8 km s^{-1}) variations on a time scale of about a year or less would account for the observed 'jitter', and might well be present in the atmospheres of these very high-luminosity objects.

Two stars (vZ 764 and vZ 911) stand at 3.5 and 4.5 times the velocity dispersion away from the mean. They have been excluded from the analysis but are definitely cluster members; we will discuss them later.

The data support one's visual impression that the cluster is quite round; the rotation, fitted to a simple rotating model, has a maximum amplitude of about 1 km s^{-1} and is barely *statistically*, much less *dynamically* significant. The raw velocity dispersion is 4.3 km s^{-1}, and one is struck immediately (fig. 14.1) by its rapid falloff with radius beyond about 10 core radii.

14.3 The models

Models were constructed in an attempt to understand the present dynamical state of the cluster. In the absence of a fully evolutionary scheme, 'thermal equilibrium' models, in which stars of various masses have achieved energy equipartition in the cluster core (King, 1966; Da Costa & Freeman 1976; King, p. 259, this volume) were constructed. The core relaxation times are short enough ($\sim 1 \times 10^8$ yr) that this scheme is likely to be a good

Fig. 14.1. Measured velocity dispersions with error bars in five radial bins. The heavy dot is the average of the first three points. The horizontal arrows indicate the extent of the bins. Also plotted are the fits for the best model (solid line) and a model which fits the luminosity profile quite well and has identical population and dynamical mass-to-light ratios (see text). This is the only isotropic model which does this, but the fit to the velocity dispersion data is clearly very poor.

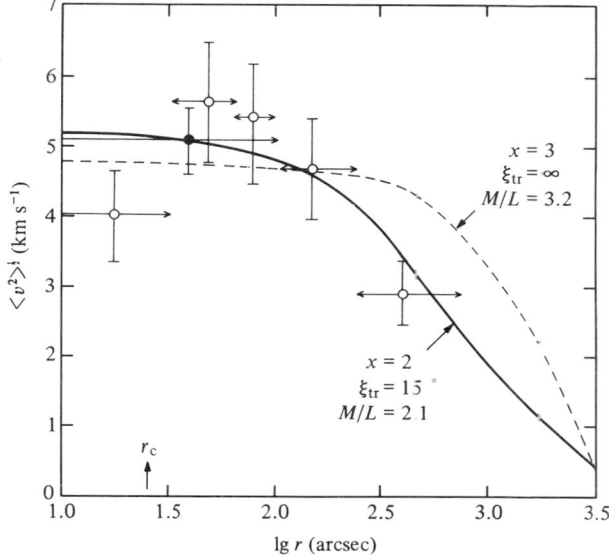

approximation for stars which spend much time in the core. In the outer parts, the relaxation times are long enough that velocity anisotropy can persist, and we have included it in our models. A distribution function, f, of the form

$$f(E, J, m) = C_m e^{-\beta J^2/m} (e^{-\alpha E} - 1), \tag{1}$$

where, as usual, E and J represent energy and angular momentum and m is the stellar mass, with α and β constant for all masses, and with the zero of energy defined by a star at rest at the cluster edge, has been used. This form has a transition radius for anisotropy which is independent of mass, and one can make reasonable arguments that this assumption is not far wrong. The energy dependence is roughly that predicted by solutions of the Fokker–Planck equation (King 1965).

The structure equations were cast in dimensionless form in the standard way (King 1966), which results in the first-order system

$$\xi^2 \frac{dW}{d\xi} = -U,$$
$$\frac{1}{\xi^2} \frac{dU}{d\xi} = -9 \sum_j \sigma_j(\xi, W), \tag{2}$$

where $-W$ is the potential in units of the central mass-weighted velocity dispersion v_0^2, ξ is the radius in units of the core radius r_c, U is the contained mass in units of $v_0^2 r_c/G$, and σ_j is the density of the jth species in units of the total central density. The σ_j are algebraic functions of the scale radius and W:

$$\sigma_j(\xi, W) = \sigma_j I_1(\mu_j W, \xi/\xi_{tr})/I_1(\mu_j W_0, 0). \tag{3}$$

Here μ_j is the mass of the stars in the jth component in units of the central mean mass, and ξ_{tr} is the scale anisotropy transition radius. The function $I_1(x, y)$ and similar functions $I_2(x, y)$ and $I_3(x, y)$ for the radial and tangential velocity dispersions can be evaluated from the form of f (eq. (1)) by quadratures over velocity space once and for all and tabulated.

The construction of a model then consists of specifying the transition radius, ξ_{tr}, the central density function α_j for each component and the central potential W_0, and integrating (2) out from the centre until the density (or equivalently, the potential) falls to zero. In practice it is not the central mass function (which the α_j clearly represent) but the global one which one specifies on astrophysical (or aesthetic) grounds. A straightforward iteration scheme for the αs accommodates this, and models with reasonable mass functions converge satisfactorily in fewer than 10 iterations with 7 components.

14.4 Astrophysical input parameters: fitting

To compare one of the models with a real cluster, one must have a mass function (clearly only relative masses are important until and if one considers relaxation processes), and one must fit to the dimensional scales r_c and v_0. All other physical quantities are then specified in terms of these and the dimensionless model scale parameters.

We have used Böhm-Vitense & Szkody's (1973) value of $y = 0.25$ for the helium abundance. With P. Demarque's (private communication) evolutionary models (kindly lent to us prior to publication) and Böhm-Vitense's (1973) bolometric corrections, an excellent fit to the colour–magnitude diagram of M3 (Sandage 1970) is obtained for a cluster age of 14 billion years, a metal abundance $Z = 0.004$, and a distance modulus of 15.08 mag, which we have adopted for all our dynamical comparisons. The turnoff mass is 0.79 $M_\odot$ for these parameters.

The mass function is a very difficult thing to get a reasonable estimate for. Sandage's (1954) counts in M3, when combined with a mass–luminosity relation, define the mass function rather imprecisely over a small range in masses near the turnoff, but we must rely on the properties of the models themselves to define it at lower masses. We have used the giant bolometric corrections of Böhm-Vitense (1973), the lower main-sequence bolometric correction of Veeder (1974), and models of P. Demarque (private communication) and Copeland, Jensen & Jørgensen (1970), to arrive at our adopted mass–luminosity relation. The Sandage counts then define the mass function from $\lg m = -0.13$ to -0.23. If we suppose that the mass function has the form $dN/dm = cm^{-(1+x)}$ down to some cutoff where evaporation and tidal shocks have depleted it, the suggested value of x is ~ 2 from the count data. There is some uncertainty in this value; see Gunn & Griffin (1979) for a fuller discussion. We have left x as a free parameter in our models, and it is clear that a value near 2 is preferred dynamically. The cutoff mass is also unknown, but we shall see that it does not have a large effect on the observable properties of the models.

The surface-brightness profile of the cluster has been discussed by Da Costa & Freeman (1976), who have also made dynamical models of the sort we are dealing with here. (They had no dynamical data, however, so it is not surprising that their model is not a viable one.) We must fit this surface-brightness profile and the velocity dispersion–radius relation we obtain from the radial-velocity data.

A complication in all this is the question of remnants. We have taken the model mass function (some value of x), and computed the number of dead stars. We have assumed that stars less massive than 5 $M_\odot$ make white dwarfs, and have kept an open mind about what more massive stars do. For most mass functions of interest, there are so few stars above 5 $M_\odot$ that it does not matter what they do; we shall see that no very heavy remnants are likely to be left in the cluster.

The white dwarfs we have assumed are evenly distributed in mass between about 0.8 and 0.4 $M_\odot$, with a free number (subject to the calculated total number) of 1.2 $M_\odot$ ones. Thus the mass function is described by three parameters, viz. $x, f_{1.2}$ (the fraction of the white dwarfs in the 1.2 $M_\odot$ bin), and m_0, the lower cutoff mass – actually, we have cut off the mass function smoothly, and m_0 is the mass at which it peaks.

The dimensionless models have two additional parameters, the central potential W_0 and the transition radius ξ_{tr}. Thus five dimensionless parameters specify a model. Note that the mass function chosen has a value of the visual population mass-to-light ratio $(M/L)_{pop}$ associated with it, which can be easily calculated.

The observations of the luminosity profile contain essentially 3 dimensionless pieces of information: a parameter which specifies the *core shape* and which is affected solely by the proportion of heavy remnants; the slope of the lg I versus lg r curve outside the core; and the ratio of the 'tidal radius' r_t to the core radius. The dynamical data contain really only one: the ratio of the central velocity dispersion to that at some radius well outside the core radius.

Thus there should be a one-parameter family of fits. The observations also determine the dimensional parameters v_0 and r_c, and hence the run of all physical quantities in the cluster model. In particular the total mass of the cluster is determined. From the mass and total visual luminosity, a visual dynamical mass-to-light ratio $(M/L)_{dyn}$ is calculated. Forcing the agreement of $(M/L)_{dyn}$ with $(M/L)_{pop}$ then determines the fit.

14.5 The fit to M3: discussion

It was at once apparent that isotropic models could not adequately reproduce both the extent of the cluster and the rapid fall-off of the velocity dispersion with radius. The observed core shape, very nearly the classical King $(1+\xi^2)^{-1}$ profile, puts a stringent upper limit on the mass present in 1.2 $M_\odot$ or heavier remnants, and the rather high dynamical M/L derived demands that x be greater than the solar neighbourhood (~ 1.3) value, but not so high ($x \sim 3$) as to preclude a proper fit to the run of velocity dispersion with radius. Quite good fits are observed with $x \sim 2$, $m_0 \sim 0.3$ $M_\odot$, about one-eighth of the total white dwarf mass ($\sim \frac{1}{16}$ by number) in 1.2 $M_\odot$ objects, and a transition radius of about 15 core radii. The adopted mass function and the model parameters are outlined in tables 14.1 and 14.2. Fig. 14.2 shows the projected density distribution of each of the seven mass classes, and the fit of the projected luminosity distribution (Da Costa & Freeman 1976) with mass class 2 (from which essentially all the light comes). Fig. 14.1 shows the fit to the velocity dispersion of the chosen model and the 'best' isotropic one which satisfies all the other constraints. It is a very light-star dominated one ($x = 3$) and has a very slow falloff of radial velocity dispersion with radius.

It is interesting that the relaxation time for deflections is about 10^{10} yr at about half the transition radius; thus the model in some sense is almost as anisotropic as it can be.

Table 14.1. *The mass function*

Mass class	$M/M_\odot$	Luminous mass fraction	L/M	Remnant mass fraction
1	1.2	0	—	0.015
2	0.72	0.071	6.33	0.030
3	0.53	0.152	0.38	0.030
4	0.38	0.214	0.04	0.030
5	0.27	0.232	0.01	0
6	0.19	0.161	0.01	0
7	0.13	0.062	0.00	0

Table 14.2. *Characteristics of the best model*

Central potential W_0	13
Transition radius	15 core radii
Scale velocity v_0	4.2 km/s^{-1}
Core radius r_c	1.29 pc, 25.6 arcsec
Limiting radius r_t	128 pc
Half-mass radius	13 pc
Luminosity	2.7×10^5 $L_\odot$ (V)
Mass	5.7×10^5 $M_\odot$
M/L (V) (total)	2.1
M/L (V) (core)	0.67
Average stellar mass in core	0.822 $M_\odot$
Escape velocity (to r_t)	24.0 km/s^{-1}
Central relaxation time	9.6×10^7 yr

The two high-velocity stars have velocities which are well below the escape velocity of the cluster. With our assumed form of the distribution function, there should be no such stars in a sample of 100; the probability of two such is less than one part in 10^6. It thus seems likely that the high-energy end of our assumed distribution function is much too small. Such high-energy stars, even in the nearly-radial orbits that these have (they are both projected on the core of the cluster) can reasonably have survived since the formation of the cluster, though the slower of them will succumb to dynamical friction in about 10^9 yr. It is thus not surprising that thermal-equilibrium distribution functions do not adequately describe the high-energy tail. The situation almost certainly has a bearing on another rather embarrassing feature of our best-fitting model, viz, that its limiting radius is very nearly the same as its Roche radius *now*. Since the extreme wings of the cluster are populated by very high-energy stars, and since increasing the population of high energy stars for a given energy cutoff makes the cluster cut off more sharply (King 1965), we are led again to the same conclusion about the high-energy end of the distribution function.

Fig. 14.2. Run of projected density with radius for each of the seven mass classes (see table 14.1) for the best-fitting model. Also plotted (filled triangles) are the photometric data from the compilation of Da Costa & Freeman (1976). It should (and does) fit the run of component 2, the giants and turnoff stars.

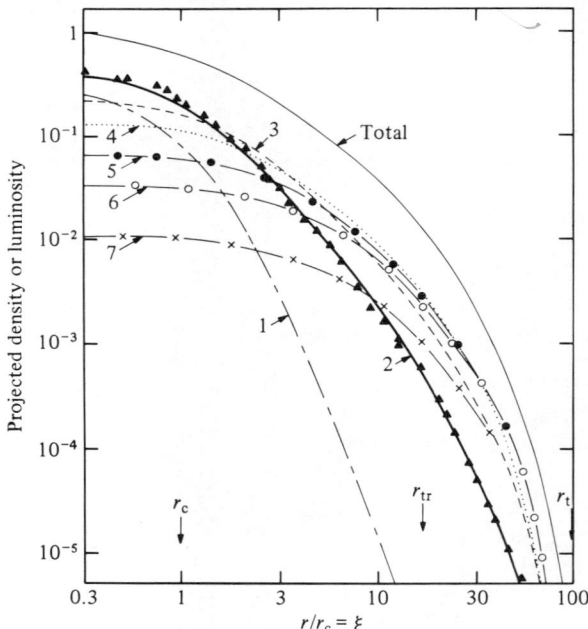

It is gratifying that these simple models *do* adequately describe the cluster with quite reasonable parameters; most noteworthy is that there need be no appeal to any dark mass not accounted for by normal stellar evolution (see King, p. 265, this volume). We certainly do not regard these models as the best which can be done, but hope that they and our data will motivate a proper dynamical-evolutionary treatment by the techniques of Spitzer or Hénon or some similar one, properly including stellar evolution and stellar mass loss.

This work was supported in part by research grants from the United States National Science Foundation and the United Kingdom Science Research Council. I would like to acknowledge the hospitality of the Cambridge Institute of Astronomy and the University of Chicago, where much of the theoretical work was done.

References

Böhm-Vitense, E. (1973). *Astron. Astrophys.* **24**, 447.
Böhm-Vitense, E. & Szkody, P. (1973). *Astrophys. J.* **184**, 211.
Copeland, H., Jensen, J. O. & Jørgensen, H. E. (1970). *Astron. Astrophys.* **5**, 12.
Da Costa, G. S. & Freeman, K. C. (1976). *Astrophys. J.* **206**, 128.
Da Costa, G. S. (1977). Thesis, Australian National University.
Griffin, R. F. & Gunn, J. E. (1974). *Astrophys. J.* **191**, 545.
Gunn, J. E. & Griffin, R. F. (1979). *Astron. J.* **84**, 752.
Hills, J. G. (1975). *Astron. J.* **80**, 809.
King, I. R. (1965). *Astron. J.* **70**, 376.
King, I. R. (1966). *Astron. J.* **71**, 64.
Sandage, A. R. (1954). *Astron. J.* **59**, 162.
Sandage, A. R. (1970). *Astrophys. J.* **162**, 841.
Veeder, G. J. (1974). *Astron. J.* **79**, 1056.

15
Dynamical theory of binaries in clusters

DOUGLAS C. HEGGIE

15.1 Introduction

There are three main reasons for studying the dynamical role of binaries in star clusters. First, there is interesting evidence from the study of comparatively small systems ($N \leqslant 500$: Aarseth 1972) that binaries can absorb a large fraction of the binding energy of a cluster and can influence various aspects of its dynamical evolution, such as the escape rate. It is important to know whether similar processes can occur in much larger systems, such as globular clusters. Secondly, even if binaries cannot form dynamically in significant numbers in large systems, it seems possible that they might be present initially in numbers sufficient to alter radically the predicted dynamical evolution of a cluster (Hills 1975b). Finally, the variable X-ray sources which have been found in a few globular clusters might be binary stars (Clark 1975; Lewin, p. 339, this volume), and a study of this hypothesis requires some knowledge of the dynamical behaviour of binaries.

The basic information on which these three investigations largely depend is a knowledge of the outcome of encounters between binary stars and single stars, i.e. a type of three-body encounter. The first half of this review, consisting of sections 15.2 and 15.3, is devoted to a summary of some important established results on this process, and also on three-body encounters leading to the formation of binary stars. We consider both theoretical and numerical investigations, but concentrate heavily on purely gravitational aspects of these events. In the last section of this chapter we consider the effect of such encounters in the context of the dynamics of star clusters, i.e. their effect on the numbers of binaries and on the distributions of their orbital parameters, and also the effect of the binaries on the dynamical behaviour of the star cluster itself. We shall not attempt to carry out any systematic dynamical discussion of the hypothesis that X-ray sources in globulars are binaries, though we shall make occasional remarks which are of specific relevance to this problem.

15.2 Theory of three-body encounters

It turns out that, in statistical terms, the outcome of an encounter between a binary and a single star depends radically on the ratio of the binding energy of the binary, ϵ, to the energy of relative motion of the single star and the centre of mass of the binary. Let us consider first a system in which all stars have the same mass m and a distribution of velocities v from which we define

$$\beta^{-1} \equiv \tfrac{1}{3}\langle mv^2 \rangle.$$

Thus $\beta^{-1} = kT$ in the usual notation. Then we define a binary to be 'soft' if

$$\beta\epsilon < 1 \tag{1}$$

and 'hard' if $\beta\epsilon > 1$. Actually, in what follows we shall often concentrate on the behaviour of *very* soft and *very* hard binaries, and there will be some region near $\beta\epsilon = 1$ where the properties are transitional between those of the extremes (see eq. (13) below).

15.2.1 Soft binaries

The theory in this case is comparatively simple, since encounters are approximately impulsive. At impact parameters much greater than the semi-major axis, i.e. $p \gg a$, the encounter is tidal, and the typical change in binding energy is

$$|\Delta\epsilon| \sim \epsilon \frac{a^2}{p^2}(\beta\epsilon)^{\frac{1}{2}} \quad (p \gg a,\ \beta\epsilon \ll 1), \tag{2}$$

except at impact parameters so large that the encounter is no longer impulsive. If the impact parameter to either component of the binary is much less than a, however, the result is

$$|\Delta\epsilon| \sim \begin{cases} \epsilon\dfrac{a}{p}(\beta\epsilon)^{\frac{1}{2}} & \text{if } a(\beta\epsilon)^{\frac{1}{2}} \ll p \ll a, \\[6pt] \epsilon\dfrac{a^2}{p^2}(\beta\epsilon) & \text{if } p \ll a(\beta\epsilon)^{\frac{1}{2}}, \end{cases} \tag{3}$$

provided that p is not much less than the 90° deflection distance, $\sim (\beta\epsilon)a$. (For a derivation of results similar to these, see Heggie 1977b.)

In physical terms, the net result of many encounters can be understood by an equipartition argument (Gurevich & Levin 1950). The 'internal' kinetic energy is much less than the kinetic energy of relative motion of the single star and the centre of mass of the binary, and so kinetic energy tends to be transferred to the internal degrees of freedom. This results in a decrease in ϵ. More thorough calculations lead to the asymptotic result

$$\langle \Delta \epsilon \rangle \approx 4\pi^{\frac{1}{2}} G^2 n m^{\frac{7}{2}} \beta^{\frac{1}{2}} \ln \beta\epsilon \quad (\beta\epsilon \ll 1) \tag{4}$$

for the average change per unit time, where n is the number-density of single stars, the eccentricity e has the distribution (7) below, the mean square velocity of the centres of mass of the binaries equals that of the single stars, and the distributions are Maxwellian (Heggie 1975a, eq. 4.30). This result includes disruptive encounters; if these are excluded, $\langle \Delta \epsilon \rangle$ is reduced by a factor of two.

Results (2) and (3) imply that a very soft binary may be disrupted only by a close approach to one of its components. Thus such a binary may well survive many encounters at an impact parameter of order a or less. In fact the rate of destruction by a *single* encounter (see section 15.4.1 below) is approximately

$$\frac{10\pi^{\frac{1}{2}}}{3} G^2 n m^{\frac{7}{2}} \beta^{\frac{1}{2}} \epsilon^{-1} \quad (\beta\epsilon \ll 1) \tag{5}$$

under the same assumptions as (4) (Heggie 1975a, eq. 4.13). It is approximately independent of eccentricity. The inverse process, i.e. formation of soft binaries by three-body encounters, occurs at the rate

$$5\sqrt{6\pi^2} G^5 n_1 n_2 n_3 (m_1 m_2)^{\frac{5}{2}} (m_1 + m_2)^{\frac{1}{2}} m_3^2 \frac{1}{\langle v^2 \rangle^2} \epsilon^{-\frac{7}{2}} \tag{6}$$

per unit volume per unit range of energy (Heggie 1975a, eq. 4.14); here we are considering the rate of formation of binaries whose components have masses m_1, m_2, by three-body encounters involving a third star of mass m_3; the number-densities of the three species are n_1, n_2, n_3, and we assume they have Maxwellian distributions of velocities with the same mean square velocity $\langle v^2 \rangle$. Note the fact that the formation rate appears to be strongly dependent on the masses of the components, which can be understood in terms of a phase-space volume argument (Miller 1975). However, to some extent this dependence is artificial. If we calculate the formation rate per unit range of semi-major axis, for example, it depends on m_1 and m_2 only as $(m_1 + m_2)^{\frac{1}{2}}$. The distribution of eccentricities among newly formed very soft binaries is approximately

$$f(e) = 2e. \tag{7}$$

15.2.2 Hard binaries

The theory of encounters between a very hard binary ($\beta\epsilon \gg 1$) and a single star is comparatively straightforward for distant encounters such that $p \gg (\beta\epsilon)^{\frac{1}{2}}a$, because then the closest approach occurs at a distance much exceeding a. Then the typical changes in energy and eccentricity are

$$|\Delta\epsilon| \sim \epsilon \left[\frac{p}{a(\beta\epsilon)^{\frac{1}{2}}}\right]^{\frac{3}{2}} exp-\left[\frac{p}{a(\beta\epsilon)^{\frac{1}{2}}}\right]^3 \tag{8a}$$

and
$$|\Delta e| \sim \left[\frac{a(\beta\epsilon)^{\frac{1}{2}}}{p}\right]^3 \tag{8b}$$

if $(\beta\epsilon)^{\frac{1}{2}}a \ll p \ll (\beta\epsilon)a$, and $\beta\epsilon \gg 1$ (Heggie 1975a, eqs. 5.40 and 5.66). Note that $\Delta\epsilon$ drops off much more rapidly with increasing impact parameter than does Δe.

Let us consider the case of equal masses. If $\Delta\epsilon < -\frac{1}{3}mV^2$, where V is the relative velocity of the single star and the centre of mass of the binary long before the encounter, the third body is captured into a highly eccentric orbit about the binary, forming a hierarchical triple system. By (8a) this typically happens only if $p \lesssim a(\beta\epsilon)^{\frac{1}{2}}(\ln \beta\epsilon)^{\frac{1}{3}}$. At subsequent close approaches, the binding energy of the third body relative to the binary performs a random walk, the changes typically being comparable with that which first led to its capture, and eventually it will escape again.

The theory of close approaches is difficult. Typically $|\Delta\epsilon| \sim \epsilon$ but, as with distant encounters, the third body may be captured. In fact three outcomes are possible:

(i) If $\Delta\epsilon > -\frac{1}{3}mV^2$, the third body recedes afterwards, and the encounter can be described as a 'flyby'. Typically the resulting change in energy is $\Delta\epsilon \sim \epsilon$ (and positive) since we are considering very hard binaries for which $\epsilon \gg mV^2$. An approximate theory of such encounters is possible because the initial acceleration of the third body towards the binary makes an impulsive calculation correct to order of magnitude.

(ii) If $-\epsilon < \Delta\epsilon < -\frac{1}{3}mV^2$, the third body becomes bound but the binary is not disrupted. As in the similar situation with distant encounters, a bound triple system is formed, but it is not hierarchical. Typically, the orbits of the three stars look chaotic, and the breakup of the system can be predicted with some success by statistical arguments (Heggie 1975a, § 5.3; Monaghan 1976). For example, the probability that a triple system of energy ϵ_3 breaks up to yield a binary of energy ϵ is

$$\tfrac{7}{2}\epsilon_3^{\frac{7}{2}}\epsilon^{-\frac{9}{2}} \quad (\epsilon > \epsilon_3)$$

(Heggie 1975a, eq. 5.24), so that the mean is $\bar{\epsilon} = \frac{7}{5}\epsilon_3$. Since ϵ_3 is approximately the original energy of the binary if this was very hard, the net result is the appearance of a binary typically harder than the original binary by about $\frac{2}{5}\epsilon$, and again the net change in energy is positive. Encounters of type (ii) are called 'resonances', and in the case of equal masses appear to be the most frequent of the three types.

(iii) If $\Delta\epsilon < \epsilon$ the binary is disrupted, but the former third body must become bound to one of its components, forming a new binary. Since $|\Delta\epsilon| \sim \epsilon$ typically, the new third body usually recedes with an energy (relative to the new binary) of order ϵ, which typically much exceeds the relative energy of the old third body and the old binary. Thus the *net* change in binding energy is typically positive and of order ϵ. An approximate theory of such encounters can be attempted on the same basis as those of type (i).

Encounters of type (iii) are called 'exchanges', though this term is also often applied to *any* encounter in which the identity of the components of the binary changes, including some encounters of type (ii). The probability of exchange depends on the masses of the stars, though this has been obtained theoretically only for exchange encounters of type (ii). Consider a triple system with components of masses m_1, m_2, m_3, which is not hierarchical. Then the probability that this breaks up so that the star of mass m_3 escapes, leaving a binary with components of masses m_1 and m_2, has been given as proportional to $m_3^{-\frac{3}{2}}(m_1+m_2)^{-\frac{1}{2}}$ by Heggie (1975a, §5.3), and to m_3^{-2} by Monaghan (1976, 1977). In either case the lightest star tends to be expelled, but there is some numerical evidence (Anosova 1969) that the dependence is steeper even than Monaghan's result would predict.

As shown qualitatively above, all three types of close encounter lead typically to positive net values of $\Delta\epsilon$, and the same is true on average for wide encounters, whether or not a hierarchical three-body system is formed. Detailed theory leads to the order-of-magnitude result

$$\langle \Delta\epsilon \rangle \sim G^2 n m^{\frac{7}{2}} \beta^{\frac{1}{2}} \quad (\beta\epsilon \gg 1) \tag{9}$$

for the average rate of change (Heggie 1975b). This is independent of ϵ, though for large ϵ, typical changes in ϵ are large but take place very infrequently. In fact the rate of occurrence of close encounters, such that the minimum separation of the third body and the centre of mass of the binary is less than the semi-major axis of the binary, is approximately

$$3(3\pi)^{\frac{1}{2}} \frac{G^2 m^3 n}{\epsilon \langle v^2 \rangle^{\frac{3}{2}}} \quad (\beta\epsilon \gg 1), \tag{10}$$

where we have assumed Maxwellian distributions with the same mean square velocity for both single stars and binaries. These encounters lead typically to changes of order ϵ.

The rate of destruction of very hard binaries is very small, because it requires a suitable encounter with a star of very high velocity, and these are rare in real systems or on the assumption of Maxwellian distributions. In the latter case the destruction rate is of order

$$G^2 m^{\frac{7}{2}} n \beta^{-\frac{1}{2}} \epsilon^{-2} e^{-\frac{3}{4}\beta\epsilon} \quad (\beta\epsilon \gg 1)$$

on the same assumption as (10) (Heggie 1975a, eq. 5.5). The rate of formation due to three-body encounters is of order

$$G^5 n^3 m^{\frac{19}{2}} \beta \epsilon^{-\frac{9}{2}} \quad (\beta\epsilon \gg 1) \tag{11}$$

per unit density per unit range of binding energy, though hard binaries may also form by the hardening of soft binaries and by dissipative processes (see section 15.4.2 below). We deduce from (11) that most newly formed hard binaries are only slightly hard. Also, it can be argued that the distribution of their eccentricities will be very roughly as given by (7).

15.3 Numerical experiments on three-body encounters

Many computational studies of the three-body problem have been described in the literature, but we shall concentrate on those which are of significance in the present context.

15.3.1 Soft binaries

It has been shown numerically that the cross-sections derived from the impulsive theory are approximately correct for very soft pairs, and it has been verified that the rate of destruction is approximately independent of eccentricity (Heggie 1975a, §4.4). Thus the theoretically determined average rate of change of energy, and the rates of destruction and formation, whose asymptotic forms are (4), (5) and (6) respectively, should be approximately correct.

The most extensive useful series of computations on this problem is Hills' (1975a), the one specialization being that he took $e = 0$ initially. Comparison with results of the analytic impulsive theory indicate that his value of the cross-section for disruption is smaller by about 30 per cent in the limit of very soft binaries, in the case of equal masses. On the other hand the average rate of change of binding energy agrees within the limits of statistical errors, despite the different assumed eccentricity distributions.

Hills also gives information on the probability of exchange encounters, and this drops sharply to zero in the limit of very soft binaries, as the impulsive theory would predict. The outcome of this comparison is that the impulsive theory seems reasonably satisfactory in the limit of very soft pairs, and its use is probably preferable to the use of numerical results, because of the convenience of analytic forms and the absence of a restriction to certain masses.

15.3.2 Hard binaries

In the case of equal masses, the theoretical cross-sections for resonances and exchanges (encounters of types (ii) and (iii), respectively) have been verified numerically to be correct within a factor of two or so (Heggie 1975a). However, the mass-dependence of these processes has not yet been established theoretically, and it seems best to place most reliance on numerical results. The best of these are again Hills', though the restriction to initially circular orbits may be more important here.

The average rate of change of energy, in the case of equal masses and the limit for a very hard binary, is given by Hills (1975a, eqs. 9, 10, 15) as

$$\langle \Delta\epsilon \rangle \approx 3.1 \frac{G^2 m^3 n}{V} \quad (\beta\epsilon \gg 1)$$

in our notation, where V is the initial relative velocity between the single star and the centre of mass of the binary. If the velocities of single stars and binaries are Maxwellian with the same dispersion $\langle v^2 \rangle$, we find that

$$\langle \Delta\epsilon \rangle \approx 3.0 \frac{G^2 m^3 n}{\langle v^2 \rangle^{\frac{1}{2}}} \quad (\beta\epsilon \gg 1). \tag{12}$$

This is smaller, by a factor exceeding 3, than would be expected from existing theory (Heggie 1975b, eq. 20 and discussion). Some of this discrepancy could be due to Hills' assumption of initially circular orbits, since this depresses the frequency of ejection at high velocity (Valtonen 1976). Hills also gives formulae for the rate of occurrence of exchange encounters, without distinguishing between cases (ii) and (iii) described in section 15.2.2 above. In the limit of very hard pairs the rate is approximately

$$10 \frac{Gmna}{\langle v^2 \rangle^{\frac{1}{2}}} \quad (\beta\epsilon \gg 1).$$

This is within a factor of 2 of what would be expected from theory (Heggie 1975a, eq. 5.22) if we assume that resonant exchanges dominate.

A very extensive set of calculations by Valtonen, reported in several papers (e.g. Valtonen 1974), should be mentioned. The initial conditions were selected in a variety of ways, but these are not suitable for determining quantitatively the average rate of change of energy of a binary in a cluster, for example. However, their scope makes them very useful for giving qualitative information on how the outcome of an encounter depends on the masses, the relative inclination of the orbits, their eccentricities, and the ratio of a to the distance of closest approach of the third body.

Finally we should mention a few sets of experiments on the breakup of bound triple systems, since this is relevant to the outcome of resonant encounters with hard binaries. Those by Agekian & Anosova (1967, 1968) dealt with the case of equal masses, but a later paper considered the breakup of triple systems with a small variety of different masses (Anosova 1969).

15.3.3 Future work

Numerical calculations are particularly necessary in the case of hard binaries, and there are still important results on which there is inadequate information or even none at all. From the point of view of a Fokker–Planck description of the dynamics of a star cluster, one would like to know certain moments of the changes per unit time in the velocity $\mathbf{v} = (v_1, v_2, v_3)$ of a star which moves through a medium consisting partly of single stars and partly of binaries. In particular it is necessary to find $\langle \Delta v_i \rangle$ and $\langle \Delta v_i \Delta v_j \rangle$, even though the changes due to close encounters with hard binaries are liable to be large but infrequent, in contrast to the frequent occurrence of small changes, which is usually required to justify the Fokker–Planck approach. Only rough semi-theoretical results on these moments are available (Heggie, 1979).

Recently a number of studies (Hills 1975b, Heggie, 1979; Spitzer 1978) have been devoted to the investigation of clusters containing initially many binaries. The occurrence of encounters between pairs of binaries, where it has not been neglected, has been treated only rather intuitively on the basis of already rather incomplete information about three-body encounters. Therefore numerical work on four-body encounters (i.e. encounters between binaries) would also be justified.

More work should perhaps be done on the distinction between hard and soft binaries. The criterion $\beta\epsilon \gtrsim 1$ is rather artificial, and one needs to identify a suitable energy where the properties of very soft binaries go over to the very different properties of very hard binaries. Perhaps the most useful and convenient criterion would be the energy at which $\langle \Delta\epsilon \rangle = 0$. Hills' work (1975a) shows that, in the case of equal masses, this energy

is approximately $\epsilon \approx \tfrac{1}{2} m V^2$, or
$$\beta \epsilon \approx 3 \tag{13}$$
on the same assumptions as (12). However, work needs to be done on what this energy is when a spectrum of masses is present.

Even in the case of equal masses we do not have either an accurate theoretical prediction or an accurate numerical measurement of the rate of formation of hard binaries (in the sense of those that will survive further encounters), and there are several reasons for supposing that this is a quantity of some importance.

15.4 Binaries in clusters

In the previous two sections we summarized some important facts about the effect of encounters between binaries and single stars. Now we shall use this information to study the implications for the dynamics of star clusters and the binaries within them. As we have seen, the most important distinction is between hard and soft binaries, which occurs at the energy given by (13), i.e. when

$$a \approx 4 \left(\frac{m}{M_\odot}\right)\left(\frac{100 \text{ km}^2 \text{ s}^{-2}}{\langle v^2 \rangle}\right) \text{AU} \tag{14}$$

if all masses are equal. Here, as in what follows, we scale various quantities in a way appropriate for equal-mass models of the cores of dense and concentrated globular clusters (Peterson & King 1975).

15.4.1. Theory of soft binaries

Soft binaries tend to be destroyed by encounters at a rate given by (5) for destruction by a single encounter. It is also possible to disrupt a soft binary by successive non-disruptive encounters, and the rate for this process may be estimated to be $|\langle \Delta\epsilon \rangle|/\epsilon$, using (4). The result is shorter by only a logarithmic factor, and so at worst we slightly underestimate the destruction rate by using (5). The resulting lifetime is of order $(\beta\epsilon)(\lg N)t_r$, where N is the total number of stars and t_r is an appropriate two-body relaxation time, or approximately

$$10^9 \left(\frac{\langle v^2 \rangle^{\frac{1}{2}}}{10 \text{ km s}^{-1}}\right)\left(\frac{10^4 \text{ pc}^{-3}}{n}\right)\left(\frac{M_\odot}{m}\right)\left(\frac{1 \text{ AU}}{a}\right) \text{ yr}. \tag{15}$$

This would be of the order of the relaxation time for a binary of semi-major axis given by (14), and proportionately shorter for a softer binary (see also

Heggie 1975b). While this is just an upper limit to the time scale for the disruption of a soft binary, the detailed evolution of the distribution of binding energies of soft pairs has been studied by King (1977).

On the time scale (15), the disruption of soft binaries will achieve equilibrium with their formation by three-body encounters. For a cluster in virial equilibrium, the resulting number-distribution by energy is of the form $n(\epsilon)/n \propto N^{-2}(\beta\epsilon)^{-5/2}\beta$ (see Heggie 1975b, eq. 4), or approximately

$$\frac{n(a)}{n} \approx 10^{-9}\left(\frac{n}{10^4 \text{ pc}^{-3}}\right)\left(\frac{100 \text{ km}^2 \text{ s}^{-2}}{\langle v^2 \rangle}\right)^{3/2}\left(\frac{m}{M_\odot}\right)^{3/2}\left(\frac{a}{1 \text{ AU}}\right)^{1/2}\text{AU}^{-1} \quad (16)$$

by semi-major axis, where we have expressed both results as a fraction of the number-density of single stars. This is a minute fraction, even up to separations of several hundred AU. One would also expect the eccentricities of soft binaries to be randomized to the distribution (7) on the time scale (15).

15.4.2 Theory of hard binaries

It is possible to write down a Boltzmann distribution for hard pairs, but it rises exponentially with decreasing semi-major axis and is quite unphysical. In fact their distribution is the combined result of their formation at a rate given in order of magnitude by (11), and their hardening at a rate given roughly by (9). It has been argued that the resulting distribution of energies is flat as far as the energy to which binaries have had time to harden (Lightman & Fall 1978; but see Retterer 1980). Actually, hard binaries also form by hardening of soft binaries, but if these are in equilibrium the rate is of the same form as (11) (Heggie 1975b). We now discuss these rates in more detail.

The rate at which hard binaries form is of order

$$\frac{1}{N \lg N} t_r^{-1} \quad (17)$$

for the whole cluster (Spitzer & Hart 1971). More precisely, the rate of formation of binaries with energies exceeding (13) is roughly

$$10^2 \frac{n^3 G^5 m^5}{\langle v^2 \rangle^{9/2}} \quad (18)$$

per unit volume (Heggie 1975b, eq. 21), or

$$2 \times 10^{-13}\left(\frac{n}{10^4 \text{ pc}^{-3}}\right)^3\left(\frac{m}{M_\odot}\right)^5\left(\frac{100 \text{ km}^2 \text{ s}^{-2}}{\langle v^2 \rangle}\right)^{9/2} \text{pc}^{-3} \text{ yr}^{-1}. \quad (19)$$

(The numerical coefficients in these last two results are not well established.) Most new hard binaries will have semi-major axes of order (14) or a little less, and because of the large dependence of the formation rate on density, almost all form within the core of the cluster. In systems with a spectrum of masses, hard binaries should form preferentially from the largest masses, an effect enhanced by mass-segregation caused by two-body encounters.

The above considerations are concerned with formation of hard pairs by purely gravitational three-body encounters. They may also form by dissipative two-body encounters, and the corresponding rate is approximately

$$2 \times 10^{-8} \left(\frac{n}{10^4 \text{ pc}^{-3}}\right)^2 \left(\frac{m}{M_\odot}\right)^{1.1} \left(\frac{\langle v^2 \rangle}{100 \text{ km}^2 \text{ s}^{-2}}\right)^{-0.6} \left(\frac{R}{R_\odot}\right)^{0.9} \text{pc}^{-3} \text{ yr}^{-1},$$

where R is the stellar radius (Press & Teukolsky 1977). This rate generally much exceeds (19).

Having formed, a new hard binary has a tendency to harden. From (12) we deduce that the mean rate of hardening by purely gravitational processes is of order

$$\langle \Delta \epsilon \rangle \sim \beta^{-1} \frac{1}{\lg N} t_r^{-1}$$

for a cluster in equilibrium, and

$$\left\langle \Delta\left(\frac{1}{a}\right)\right\rangle \approx 1.3 \times 10^{-10} \left(\frac{m}{M_\odot}\right)\left(\frac{n}{10^4 \text{ pc}^{-3}}\right)\left(\frac{10 \text{ km s}^{-1}}{\langle v^2 \rangle^{\frac{1}{2}}}\right) \text{AU}^{-1} \text{ yr}^{-1}. \quad (20)$$

It is important to remember that the hardening of a binary is a stochastic process, but if we treat (20) as a differential equation, i.e. we set $\langle \Delta(1/a) \rangle = -\dot{a}/a^2$, we find that

$$a = \frac{a_0}{1 + \frac{t}{T}}, \quad (21)$$

where $a = a_0$ when $t = 0$ and

$$T \approx 8 \times 10^9 \left(\frac{M_\odot}{m}\right)\left(\frac{10^4 \text{ pc}^{-3}}{n}\right)\left(\frac{1 \text{ AU}}{a_0}\right)\left(\frac{\langle v^2 \rangle^{\frac{1}{2}}}{10 \text{ km s}^{-1}}\right) \text{ yr}. \quad (22)$$

The average rate at which close encounters occur is of order $(\beta \epsilon \lg N)^{-1} t_r^{-1}$ by (10), or approximately

$$4 \times 10^{-10} \left(\frac{m}{M_\odot}\right)\left(\frac{a}{1 \text{ AU}}\right)\left(\frac{n}{10^4 \text{ pc}^{-3}}\right)\left(\frac{10 \text{ km s}^{-1}}{\langle v^2 \rangle^{\frac{1}{2}}}\right) \text{ yr}^{-1}. \quad (23)$$

For sufficiently hard binaries, then, say $a < 0.1$ AU, close encounters are rare even in the cores of dense clusters. However, the eccentricities of hard binaries are likely to be randomized to (7) on a time scale shorter than the interval between successive close encounters. The rate at which exchange encounters may be expected to occur is another quantity of interest, and it has been considered by Hills (1977); the rates are within a factor two of (23).

The average change in the binding energy of the binary in a close encounter is of order $\frac{1}{2}\epsilon$. Typically, then, the velocity with which the third body emerges from the encounter is about

$$20\left(\frac{m}{M_\odot}\right)^{\frac{1}{2}}\left(\frac{1 \text{ AU}}{a}\right)^{\frac{1}{2}} \text{ km s}^{-1} \qquad (24)$$

and it is possible for this to exceed the escape velocity (Aarseth 1977). The ejection velocity of the binary will be about half of (24), but this may still be enough to move the binary out of the core. The effect of binaries on the rate of escape from clusters has been considered by Heggie (1975b).

Let us summarize, then, for a star cluster with $m = 0.8$ $M_\odot$, $n = 4 \times 10^4$ pc^{-3} in a core of volume 1 pc^3, $\langle v^2 \rangle^{\frac{1}{2}} = 10$ km s^{-1} and a lifetime so far of 10^{10} yr. (Several dense globular clusters may have parameters not far from these values, see Peterson & King (1975).) Soft binaries are those with $a > 3$ AU, by (14). The time scale for their disruption is at most about 10^8 yr, by (15), and so we expect that these should have achieved equilibrium, but (16) implies that their numbers are negligible. By (19) it is unlikely that any hard binaries will have formed by three-body processes.

It is important not to take the above example as typical. The rate of formation of binaries depends sensitively on the central density, which may vary greatly between equal-mass models and those with a spectrum of masses (compare Peterson & King 1975 with Illingworth & King 1977). Furthermore, we do not know the dynamical histories of globular clusters. However, dynamical theory predicts that all clusters must sooner or later pass through a phase of high central density, towards the end of core collapse and perhaps after. It is therefore interesting to compute the numbers and properties of hard binaries formed in this process.

There are theoretical arguments for supposing that the energy of the collapsing core remains approximately constant, while decreasing in mass (Lightman & Shapiro 1978). Hence the process could end with the formation of a single very energetic binary (Aarseth 1972). More detailed calculation, however, suggests that the result actually depends on the precise form of core collapse, as follows. Suppose the central density and velocity dispersion vary like $n_c \propto \tau^m$, $\langle v_c^2 \rangle \propto \tau^n$, where τ is the time until complete collapse, and m, n are constants which have been determined for various computed and analytic models by Davoust (1977). Then the total number of hard binaries expected to form during core collapse is proportional to N^p, where

$$p = \frac{3n - 2m - 2}{3n - m}$$

(Heggie 1975b), and N is the total number of stars in the system. Now $p = 0$ on the evaporative model of core collapse, which would imply that the total number of binaries formed during core collapse should be independent of N. However, from Davoust's results for Monte Carlo models (which were not available when the previous treatment was written), values between 0.4 and 0.9 occur. This suggests that if one hard binary can form in the evolution of a small cluster then very large numbers may form in the collapse of the core of a globular cluster. (This is contrary to what might be expected from (17), but the latter takes no account of the evolution of the cluster.) Of course the formation of a large number of binaries might greatly change the assumed form of core collapse.

Virtually nothing is known of circumstances after core collapse. The presence of newly formed hard binaries may tend to retard further collapse, unless the binaries move out of the core, or become too hard to interact on a short enough time scale. However, it may be wrong to think that the cluster now evolves at the bidding of the hard binaries in its core. Hénon (1975) has argued ingeniously that the evolution of existing binaries and their further formation is controlled by a flow of energy into the core, caused by two-body relaxation.

15.4.3 Theory of initial binaries

As we have seen, the rate at which hard binaries form during core collapse is uncertain, but the known abundance of binaries in the solar neighbourhood makes it worth investigating the effect on the evolution of a cluster of a population of binaries present initially. As was stated above, soft binaries, with semi-major axes exceeding (14), will be disrupted rapidly. The effect of the initial hard binaries may be more prolonged, and we consider the effects of the energy which tends to be released in their encounters with single stars.

If the semi-major axis is sufficiently small that a close encounter is unlikely to occur throughout the life of the cluster (see eq. (23)), the binary may be regarded dynamically as a massive star. Also, if a is so small that a star encountering the binary is ejected from the cluster (see eq. (24)), then only a fraction of the energy released in the encounter will be fed to the remaining bound stars. For these reasons, only binaries with semi-major axes within a certain range are efficient producers of energy (see Heggie 1975b).

Hills, using an evaporative theory of core collapse (Hills 1975b), has shown that the presence of a sufficient fraction, f_b, of hard binaries (more exactly, of efficient hard binaries) can arrest the process of core collapse. In fact this occurs if

$$f_b \gtrsim 0.07 \lg N, \qquad (25)$$

where N may be thought of as the number of stars in the core. Heggie (1979), using a more detailed theory of core collapse, finds a similar condition, with the additional feature that the critical fraction depends quite sensitively on the velocity of escape from the centre of the cluster. (An early rough estimate of the numbers of binaries needed to influence the evolution of the cluster (Heggie 1975b) now seems definitely wrong (Heggie 1977a).)

It should be noted that both theories neglect binary–binary interactions, and the process of mass-segregation, which by itself would tend to concentrate the binaries close to the centre of the cluster. However, if the fraction of binaries is insufficient to arrest core collapse initially, N decreases and the velocity of escape increases by the usual process of core collapse. Both effects mean that the given proportion of binaries will ultimately be able to halt core collapse. Nevertheless, the halt can be only temporary, since the binaries which are supporting the core will eventually become too hard for efficient energy production, and collapse may set in again. Monte Carlo methods (Spitzer 1978), which include most of these effects, indicate that the simple theory on which results such as (25) are based is unlikely to be correct.

As an example of the hardening of an initial hard binary, let us consider again the core of a centrally condensed globular cluster with $\langle v^2 \rangle \approx 10^2$ km^2 s^{-2}, $n \approx 4 \times 10^4$ pc^{-3}, and $m \approx 0.8$ M$_\odot$. Then a binary which is slightly hard corresponds to $a \approx 2$ AU. By (21) and (22) this will have had time to harden to about $a \sim 0.3$ AU. By (23) it will be undergoing close encounters every 3×10^9 yr or so. Incidentally, together with the results for soft binaries, these results may imply that there should be virtually no binaries with $a \gtrsim 0.3$ AU in the dense cores of such clusters, if they have been as dense as this for much of their lifetimes. This conclusion would be strengthened by inclusion of dissipative effects, especially for giants.

15.4.4 Numerical work on binaries in clusters

The formation of binaries in the computed evolution of N-body systems has been noted many times since the earliest work of von Hoerner (1963). Since we are concerned mainly with globular clusters, the relevance of some of these results to large systems often requires the intervention of some theory. Furthermore, numerical calculations demonstrate the importance of the heaviest stars, and the effects of mass loss in their evolution may greatly modify the results. However, to some extent the heaviest stars ultimately evolve dynamically rather like an independent cluster in the core (Lightman & Fall 1978), and so the best guide to the evolution of the core in a globular cluster may be the computed N-body systems of several hundred stars with the same mass.

Cruz-González & Poveda (1972) have considered numerically the evolution of a distribution of fairly soft binaries subject to repeated encounters, though their conclusions are subject to a correction pointed out by Hénon (1972). They considered the case in which one of the components was massless, but the theory of this case is similar to that of components with non-zero masses. They compared the lifetimes with those predicted theoretically in various ways. Heggie (1975a) has analysed the soft pairs that were found in a few computed N-body systems with $N \leqslant 250$. The distributions of eccentricity and binding energy are consistent with those expected on the assumption that soft binaries are in statistical equilibrium (eqs. (7) and (16)). The rate of formation of soft binaries has also been studied by Agekian & Anosova (1971) and by Aarseth & Heggie (1976).

A series of experiments on small ($N \leqslant 24$) isolated systems was executed by van Albada (1968) with the specific intention of investigating the occurrence of binaries and their role in the evolution of systems. He found that binaries generally formed, the maximum binding energy tending to increase with time, and ultimately containing as much energy as the original system, or more. The evolution of the binaries, which took place partly by exchange encounters, was accompanied by energetic escape of stars, especially late in the evolution of the system. Even binaries may escape. Ultimately, the components of the most energetic binaries tended to be the heaviest stars in the system, in those cases where a spectrum of masses was present. Hayli (1972) has shown that some of these results are little affected by the existence of a tidal gravitational field.

In these conclusions one can recognize several phenomena which might be expected theoretically to occur also in somewhat larger systems. That this was so was shown by Aarseth (1968, 1972, 1974) in cases with $N \leq 500$. In these models there are frequently several soft binaries, and these can be remarkably persistent if formed on orbits which take them only infrequently within the core. The total number of hard binaries formed is remarkably small, though larger numbers occur if the initial structure of the system contains subclusters (Aarseth & Hills 1972). Otherwise, a certain amount of core collapse must occur before any persistent hard binaries are formed. In some respects these act as massive single stars and merely enhance two-body relaxation effects, such as escape. However, there is a marked tendency for just one of the hard binaries to become harder by encounters with single stars and any other hard binaries present. This effect is especially marked in systems with a spectrum of masses, and exchange of companions frequently ensures that its components are the most massive stars in the system.

The 'Aarseth binary' may eject single stars from the system, but most of the energy it gains leads to an expansion of the halo. Ultimately – and this tends to happen rather abruptly – it absorbs a large fraction of the energy of the system (except in the case of equal masses). At this point the radius of the core reaches a minimum, and we can consider core collapse to have come to an end. Escape due to interactions with the binary tends to decline at this stage, and the binary may move from the core. The accompanying loss of mass may contribute to a decrease in density in the core. However, it is possible for the core to begin collapsing again, albeit on a longer time scale than before. Again this may close with the formation and hardening of a binary, and often the old 'Aarseth binary' is one of its components, forming a hierarchical triple system. Sometimes this seems stable.

It is difficult to draw firm conclusions relevant to globular clusters from such results, though they are highly suggestive. Part of the difficulty is that the binary phenomena are less striking in the case of equal masses. Turning now to Monte Carlo models, which are thought to be directly relevant to globular clusters in important respects, we find that very little work on these phenomena has been attempted. However Hénon (1975) has boldly modelled the effect of the formation and hardening of an energetic binary by providing a source of energy at the centre of the cluster. Core collapse proceeds almost in the usual way, little energy being released by the source in the process. Thereafter, however, the whole cluster expands at the expense of the energy released by the central source.

If Hénon's arguments are correct, his central source should qualitatively model the effects of binaries, despite its seeming artificiality. Furthermore, his models are of great interest because they provide us with the clearest picture we have of what a cluster might look like after core collapse. Indeed they suggest that the spatial structure of a post-collapse cluster may be hard to distinguish from its structure before collapse. In his models there is no tendency for recollapse to occur.

Finally let us consider studies of N-body systems containing binaries initially. Unfortunately, the presence of any energetic binary seriously hampers the numerical integration of N-body systems even in spite of the application of sophisticated techniques. This accounts for the fact that many calculations are stopped at the appearance of a very close binary, and that very little has been done on systems containing initial binaries. One case studied by Aarseth (1975) with $N = 90$ contained about 10 per cent of hard binaries initially, with energies which should make them particularly efficient as producers of energy. However, lack of a comparison calculation makes it difficult to determine their effect on core collapse. A further calculation currently in progress may clarify this. The effect of initial hard binaries on the evolution of Monte Carlo models has been studied recently by Spitzer (1978).

15.4.5. Future work

In late stages of core collapse it is certain that non-gravitational phenomena come into play (see Lightman & Shapiro 1978). However, even in the purely stellar-dynamic problem we have only a general idea about the way in which core collapse in globular clusters ends, and what happens thereafter. Even in the absence of initial binaries, it seems that the following processes may be important:

(i) Two-body relaxation, resulting in a flow of energy into the central parts of the cluster.

(ii) The formation of hard binaries.

(iii) Elastic encounters between binaries and single stars, leading to mass segregation.

(iv) The heating effect of three- and four-body encounters on the distributions of single and binary stars.

(v) The hardening of binaries, making them less efficient sources of energy.

The essential gravitational phenomena controlling the evolution during late core collapse and after may be contained in the interplay of these processes, and the clarification of their respective roles remains a central unsolved problem in stellar dynamics.

I would like to thank the staff of the Institute of Astronomy, especially M. J. Rees, for their hospitality over a period during which this review was prepared. I am particularly grateful to S. J. Aarseth for the benefit of many useful conversations on dynamics during this period. Several valuable comments by I. R. King have been incorporated into the text, and I am grateful to him also for describing to me some unpublished work of his student Mr Retterer.

References

Aarseth, S. J. (1968). *Bull. Amer. astron. Soc.* **3**, 105.
Aarseth, S. J. (1972). In *Gravitational N-Body Problem*, ed. M. Lecar, p. 88. Dordrecht: D. Reidel.
Aarseth, S. J. (1974). *Astron. Astrophys.* **35**, 237
Aarseth, S. J. (1975). In *Dynamics of Stellar Systems*, IAU Symposium no. 69, ed. A. Hayli, p. 57. Dordrecht: D. Reidel.
Aarseth, S. J. (1977). *Rev. Mex. Astron. Astrofis.* **3**, 199.
Aarseth, S. J. & Heggie, D. C. (1976). *Astron. Astrophys.* **53**, 259.
Aarseth, S. J. & Hills, J. G. (1972). *Astron. Astrophys.* **21**, 255.
Agekian, T. A. & Anosova, Zh. P. (1967). *Astron. Zh.* **44**, 1261, transl. in *Soviet Phys.-Astron.* **11**, (1968) 1006.
Agekian, T. A. & Anosova, Zh. P. (1968). *Astrofiz.* **4**, 31, transl. in *Astrophys.* **4**, (1968) 11.
Agekian, T. A. & Anosova, Zh. P. (1971). *Astron. Zh.* **48**, 524, transl. in *Soviet Astron.* **15** (1971) 411.
Anosova, Zh. P. (1969). *Astrofiz.* **5**, 161, transl. in *Astrophys.* **5** (1969) 81.
Clark, G. W. (1975). *Astrophys. J. Lett.* **199**, L143.
Cruz-González, C. & Poveda, A. (1972). In *Gravitational N-Body Problem*, ed. M. Lecar, p. 99. Dordrecht: D. Reidel.
Davoust, E. (1977). *Astron. Astrophys.* **61**, 391.
Gurevich, L. E. & Levin, B. Yu. (1950). *Astron. Zh.* **27**, 273, transl. in *NASA TT* F-11, 541.
Hayli, A. (1972). In *Gravitational N-Body Problem*, ed. M. Lecar, p. 73. Dordrecht: D. Reidel.
Heggie, D. C. (1975a). *Mon. Not. Roy. astron. Soc.* **173**, 729.
Heggie, D. C. (1975b). In *Dynamics of Stellar Systems*, IAU Symposium no. 69, ed. A. Hayli, p. 73. Dordrecht: D. Reidel.
Heggie, D. C. (1977a). *Comm. Astrophys.* **7**, 43.
Heggie, D. C. (1977b). *Rev. Mex. Astron. Astrofis.* **3**, 169.
Heggie, D. C. (1979). *Mon. Not. Roy. astron. Soc.* **188**, 525.
Hénon, M. (1972). *Astron. Astrophys.* **19**, 488.
Hénon, M. (1975). In *Dynamics of Stellar Systems*, IAU Symposium no. 69, ed. A. Hayli, p. 133. Dordrecht: D. Reidel.
Hills, J. G. (1975a). *Astron. J.* **80**, 809.
Hills, J. G. (1975b). *Astron. J.* **80**, 1075.
Hills, J. G. (1977). *Astron. J.* **82**, 626.
Illingworth, G. & King, I. R. (1977). *Astrophys. J. Lett.* **218**, L109.
King, I. R. (1977). *Rev. Mex. Astron. Astrofis.* **3**, 167.
Lightman, A. P. & Fall, S. M. (1978). *Astrophys. J.* **221**, 567.
Lightman, A. P. & Shapiro, S. L. (1978). *Rev. Mod. Phys.* **50**, 437.
Miller, R. H. (1975). In *Dynamics of Stellar Systems*, IAU Symposium no. 69, ed. A. Hayli, p. 95. Dordrecht: D. Reidel.

Monaghan, J. J. (1976). *Mon. Not. Roy. astron. Soc.* **176**, 63.
Monaghan, J. J. (1977). *Mon. Not. Roy. astron. Soc.* **179**, 31.
Peterson, C. J. & King, I. R. (1975). *Astron. J.* **80**, 427.
Press, W. H. & Teukolsky, S. A. (1977). *Astrophys. J.* **213**, 183.
Retterer, J. M. (1980). *Astron. J.*, in press.
Spitzer, L., Jr (1978). Results presented at NATO Advanced Study Institute on Globular Clusters, Cambridge.
Spitzer, L., Jr & Hart, M. H. (1971). *Astrophys. J.* **164**, 399.
Valtonen, M. J. (1974). In *The Stability of the Solar System and of Small Stellar Systems*, ed. Y. Kozai, p. 211. Dordrecht: D. Reidel.
Valtonen, M. J. (1975). *Mem. Roy. astron. Soc.* **80**, 77.
Valtonen, M. J. (1976). *Astrophys. Space Sci.* **42**, 331.
van Albada, T. S. (1968). *Bull. astron. Inst. Netherlands*, **19**, 479.
von Hoerner, S. (1963). *Z. Astrophys.* **57**, 47.

16
Some thoughts concerning the origin of globular clusters

JAMES E. GUNN[†]

16.1 Introduction

We consider briefly some topics relevant to the origin of globular clusters by first considering the properties of protoclusters which can be inferred from the present state of clusters and the application of simple dynamical and nucleosynthetic theory. It is tentatively concluded that clusters probably do not lose a large fraction of their mass as gas, but probably do lose a substantial fraction as low-mass stars. Several arguments suggest that large-amplitude fluctuations likely to be present at the era of decoupling have nothing to do with globular clusters. A more promising scenario is afforded by condensation of cooling shocks in the material of the forming protogalaxy. The suggestion is made that the 'young globular clusters' now seen in the Magellanic Clouds have a quite different origin.

16.2 What does one need to form?

Before one can understand the physics of the formation of globular clusters, one needs to determine whether the cluster at its formation is likely to bear much resemblance to the presently seen aggregates in such important properties as total mass and binding energy (see e.g. Fall, p. 309, this volume).

One important and inevitable mass-loss process is that attending stellar evolution after most of the star formation has taken place. The amount of such loss depends on the mass function, of course, but for reasonable mass functions (x between 1 and 3, say) the loss is always less than half the total mass, and it occurs on an adiabatic time scale; i.e. one much longer than the crossing time. For such processes the adiabatic invariants are MR and Rv, the constancy of which imply that

$$R \propto 1/M, \quad v \propto M, \tag{1}$$

[†] Supported in part by the National Science Foundation (AST 76-80801 AOI and AST 75-01398 AOI).

where R is a characteristic radius, M is the cluster mass, and v a characteristic velocity.

The cluster then grows a little and the velocity dispersion drops a little during the course of stellar evolution. The two-body processes discussed by Spitzer (1975) and by Lightman & Shapiro (1978) also give rise to a slow expansion of the body of the cluster and an accompanying shrinkage of the core, but the mean energy per unit mass is little changed.

Tidal shocks remove low-mass high-energy stars effectively, as shown by Ostriker, Spitzer & Chevalier (1972), and can disrupt the cluster completely once it has lost a great deal of mass, as has been discussed by van Albada (1978). Shocks heat the cluster to some extent, but for centrally-condensed aggregates it is probable that the energy change is less important than the mass loss. Thus if one approximates the velocity dispersion as $M^{-\alpha}$, the exponent α is positive and is somewhat smaller than $\frac{1}{2}$. Tidal shocking is probably important for presently extant clusters only for the low-mass end of the mass function, about which we will say more later.

'Presently extant' is a key qualification to all the above discussion. It is quite possible, as Fall & Rees (1977) have argued, that only a narrow range of clusters formed early have survived, and the clusters we see today are the results of selective survival rather than any preferential set of initial conditions at formation. (See also Fall, p. 309, this volume.) The situation regarding this possibility is, to say the least, unclear. On the one hand is the existence of the galactic halo, dynamically and chemically essentially identical to the globular cluster population, but containing many times the total mass in the clusters. To my mind the only reasonable way to form these stars is in globular-cluster-like aggregates which are then disrupted somehow, and this suggests that the clusters we see *are* a small part of the protocluster population.

On the other hand, the fact that the radial density distribution of the clusters and the halo stars are essentially the same in the Galaxy (and in other systems) and the near-universality of the cluster luminosity function (Hanes 1977) argues that tidal shocks, at least, are not a major destruction process except, perhaps, in the very inner regions of the Galaxy. (The process should not occur at all in ellipticals.) It is not clear at present whether two-body processes really disrupt clusters at all – it is particularly unclear whether core collapse has any particularly adverse effect on the cluster.

Let us keep in mind that the objects that become the globular clusters seen today may well not be typical of the structures forming early, but until the destruction mechanism is identified we can do little except be cautious.

We have concerned ourselves so far with 'slow' processes. The cluster must survive one 'fast' process as well, namely star formation. The dynamical consequence of this may be quite severe.

Chemical inhomogeneities in clusters and dwarf spheroidals, described by Zinn (p. 206, this volume) make it seem likely that at least some of the metal abundance in clusters is from self-enrichment, i.e. successive generations of stars formed from material enriched by (presumably) supernovae from previous generations, all within the cluster itself. Indeed, it seems difficult to escape this phenomenon. The crossing times for massive clusters *now* is a few million years, comparable to the lifetimes of massive stars, and it is highly unlikely that any 'wave of star formation' could cross the cluster faster than the crossing time. Indeed, Freeman (1978) has reported that the young Magellanic Cloud globulars have age spreads of at least 10^7 yr. When a fair fraction of the mass of a cluster is in gas, the mean densities are of order 10^3 cm^{-3} or greater, and the cooling times are very short. The momentum transfer from a typical supernova event will not eject mass unless the supernova is in the low-density outer parts. Thus supernovae cannot easily eject mass from the young cluster until most of the mass has been converted to stars. The situation is rather different from that in dwarf galaxies because of the high densities and attendant efficient cooling in globular clusters.

There are two observations to be made at this point, which might be of some worth. First, the cooling times become very long below about 10^4 K for very metal-poor material even at these densities, so the survival of early clusters may be simply determined by whether they were sufficiently tightly bound to retain 10^4 K gas or not. In particular, if the sound speed at 10^4 K, about 13 km s^{-1}, is sufficiently larger than the r.m.s. stellar velocity, one energetic massive supernova can, if more than half the cluster is still gaseous at its ignition, disrupt the whole system. Thus the disruption of the majority of protoclusters may have nothing whatever to do with slow dynamical processes.

Secondly, for those clusters which can remain intact for more than one generation of star formation, there are severe problems with questions of chemical composition. Simple chemical evolution theory predicts that the metal abundance Z in a population is related to the yield y and the fraction s of the mass converted into stars by

$$Z = ys + Z_0,$$

where Z_0 is the initial abundance, as long as $s \ll 1$. The popular value for y is about the solar abundance, $y \sim 0.01$–0.02. Thus a typical massive globular cluster must stop star formation and clear itself of gas (adiabatically!) when $s \sim 0.05$–0.02. This point has been made before in connection with the dwarf spheroidals by Sandage (1965) and by Searle & Zinn (1978) and has been discussed in this volume by Zinn. The dynamical consequence of this mass loss has not been discussed and is very important. If we take a cluster now which has a mass of 5×10^5 $M_\odot$ and a half-mass radius of 15 pc with a (three-dimensional) velocity dispersion of 7 km s^{-1} (parameters appropriate to M3), its original parameters must have been as follows: mass $\sim 1.5 \times 10^7$, velocity dispersion ~ 200 km s^{-1} and half-mass radius ~ 0.5 pc! The average density in this structure is 10^9 cm^{-3}, and the electron-scattering optical depth is greater than 1000. Furthermore, its photon diffusion time is of order its free-fall time; the object is a *star*, not a protocluster, and it is clear that whatever an object like this becomes, it does not fragment to become a globular star cluster.

The assumption of instantaneous recycling which goes into the simple theory is clearly not a very good one if there are only a few generations, but the problem remains, and the trouble is clearly that the yield is so large. The suggestion is that the Population II yield is small, perhaps much smaller than the solar neighbourhood one. One straightforward way of making the yield smaller is to make the main sequence luminosity function steeper, and indeed there is evidence that this may be the case. We have seen from the dynamical models discussed in Gunn (p. 276, this volume) that x seems to be about 2 in M3; Freeman has reported (1978) that x is not constant from cluster to cluster, and may be as large as 3 in some clusters. Schmidt's (1975) counts of halo dwarfs also indicate an $x \sim 2$. Assuming that a Population II supernova produces as many metals as a Population I supernova, one can calculate easily that the yield for $x = 2$ is, roughly,

$$y \approx 0.001 \left(\frac{m_l}{0.1 \, M_\odot}\right),$$

where m_l is the lower mass cutoff. The corresponding present visual mass-to-light ratio is about

$$(M/L)_V \sim 0.7/m.$$

If the cutoff is at ~ 0.1 $M_\odot$, the yield is thus a factor of 20 lower than the Population I yield, and the mass-to-light ratio is acceptably small.

The low-mass end of the mass function has presumably been lost from present-day clusters by the continual action of segregation and tidal shocks (Ostriker *et al.* 1972), so that the cluster need have lost only a factor of 3–5 in mass from all causes to have become the objects we see today, and in all likelihood the velocity dispersion has not changed very much since the initial formation.

It seems hardly worth saying that this chain of argument is almost laughably uncertain, but in view of our ignorance of the early phases of cluster evolution is perhaps the best one can do. Let us see whether a natural mechanism for producing objects of a few times 10^6 solar masses with velocity dispersions near 10 km s^{-1} can be found.

16.3 Are globular clusters primordial?

That is, did they form from density fluctuations present at decoupling on a time scale much shorter than the time scale for the formation of galaxies (Peebles & Dicke 1968)? Van den Bergh has reviewed the current observational evidence in chapter 10; the observed metallicity gradients in the cluster population remain the most difficult hurdle. In my opinion, the observed radial distribution of clusters is almost as damning; the infall of an initially almost homogeneous cluster population results always in a distribution which is close to $\rho \propto r^{-2}$, rather than the very steep $r^{-3} - r^{-3.5}$ observed (Gunn 1977).

There is also some theoretical evidence against the primordial hypothesis. There is evidence (see e.g. Gott & Rees 1975) that if the universe is dense ($\Omega \sim 1$) the spectrum of initial perturbations was white noise; i.e. the r.m.s. density contrast δ^+ varies like the inverse square root of the mass of a region. If Ω is nearer to 0.1, the spectrum is probably more heavily weighted to larger mass scales, and a variety of arguments suggest that $\delta^+ \propto M^{-\frac{3}{2}}$. In the former case, the characteristic mass when the amplitude is unity (and it is probably *this* mass scale which sets the 'characteristic mass' at decoupling, not the Jeans mass – provided, of course, that the Jeans mass is not the larger) is about 3×10^7 M$_\odot$. For low-density universes, the number is about 5×10^6 M$_\odot$. Both are rather high for globular cluster masses, the $\Omega = 1$ value probably prohibitively so. Worse are the binding energies. If there is no dissipation, the velocity dispersions are of order 150 km s^{-1} and 40 km s^{-1} for the two cases, but dissipation is almost certainly very important for such small, dense aggregates. Dissipation, of course, *increases* the velocity dispersion. Thus, if globular clusters are primordial, they must be the loosely bound tail of some much more energetic initial distribution. The other point, of course, is that one expects the formation of these initial objects from the associated high-amplitude perturbations to be very efficient. It is this efficiency which makes them promising candidates for the formation of 'Population III', the hidden mass in galactic halos and great clusters. But whatever this stuff is, it must have a mass-to-light ratio two orders of magnitude larger than that of globular clusters. All in all, the primordial hypothesis is rather unattractive, but the uncertainties are such that it probably cannot be completely ruled out.

16.4 What, then?

The same evidence on radial distribution and composition gradients makes it seem likely that the clusters were formed somehow in the process of the early collapse of the galaxy (Eggen, Lynden-Bell & Sandage 1962). Let us explore this possibility.

Subcondensations should form as the galaxy collapses (either as a large-scale object or as a hierarchy of smaller pieces) when those subcondensations are more massive than the Jeans mass for the relevant physical conditions. The Jeans mass is approximately

$$M_J \sim 1.6 \times 10^8 \, n^{-\frac{1}{2}} T_4^{\frac{3}{2}} \, M_\odot,$$

where n is the number-density (cm^{-3}) and T_4 is the temperature in units of 10^4 K. The average density in the protogalaxy must have been near 1 cm^{-3}, so the Jeans mass is very large (the temperature is likely to have been about 10^4, or perhaps a bit higher).

In a chaotic collapse, however, there will be strong shocks, and since the cooling time for a large mass of metal-poor material with $n \sim 1$, $T_4 \sim 1$ is about 3×10^5 yr, while the sound travel time across the Jeans mass is two orders of magnitude greater, the shocks will be isothermal ones. The density increase following an isothermal shock is roughly

$$n_f/n_o \sim 0.04 \, V_s^2/T_{4,\,f},$$

where V_s is the shock velocity in km s^{-1} and $T_{4,\,f}$ is the postshock temperature. For $V_s \sim 300$ km s^{-1}, a characteristic size of the mass motions which must have been present during the collapse, the postshock density is about 3000, and the Jeans mass (again $T_4 \sim 1$) is about 3×10^6 $M_\odot$, just about right for our hypothetical protocluster. If fragmentation occurs shortly after this without too much dissipation, the velocity dispersion will be about the sound speed at 10^4 K, again about right. For $Z/Z_\odot \sim 10^{-2}$, the cooling function below 10^4 K is in the range 10^{-27}–10^{-28} erg cm^3 s^{-1}, which at these densities corresponds again to cooling times of order 10^5–10^6 yr. These times are a bit shorter than the dynamical time, so fragmentation can proceed. Since the cooling curves (see Dalgarno & McCray 1972) do not fall as rapidly as the temperature, as the temperature is lowered, the fragmentation can proceed a long way, perhaps even to stars in one step.

If there are *no* metals, it seems likely that the protocluster will not fragment immediately but will evolve slowly at 10^4 K into something much more interesting, which perhaps makes the first metals.

Freeman (1978) has suggested that the young Magellanic Cloud globular clusters, objects of mass $\sim 10^4$ $M_\odot$, have arisen from interactions of the Clouds with the gaseous halo of the Galaxy. The arguments presented here suggest that if that is so, the detailed mechanisms are very different from the ones suggested for the old galactic globulars.

If the Cloud orbital velocity is ~ 200 km s^{-1}, the bow shock ram pressure $\rho V^2/k \sim 1 \times 10^4$ K cm^{-3} if the halo number density is about 10^{-3}. If the material behind the shock is at 10^4 K, $n \sim 1$ and the Jeans mass is 2×10^8 $M_\odot$; if, however, the material can cool to 100 K – which it can, since the metal abundance is near solar – the Jeans mass is about 2×10^4 $M_\odot$. The clusters could also arise in the $n \sim 0.1$ cm^{-3} cool medium surrounding the Clouds through the intervention of a ~ 50 km s^{-1} shock excited by interaction between the Small and Large Clouds, again if cooling to 100 K is considered.

References

Dalgarno, A. & McCray R. A. (1972). *Ann. Rev. Astron. Astrophys.* **10**, 375.
Eggen, O. J., Lynden-Bell, D. & Sandage, A. R. (1962). *Astrophys. J.* **136**, 748.
Fall, S. M. & Rees, M. J. (1977). *Mon. Not. Roy. astron. Soc.* **181**, 37P.
Freeman, K. C. (1978). Results reported at NATO Advanced Study Institute on Globular Clusters, Cambridge.
Gott, J. R. & Rees, M. J. (1975). *Astron. Astrophys.* **45**, 365.
Gunn, J. E. (1977). *Astrophys. J.* **218**, 592.
Hanes, D. A. (1977). *Mon. Not. Roy. astron. Soc.* **180**, 309.
Lightman, A. & Shapiro, S. (1978). *Rev. Mod. Phys.* **50**, 437.
Ostriker, J. P., Spitzer, L. & Chevalier, R. A. (1972). *Astrophys. J. Lett.* **176**, L51.
Peebles, P. J. E. & Dicke, R. H. (1968). *Astrophys. J.* **154**, 891.
Sandage, A. (1965). In *The Structure and Evolution of Galaxies*. London: Interscience.
Schmidt, M. (1975). *Astrophys. J.* **202**, 22.
Searle, L. & Zinn, R. (1978). *Astrophys. J.* **225**, 357.
Spitzer, L. (1975). In *Dynamics of Stellar Systems*, IAU Symposium no. 69, ed. A. Hayli, p. 3. Dordrecht: D. Reidel.
van Albada, T. S. (1978). Results reported at NATO Advanced Study Institute on Globular Clusters, Cambridge.

17
Globular clusters as survivors

S. MICHAEL FALL

By astronomical standards, globular clusters are remarkably similar to one another. In particular, most of them have masses and radii that fall within fairly narrow limits (see fig. 17.1). This is often taken to imply some mechanism that would have given these objects their characteristic properties at the time of formation. A well-known example of this kind of mechanism is the Jeans instability amplifying primordial isothermal density fluctuations just after recombination (Peebles & Dicke 1968). It may well be that, by this mechanism or some other, globular clusters did

Fig. 17.1. Survival triangle for galactic substructure. The sides of the triangle are set by the three stellar-dynamical processes that might limit the present gross structural properties of substructure. A power-law spectrum of initial substructure is indicated by the stippling but any spectrum that passes through the observed globular clusters and avoids the large mass side of the triangle will do.

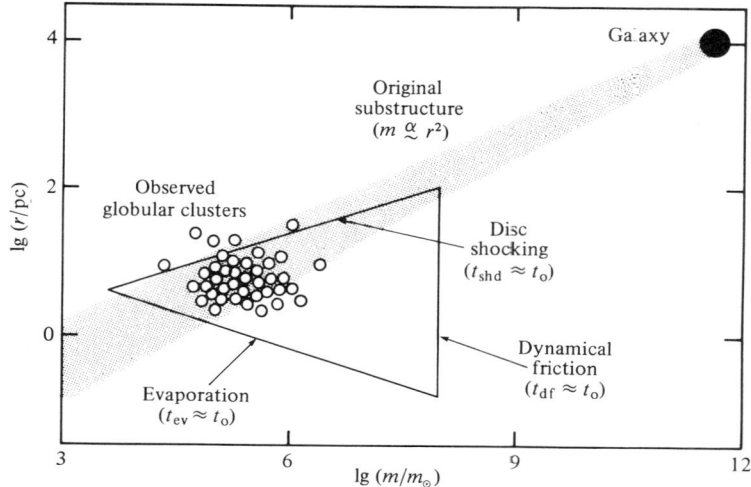

309

form with only the properties they now have. But in this short chapter I want to stress another possibility. It appears that a larger spectrum of subgalactic structure could have been broken down in such a way that globular clusters, as we see them now, would have been the only survivors (Fall & Rees 1977).

At present, there are three stellar-dynamical mechanisms limiting the positions of galactic substructures in the mass (m) versus radius (r) plane. They are: the evaporation of stars from clusters (Spitzer 1975), the shocking of clusters during passage through the galactic disc (Ostriker, Spitzer & Chevalier 1972) and the dragging of clusters into the galactic centre by dynamical friction (Tremaine, Ostriker & Spitzer 1975). The time scales for these processes are given approximately by the following formulae:

evaporation (mean stellar mass ≈ 0.3 m$_\odot$)

$$t_{\rm ev} \approx 2 \times 10^7 (m/m_\odot)^{\frac{1}{2}} (r/{\rm pc})^{\frac{3}{2}} \text{ yr}, \tag{1}$$

disc shocking (solar neighbourhood)

$$t_{\rm shd} \approx 1 \times 10^8 (m/m_\odot) (r/{\rm pc})^{-3} \text{ yr}, \tag{2}$$

dynamical friction (galactocentric distance ≈ 7 kpc)

$$t_{\rm df} \approx 1 \times 10^{18} (m/m_\odot)^{-1} \text{ yr}. \tag{3}$$

(See Fall & Rees 1977 for further details.) When these time scales are set equal to the Hubble time scale, $t_0 \approx 10^{10}$ yr, they define a 'survival triangle' in the lg m–lg r plane (Tremaine 1975; Fall & Rees 1977; see also fig. 17.1). Galactic substructure that formed outside the triangle would have been destroyed by one of these mechanisms.

The triangle has been shrinking and will continue to shrink until eventually no clusters remain. In view of this fact, it seems unlikely that some galactic substructure has not already been destroyed. Indeed, a wide range of substructure arises naturally in many pictures of galaxy formation: either by hierarchical clustering (Layzer 1977 and references therein) or by hierarchical fragmentation (Rees & Ostriker 1977 and references therein). Notice, however, that the triangle is not completely filled by the observed globular clusters, especially at the large mass side. Dynamical friction may have destroyed some massive clusters ($m \geqslant 10^8$ m$_\odot$) and it may have destroyed a few small ones that started out near the galactic centre; but for the most part it has not been important in setting the maximum masses of surviving substructure. This suggests that the original mass–radius relation of substructure somehow avoided the high-mass side of the triangle. One possibility is indicated in fig. 17.1. (See also Fall & Rees 1977.)

If the protogalaxy was originally filled with a wide spectrum of substructure, encounters would have been important in disrupting the more massive objects. In the case that these objects had already formed stars and had lost most of their gas, indirect impulsive encounters would have transferred orbital energy to internal energy on the approximate time scale (Spitzer 1958; Fall & Rees 1977):

cluster shocking (structures not bound within each other)

$$t_{\rm shc} \approx 3 \times 10^{10} (fr/{\rm pc})^{-1} \text{ yr}, \tag{4}$$

where f is the fraction of mass in undisrupted clusters. Collisions between objects in the gaseous state would have been even more effective in disrupting them (unless the specific binding energy decreases with mass). Moreover, the increased rate of encounters in a hierarchy might also make (4) an over-estimate of the time scale for disruption. (See also Rees 1977; White & Rees 1978; Tinsley & Larson 1979.)

Including encounters leads to a picture of the disruption of substructure such as that shown in fig. 17.2, which is based on (1), (2) and (4) with masses

Fig. 17.2. Dependence of disruptive processes on the index α of a power-law initial mass–radius relation ($m/r^3 \propto m^{-3\alpha}$). The reduction of an initial mass spectrum to one like the observed globular cluster mass spectrum is indicated schematically in the side panels of the diagram.

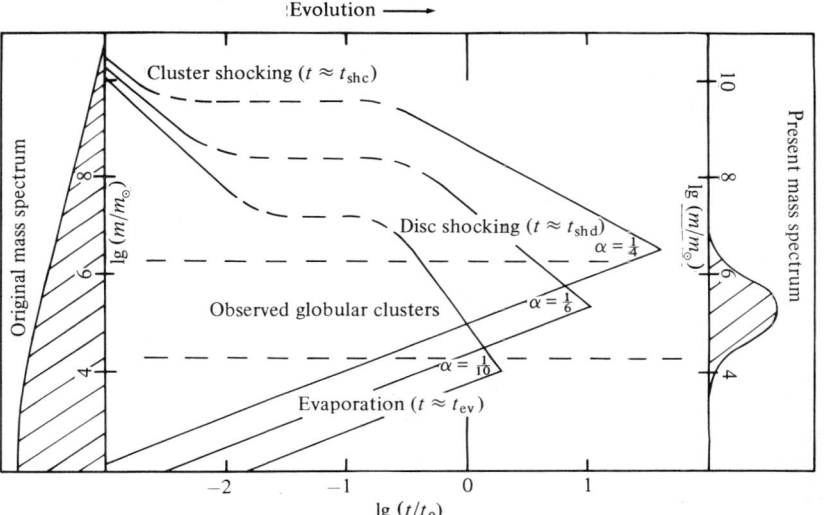

and radii following the relation $m/r^3 \propto m^{-3\alpha}$. Initially, the more massive objects are destroyed by encounters and the less massive ones by evaporation (for reasonable values of α). But once most of the massive clusters have been destroyed ($f \ll 1$), disc shocking and evaporation become the dominant mechanisms (as in fig. 17.1). There is at least the possibility that a fairly arbitrary initial mass spectrum could have been carved down to one resembling the mass spectrum of observed globular clusters, especially with $\alpha \approx \frac{1}{6}$ (Fall & Rees 1977). One can show that the mass range of surviving clusters is not very sensitive to the assumed masses and radii of the parent galaxy, but detailed calculations have not been done and may not even be warranted at this stage. An unambiguous prediction of the model, however, is that elliptical galaxies ought to have somewhat more massive globular clusters than comparable disc galaxies.

An important question hanging over this kind of speculation is the extent to which disrupted substructure could have contributed to the present galactic halo. This is a problem because the mass-to-light ratios of globular clusters are of order unity whereas the mass-to-light ratio of the halo, if it is 'heavy', may be as high as 10^2. Thus, one might suppose that disrupted

Fig. 17.3. A stellar-dynamical scheme for producing low M/L globular clusters and a high M/L halo from a protogalaxy with hierarchical structure and a broad stellar mass function. The hierarchy ensures that massive stars (~ 1 m$_\odot$) settle into aggregates with masses comparable to the masses of observed globular clusters ($\sim 10^5$ m$_\odot$) in the available time ($\sim 10^{10}$ yr). Estimates of the time scales are based on eqs. (1) and (3) of Fall & Rees (1977) with $\alpha = \frac{1}{6}$ and are accurate only to order of magnitude.

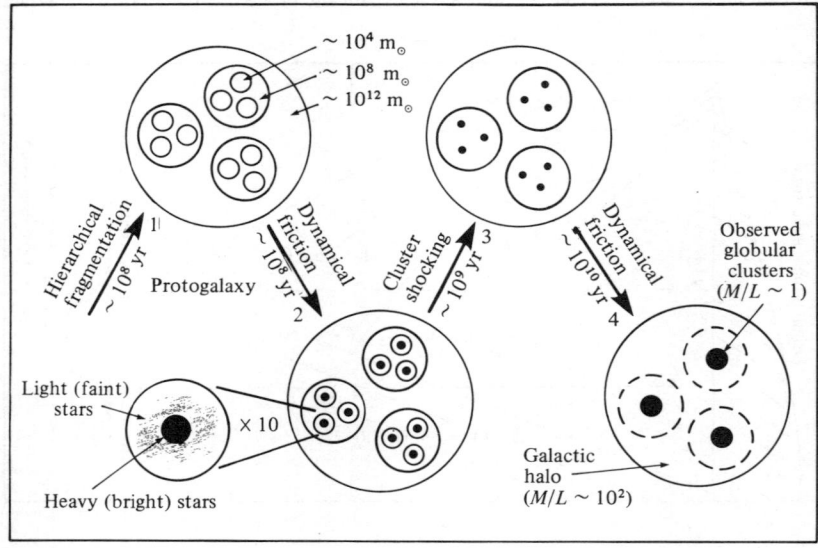

substructure could have contributed only $\sim 10^{-2}$ of the halo mass unless some mechanism ensures that low M/L material survives longest. An example of a mechanism that produces globular clusters ($M/L \sim 1$) and the halo ($M/L \sim 10^2$) from disrupted substructure is sketched in fig. 17.3. The important ingredients are: a hierarchy of initial substructure, some cluster shocking at the appropriate times and at least two stages of dynamical friction. (A one-stage picture requires dynamical friction time scales that are too long.) This scheme is either natural or contrived depending upon one's tastes. But the particular version shown in fig. 17.3 does have roughly the right time scales and does not seem to begin with an unreasonable initial spectrum of substructure, either from agglomeration or from fragmentation.

In conclusion, it is worth asking how relevant this picture is to real galaxies and real globular clusters. Stellar-dynamical processes have been emphasized here but the formation and early evolution of galactic substructure may have been dominated by gas-dynamical effects and the poorly understood process of star formation in such a setting. It is likely that the chemical abundances of globular clusters are telling us something about these processes (Searle 1977). A general problem that any survival-disruption picture will eventually have to face up to is the fact that the average properties of globular clusters depend only weakly on position in the galaxy. However, the strengths of at least some of the disrupting mechanisms depend on galactocentric distance (e.g. disc shocking) and this has not been taken into account in the schematic calculations and models discussed here. Nevertheless, stellar-dynamical effects of these kinds are likely to have been important at some level in the evolution of globular clusters.

I have benefited from pleasant conversations on this subject with Scott Tremaine, David Layzer and, especially, Martin Rees.

References

Fall, S. M. & Rees, M. J. (1977). *Mon. Not. Roy. astron. Soc.* **181**, 37P.
Layzer, D. (1977). *Int. J. Quantum Chem.* **11**, 637.
Ostriker, J. P., Spitzer, L. & Chevalier, R. A. (1972). *Astrophys. J. Lett.* **176**, L51.
Peebles, P. J. E. & Dicke, R. H. (1968). *Astrophys. J.* **154**, 891.
Rees, M. J. (1977). In *The Evolution of Galaxies and Stellar Populations*, ed. B. M. Tinsley & R. B. Larson, p. 339. New Haven: Yale University Observatory.
Rees, M. J. & Ostriker, J. P. (1977). *Mon. Not. Roy. astron. Soc.* **179**, 541.
Searle, L. (1977). In *The Evolution of Galaxies and Stellar Populations*, ed. B. M. Tinsley & R. B. Larson, p. 219. New Haven: Yale University Observatory.

Spitzer, L. (1958). *Astrophys. J.* **127**, 17.
Spitzer, L. (1975). In *Dynamics of Stellar Systems*, IAU Symposium no. 69, ed. A. Hayli, p. 3. Dordrecht: D. Reidel.
Tinsley, B. M. & Larson, R. B. (1979). *Mon. Not. Roy. astron. Soc.* **186**, 503.
Tremaine, S. D. (1975). Ph.D. Thesis, Princeton University.
Tremaine, S. D., Ostriker, J. P. & Spitzer, L. (1975). *Astrophys. J.* **196**, 407.
White, S. D. M. & Rees, M. J. (1978). *Mon. Not. Roy. astron. Soc.* **183**, 341.

18
X-ray burst sources in globular clusters and the galactic bulge†

WALTER H. G. LEWIN

18.1 Introduction

Of the approximately 150 known galactic X-ray sources, about 5 per cent are located in globular clusters. Yet the total mass of stars in globular clusters is only about 0.1 per cent that of our galaxy. Thus, the probability (per unit mass) of finding an X-ray source in a globular cluster is about two orders of magnitude higher than in the rest of our galaxy. Clearly the

Table 18.1. *X-ray sources in globular clusters*

Globular cluster	X-ray source	Burst source	References
NGC 1851	4U 0513−40	Yes	a,b,c
NGC 6441	4U 1746−37	Probably	c,d
Liller 1	MXB 1730−335	Yes	e
NGC 6624	4U 1820−30	Yes	c,f,g,h
NGC 6712	4U 1850−08	Probably	c,i,j
NGC 7078	4U 2131+11	—	c
	The following associations are uncertain:		
Kron 3	4U 0026−73	—	k
NGC 6440	MX 1746−20	—	l
Terzan 2	4U 1722−30	Perhaps	k,m

References
a Forman & Jones (1976)
b Clark & Li (1977)
c Jernigan & Clark (1979)
d Li & Clark (1977)
e Doxsey *et al.* (1978)
f Grindlay *et al.* (1976)
g Clark *et al.* (1976)
h Clark *et al.* (1977)
i Swank *et al.* (1976)
j F. K. Li & G. W. Clark (private communication)
k Grindlay (1978c)
l Forman *et al.* (1979)
m Swank *et al.* (1977)

† Work supported by the National Aeronautics and Space Administration under Contract NAS5-11450.

Table 18.2. 'Steady' X-ray sources from which Type 1 bursts have been observed

Burst[a] source	Steady[a] source	Positions of steady sources from SAS-3[b]							Error[c] circle radius (arcsec)	4U catalogue[d] intensity	
		RA (1950)			Dec (1950)					Max U counts	Max/Min
		h	m	s	o	′	″				
MXB 0512−40 (in NGC 1851)	2S 0512−400 2A 0512−399 4U 0513−40	05	12	28.7	−40	05	53		20	18	3
MXB 1636−53 (blue star)	2S 1636−536 4U 1636−53	16	36	57.6	−53	39	21		20	250	2
MXB 1728−34	2S 1728−337 4U 1728−33	16 17	36 28	56.2 39.6	−53 −33	39 47	15 52		3 30	150	5
MXB 1735−44 (blue star)	2S 1735−444 4U 1735−44	17	35	19.5	−44	25	22		20	210	1.7
MXB 1820−30 (in NGC 6624)	2S 1820−303 4U 1820−30	17 18	35 20	19.0 28.4	−44 −30	25 23	19 14		3 20	320	3
MXB 1837+05 (blue star)	2S 1837+049 4U 1837+04	18	37	29.8	04	59	23		20	280	2
MXB 1906+00	2S 1905+000 A 1905+00 4U 1857+01	18 19	37 05	29.6 54.9	04 00	59 05	21 37		5 35	4.05 ± 1.1[e]	
MXB 1916−05	2S 1916−053 4U 1915−05	19	16	08.5	−05	19	51		20	20	2

	from HEAO-A3[f]							from SAS-3	
MXB 1730−335 Rapid Burster (in Liller I)	H1730−333	17	30	07.4	−33	20	34		~50 (Type II bursts) > 10
MXB 1659−29	(4U 1704−30?) H1658−298 blue star	16 16	58 58	54.9 55.5	−29 −29	49 52	55 26	5	~80 > 15

[a] For references of burst sources and optical identifications, see text.
[b] Doxsey et al. 1977; Jernigan et al. 1977; Jernigan & Clark 1979.
[c] Approximately 90% confidence.
[d] Forman et al. 1979.
[e] Average intensity.
[f] Griffiths et al. 1978; Doxsey et al. 1978.

Table 18.3. 'Steady' X-ray sources from which Type I bursts may have been observed

Burst source		Steady source				Intensity		References
Name	Position error deg^2	Name	RA (1950) h m s	Dec (1950) ° ′ ″	Position error (arcmin)^2	Max U counts*	$\frac{\text{Max}}{\text{min}}$	
XB 1608−52	∼3	MX 1608−52				10^3	∼25	a,b,c,d,e
		4U 1608−52	16 08 51	−52 18 02	0.3			
		2S 1608−523	16 08 52.2	−52 17 43	±0.5″			
		star						
MXB 1706−43	∼0.3	4U 1705−44				280	3	f,g,h,j
		2S 1705−440	17 05 18.2	−44 02 10	∼0.4			
		or						
		4U 1702−42				30	3	
		2S 1702−429	17 02 40.4	−42 57 58	∼0.8	7		
XB 1724−30	∼0.5	4U 1722−30 (in Terzan 2?)	17 22 50	−30 31 30	41			k,m,n
MXB 1743−29	∼0.2	A 1742−289 transient	17 42 26.4	−28 59 56	∼0.6	2×10^3	>200 transient	o,p,q
MXB 1742−29	∼0.3	A 1742−294				90	>2	g,o,q,r
		2S 1742−294	17 42 53.6	−29 29 50	∼0.8			
		GCX−1						
MXB 1746−37	∼1	4U 1746−37				13		s,t
		2S 1746−370 (in NGC 6441)	17 46 48.8	−37 02 25	∼0.8			
MXB 1850−08	∼1.3	A 1850−08				4		f,t,u,v
		2S 1850−087	18 50 21.9	−8 45 54	∼0.8			
		4U 1850−08 (in NGC 6712)						
XB 06??+??	∼80	4U 0614+09	06 14 22.3	09 09 25		120	5	m,w,x,y
		star						

* Approximate values in Uhuru counts/s are given to allow for an easy comparison with Table 18.2.
References
a Grindlay & Liller (1978)
b Grindlay & Gursky (1976)
c Tananbaum et al. (1976)
d Fabbiano et al. (1978)
e Apparao et al. (1978)
f Swank et al. (1976)
g Jernigan et al. (1977)
h Marshall, Li & Rappaport (1977)
j L. Cominsky, W. Ossman & J. van Paradijs (private communication)
k Swank et al. (1977)
m Forman et al. (1979)
n Grindlay (1978c)
o Lewin et al. (1976c)
p Eyles et al. (1975)
q Proctor, Skinner & Willmore (1978)
r Cruddace et al. (1978)
s Li & Clark (1977)
t Jernigan & Clark (1979)
u Doxsey et al. (1977)
v F. K. Li & G. W. Clark (private communication)
w Swank et al. (1978)
x Murdin et al. (1974)
y Davidsen et al. (1974)

conditions in globular clusters must be very special in that they are near 'perfect' for X-ray sources (Katz 1975).

The globular cluster X-ray sources are believed to be members of the class of galactic bulge X-ray sources. They are distinctly different from the well-known heavy mass binary systems (Population I) which often show periodic pulsations (rotating neutron star), eclipses and relatively hard X-ray spectra. The galactic bulge sources on the other hand emit softer spectra, and no periodic pulsations or eclipses have been reported for any of them (Jones 1977; Markert *et al.* 1977). (The galactic bulge source group contains about 70 objects of which only the brightest 20 or so have been studied in some detail.)

All of the 25–30 known X-ray burst sources (of Type I) belong to the galactic bulge source class, and a large fraction of the 7 known X-ray sources which are located in globular clusters produce X-ray bursts. Unlike the well-known Population I type heavy binaries, the geometry of the galactic bulge source class is uncertain. Supermassive black holes have been proposed (Bahcall & Ostriker 1975; Silk & Arons 1975), as well as 'classical' binary models wherein the compact object is either a neutron star or a black hole of a few solar masses (Clark 1975; Fabian, Pringle & Rees 1975; Hills 1975; van den Heuvel 1977).

When X-ray bursts were discovered by Grindlay in 1975 (Grindlay *et al.* 1976) he and his co-workers argued that these bursts were the signatures of supermassive ($\sim 10^3$ $M_\odot$) black holes. However, the present observational data argue against this.

It is very likely that all burst sources (of Type I) are alike in that the burst mechanism is the same and the source geometry similar. Therefore if we want to unravel the geometry of the mysterious globular cluster X-ray sources, any piece of information from any of the burst sources could help, whether the sources are located inside or outside globular clusters (the latter is the case for the majority of them). Particularly, one may expect to learn a great deal from studies of the four faint ($m_V > 17$) blue stars which have been positively identified as burst sources but which are not located in globular clusters.

In this chapter I will review the situation on X-ray burst sources. I will sometimes remind you whether a particular source is located inside or outside a globular cluster. However, since all burst sources (of Type I) are probably very similar, it should not matter.

In table 18.1 I list six X-ray sources which are definitely located in globular clusters. Three are certain burst sources, with two more probable (see also tables 18.2 and 18.3). The globular clusters NGC 6440, Terzan

X-ray burst sources

2 and Kron 3 were suggested as possible X-ray sources (Forman *et al*. 1979; Grindlay 1978c), but these associations are rather uncertain at this time. They are listed in the lower part of table 18.1.

18.2 Burst classification

X-ray bursts have been classified into two distinct groups, Type I and Type II (Hoffman, Marshall & Lewin 1978b). Type I bursts have been observed from more than 25 sources; they come in intervals from hours to days and generally their spectra soften during burst decay. Type II bursts have been observed from the Rapid Burster and possibly from several binary X-ray sources and pulsars. They occur on time scales of seconds to minutes and do not show the spectral softening as distinctly as observed in Type I bursts. As I will discuss later:

- Type I burst sources (members of the GBS class) are probably compact

Fig. 18.1. Rapidly repetitive Type II bursts from the Rapid Burster. Each stretch of data is about 24 min long; occultation by the Earth occurs near the end of each stretch. The arrow indicates a Type I burst from a nearby burst source, MXB 1728−34. This figure is from Lewin (1977a).

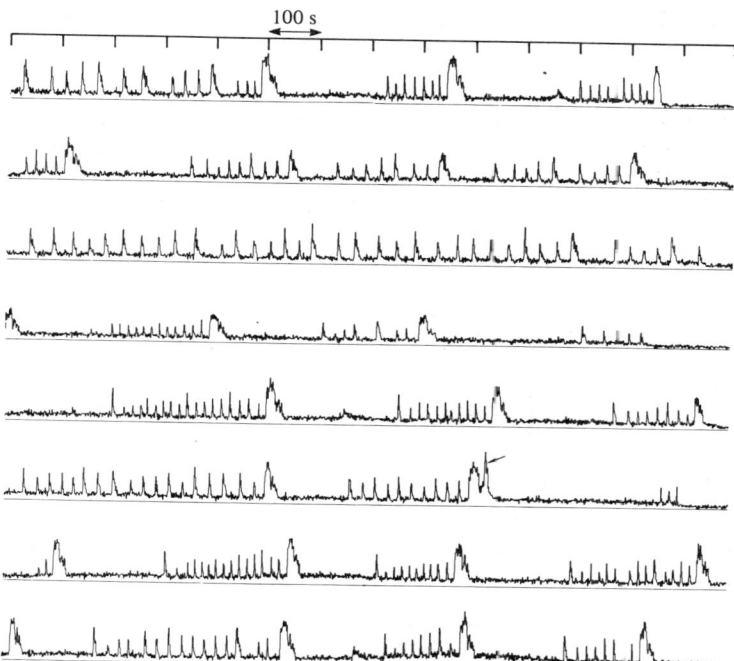

objects of solar mass. It may well be that the Type I bursts are due to thermonuclear flashes on the surface of a neutron star.

• Type II bursts are almost certainly the result of instabilities in the accretion flow onto a compact object. They have been observed from the Rapid Burster (located in a globular cluster) and possibly from several massive binary systems (Population I).

18.3 Type II burst sources

18.3.1 Type II bursts from the Rapid Burster MXB 1730−335

I will first discuss the Type II bursts as observed from the Rapid Burster, which is located in the globular cluster Liller 1. Type II bursts may also have been observed from other sources, as will be discussed in the next section.

The rapidly repetitive Type II bursts from the Rapid Burster (MXB 1730–335) are shown in fig. 18.1. These bursts do not show an appreciable amount of spectral evolution, other than a change in intensity (Lewin 1976; Lewin *et al.* 1976a; Marshall *et al.* 1979). One of my undergraduate students, Herman Marshall, made a composite of 18 Type II bursts each of about 40 s duration (Marshall *et al.* 1979). The results are shown in fig. 18.2. The composite was sliced in nine time bins. The best spectral fit during

Fig. 18.2. Composite of 18 Type II bursts from the Rapid Burster. They are all of similar duration: 33–45 s. The vertical scale is expanded on the right to show the enhancement in X-ray emission after the bursts (see text). Notice that the distinct spectral softening as seen in the 'tails' of Type I bursts (fig. 18.10) is absent here (from Marshall *et al.* 1979).

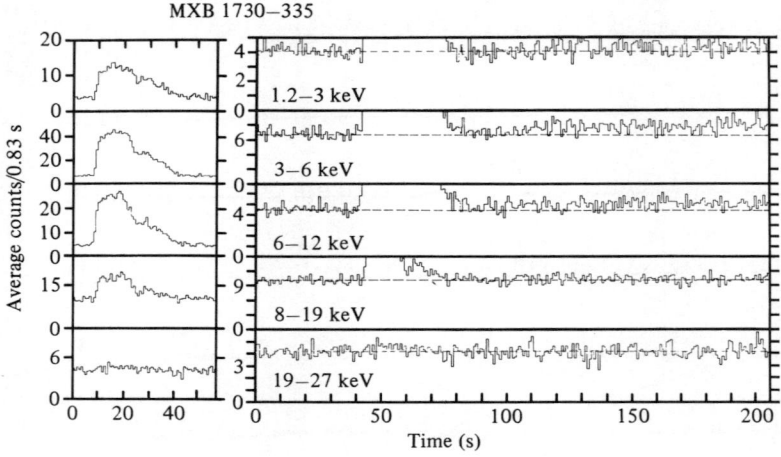

each of these time intervals is that of a black body with an approximately constant temperature of about 18×10^6 K throughout the burst. Recently, HEAO-1 data showed that some spectral softening is present in the Type II bursts; thus the temperature should decrease somewhat (Hoffman *et al.* 1979). If indeed the radiation is from a black body, then the observed decrease in all energy channels implies that the emitting region is shrinking in time. For spherical symmetry, at a source distance of 10 kpc, the radius is 16 ± 2 km during the first 15 s, decreases to 11 ± 2 km between 20 and 30 s after burst onset, and continues decreasing (the size decreases as the square root of the observed burst flux). These sizes suggest that the Rapid Burster is a neutron star or a black hole of approximately solar mass (see also Brecher, Morrison & Sadun 1977).

Notice the enhanced emission which is visible in the right-hand part of

Fig. 18.3. HEAO-1 data of the Rapid Burster (MXB 1730−335). Notice the enhanced 'steady' X-ray emission (A2 data) that builds up after the large Type II bursts and decays before the onset of the first of a series of relatively small bursts (from Hoffman *et al.* 1979).

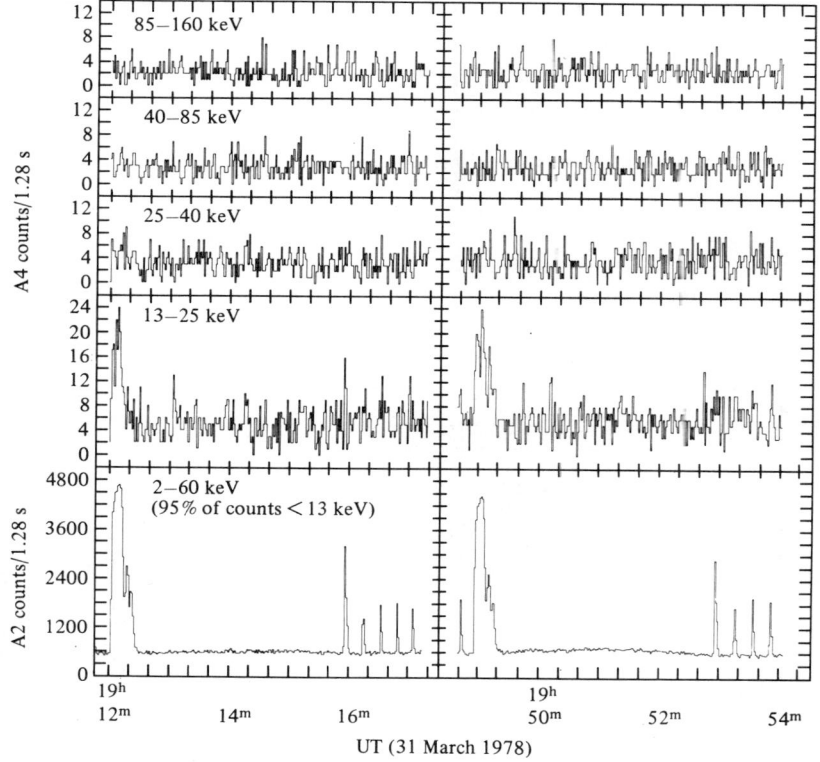

fig. 18.2. The emission (most visible in the 3–12 keV range) rises slowly after the large Type II composite burst. It was found (W. H. G. Lewin et al., in preparation) that this 'steady' emission terminates before the onset of the first of a series of small bursts. The presence of this 'steady' emission and its relation to the Type II bursts strongly suggests that the Type II bursts are instabilities in the accretion flow. Fig. 18.3 shows recent results obtained with HEAO-1 (Hoffman et al. 1979). They too show the 'steady' emission after each of two large Type II bursts (the emission is stronger after the second large burst).

Fig. 18.4 and 18.5 show some recent results from Marshall et al. (1979). Each point in fig. 18.4 represents one burst; the integrated burst energy is plotted against the time of Type II burst occurrence. One recognizes immediately two burst patterns. We have designated them Mode I and Mode II (it would have been less confusing if we had named them Mode A and Mode B!). In Mode I, the burst size distribution is double-peaked and the bursts vary by about two orders of magnitude in size (total burst energy). In Mode II, the difference is not more than about a factor of ten. As one can see in fig. 18.4, in 1976 the change from Mode I to Mode II occurred in less than a few hours. Based on many observations by SAS-3

Fig. 18.4. The total energy in each Type II burst from the Rapid Burster is plotted (abscissa) versus the UT of its occurrence (ordinate). Each dot represents one burst. The two different burst patterns (Mode I and Mode II) are clearly distinguishable (from Marshall et al. 1979).

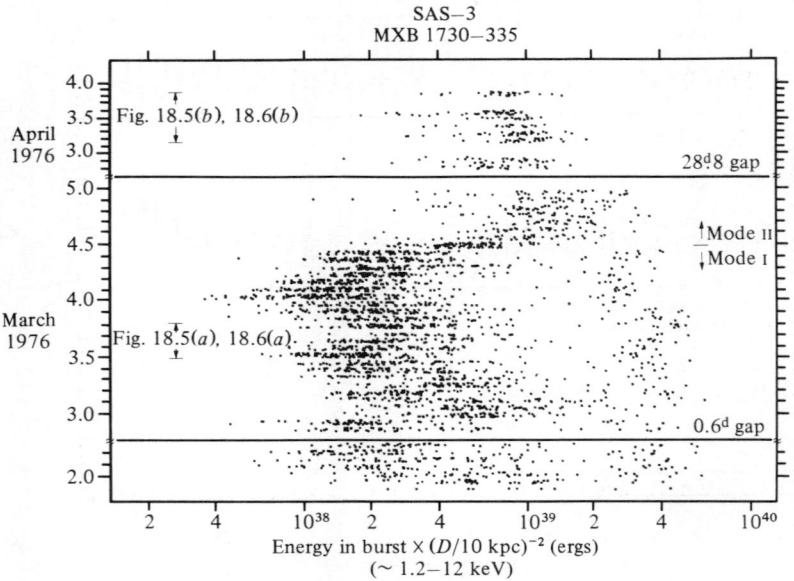

and Ariel V (Hoffman *et al.* 1978b; Lewin 1976; Lewin *et al.* 1976a; Mason, Bell-Burnell & White 1976; Ulmer *et al.* 1977; Lewin 1977a; White, Mason, Carpenter & Skinner 1978; Jernigan *et al.* 1978; Marshall & Lewin 1978; Lewin, Marshall & Cominsky 1978b), we believe that the Rapid Burster starts its burst active periods in Mode I and dies in Mode II. The Rapid Burster shows active periods of 2–6 weeks at intervals of about 6 months (Hoffman *et al.* 1978b; Lewin 1976; Lewin *et al.* 1976a; Marshall *et al.* 1979; Mason *et al.* 1976; Ulmer *et al.* 1977; Lewin 1977a; White *et al.* 1978;

Fig. 18.5. Distribution of Type II burst energies from the Rapid Burster. (*a*) Mode I data (time interval indicated in fig. 18.4). Bursts vary in energy by a factor of about 100. (*b*) Mode II data. Burst energies range by a factor of ~ 10 only (from Marshall *et al.* 1979).

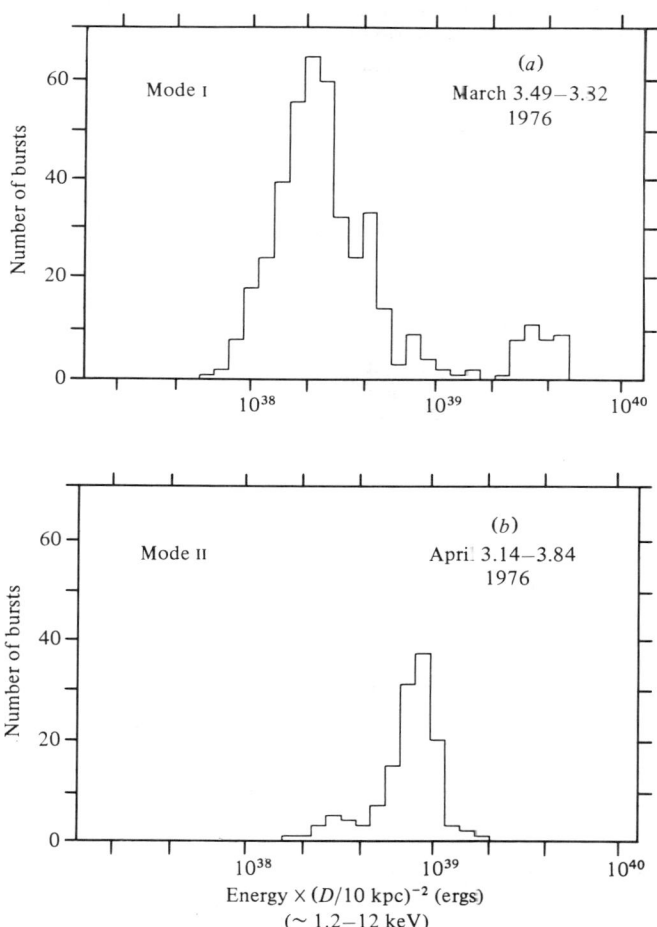

Jernigan *et al.* 1978; Marshall & Lewin 1978; Lewin *et al.* 1978b; Lewin 1977b; Grindlay & Gursky 1977).

Fig. 18.6 shows the relation between burst energy and the 'waiting' time, Δt, for the next burst to occur. As I pointed out during the Texas symposium in Boston in December 1976 (Lewin 1977a), the relation is far from linear for low values of the burst energy (see also White *et al.* 1978).

Before I start to discuss the Type II bursts from other sources, I summarize:
- The Rapid Burster is probably a neutron star or a black hole of approximately solar mass (see section 18.4.6).

Fig. 18.6. E–Δt plots for the data shown in fig. 18.5. (*a*) Mode I (same data base as used in fig. 18.5(*a*)). For low values of E, the corresponding values of Δt are considerably larger than expected from a linear relation. (*b*) Mode II (data from fig. 18.5(*b*)) (from Marshall *et al.* 1979).

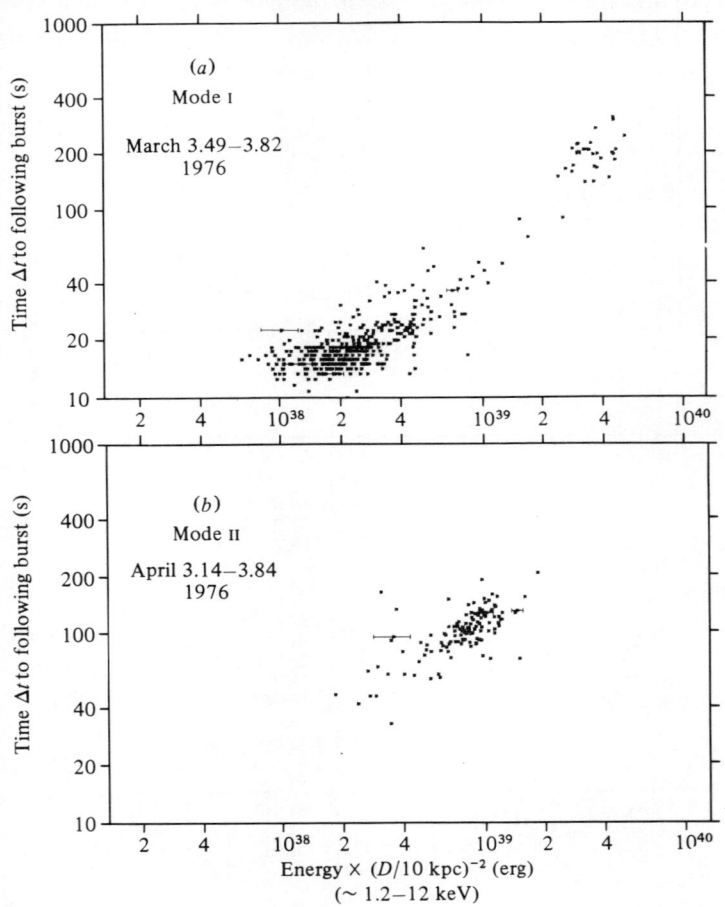

- Type II bursts from the Rapid Burster are almost certainly chopped up 'steady' emission due to instabilities in the accretion flow (see also section 18.3.2).
- The Rapid Burster is a recurrent 'transient' X-ray source. Periods between turn-ons are about 6 months.

18.3.2 Type II bursts from Population I type objects?

Type II bursts distinguish themselves from Type I bursts in two ways: (i) the time scales of recurrence are two to three orders of magnitude shorter; and (ii) there is no strong, dominating spectral softening in the burst decay. It was first noted by Hoffman et al. (1978b) that the 'flare-burst-like' events as observed from the pulsar GX 304−1 (McClintock, Rappaport, Nugent & Li 1977b) and from the binaries Cygnus X-1 (Canizares & Oda 1977) and LMC X-4 (Epstein et al. 1977) meet the phenomenological definition for Type II bursts, and that they may well be caused by the same mechanism as the Type II bursts in the Rapid Burster. In any case, it is very likely that the 'burst-like' events from these three sources and the Type II bursts from the Rapid Burster are all due to instabilities in the accretion flow.

Recently, Kelley & Bradt (1978) observed a strong burst from the pulsar GX 301−2 (Population I, massive binary). The observation was made with the Horizontal Tube system of SAS-3 (Lewin et al. 1976d. Buff et al. 1977) when it was moving over the source (SAS-3 was spinning). The burst was 'classic' in that it rose in less than one second and decayed in $\sim$ 15 s. There was no obvious spectral softening in the burst and I suggested the possibility that this was a Type II burst. This, of course, could only be verified by pointing SAS-3 at the source. This was done and indeed the source was highly active and produced 'flare-burst-like' events on time scales of seconds and minutes (R. Kelley & H. Bradt, private communication). The activity continued for several hours.

If the detection of the single burst from GX 301−2 had been found a few years ago in old SAS-3 data (taken before December 1975 when X-ray bursts were discovered by Grindlay et al. 1976), it would have had a fantastic impact. One would almost certainly have jumped to the conclusion that the nature of burst sources had been unravelled and that most, perhaps all burst sources, and thus the X-ray sources in globular clusters and the galactic bulge sources, are neutron star binary systems. This conclusion may well have been correct, but obviously for the wrong reason. Maybe it was a good thing that the burst from GX 301−2 was observed only recently after the distinction between Type I and Type II bursts was recognized (Hoffman et al. 1978b).

Clearly there is no certainty that the GX 301 − 2 events and the 'hiccups' in GX 304 − 1, Cygnus X-1 and LMC X-4 are Type II bursts in the sense that the mechanism is the same as that in the Rapid Burster. It is an area that should be studied in more detail. In any event, I have not included any of these four sources in the sky map for Type I burst sources (fig. 18.7).

18.4 Type I burst sources

18.4.1 Sky distribution − globular clusters

Fig. 18.7 shows a sky map of burst sources. The map contains twenty-five sources. I know of the existence of at least six more but their positions, probably all within $\sim 30°$ of the galactic centre, are very uncertain; they are therefore not shown in the map. It is possible that two (maybe three) of the sources shown in fig. 18.7 are not of Type I. As an example, we detected a total of only three bursts from MXB 1743 − 28 (one of the three sources in GCX) and these bursts all occurred in a 21-minute period (Lewin, *et al.* 1976c). Therefore, this source is suspect.

The sources from which only one burst was observed (there are six in

Fig. 18.7. Sky map (galactic coordinates) of 25 X-ray burst sources. Only one burst was observed from each of the 6 sources indicated with a dashed circle or a horizontal bar. There are at least 6 more burst sources, probably all within 30° of the galactic centre. Their positions are poorly known and they are not shown here. At least 21, but possibly all sources shown are of Type I (see text).

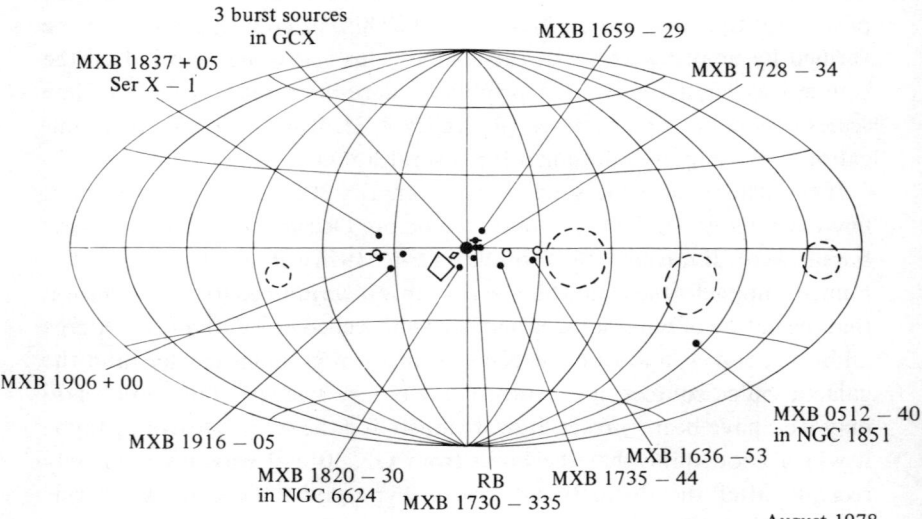

August 1978

fig. 18.7) are not necessarily suspect. For instance, the one burst from $l \sim 356^\circ.4$, $b \sim 2^\circ.3$ (Swank et al. 1977) showed the very characteristic softening in the tail (~ 100 s long) and is very likely a Type I burst. The one burst from the area near $l \sim 270°$, $b \sim -14°$ (Doty 1976) also showed the softening in the tail. In addition, in a $\sim$ 20-minute period that SAS-3 was continuously observing this area, only one burst was observed. If this burst had been of Type II, several might have been expected in this time. However, the source near $l \sim 80°$, $b \sim -15°$ is suspect (J. H. Swank, private communication).

We earlier reported a source located in a $\sim 2 \times 120$ deg² box (Lewin et al. 1977a); only one burst was recorded. I have deleted it from the map in fig. 18.7 since the one burst recorded may well have been of Type II. The burst source near $l \sim 315°$ was recently discovered by K. M. V. Apparao, S. Naranan & R. L. Kelley (private communication); it has not been reported previously.

The distribution shown in fig. 18.7 resembles that of the galactic bulge Population II type X-ray sources (Lewin 1977a; Ostriker 1977; Lewin et al. 1977a). At least three, but probably five, are located in globular clusters (see tables 18.1, 18.2, and 18.3). Certain are MXB 1820−30 in NGC 6624 (Grindlay et al. 1976; Clark et al. 1976; Clark et al. 1977) and the Rapid Burster (MXB 1730−335) in Liller 1; almost certain is MXB 0512−40 in NGC 1851 (Forman & Jones 1976; Clark & Li 1977); and probable are MXB 1746−37 in NGC 6441 (Li & Clark 1977) and MXB 1850−08 in NGC 6712 (Swank et al. 1976; F. K. Li & G. W. Clark, private communication). It has been suggested that XB 1724−30 (Swank et al. 1977) may be located in the globular cluster Terzan 2 (Grindlay 1978c) but this association is uncertain.

18.4.2 Identifications

Four burst sources have been identified with faint ($m_B > 17$) blue stars: MXB 1735−44 (McClintock's star) (McClintock et al. 1977a; McClintock, Canizares & Backman 1978), MXB 1636−53 (McClintock 1977), MXB 1659−29 (Grindlay 1978a), and MXB 1837+05 (Thorstensen, Charles & Bowyer 1978). The latter object is a very faint ($m_B \approx 19.2$) blue star and is located only ~ 2.2 arcsec from the star previously believed to be the X-ray source (Davidsen 1975; van Paradijs 1978a; Margon, Kwitter & Parkes 1978). The highly variable source MX 1608−52 (4U 1608−52) has been identified optically with a faint star ($m_I \sim 19$) (Grindlay & Liller 1978). This may well be a burst source but that is still somewhat uncertain (see table 18.3).

18.4.3 Time intervals between Type I bursts

The time intervals between Type I bursts range from ~ 1 to 50 hours (and quite possibly much longer). They can be regular as well as erratic. The most regular series was observed from MXB 1659−29 (Lewin 1977a; Lewin, Hoffman & Doty 1976b). The most erratic series were observed from MXB 1837+05 (Li *et al.* 1977) and from MXB 1735−44 (Lewin & Joss 1977; Lewin *et al.* 1977b). However, at times the bursts from MXB 1837+05 as well as from MXB 1735−44 can also be regular (Li *et al.* 1977; W. H. G. Lewin, L. Cominsky & J. van Paradijs, unpublished SAS-3 results).

18.4.4 MXB 1659−29

MXB 1659−29 is special in several ways. No 'steady' source was observed by SAS-3 when bursts from this source were discovered in October 1976 (Lewin *et al.* 1976b). The *upper limit* derived for the ratio of energy released in 'steady' emission to that in bursts is 25 (Lewin & Joss 1977). This is lower than for any other source and is perhaps a bit low to satisfy simple thermonuclear flash models (Joss 1977; Lamb & Lamb 1978; Joss 1978); see section 18.5.

In March 1978, Jeff Hoffman and I attempted to assist HEAO-1 in observing a Type I burst. At that time, the HEAO spacecraft was highly restricted in its pointing capability. It could only be pointed for a few hours continuously and only once or twice a week. The plan was (i) to observe continuously a burst source with SAS-3, (ii) to establish a burst sequence and regularity, (iii) to predict a few hours ahead of time the arrival of a burst, and (iv) to point HEAO-1 at the source in time (our strategy worked for MXB 1728−34; see Hoffman *et al.* 1978a). Because of its 'precise' regularity, MXB 1659−29 was selected first. However, to our great surprise, instead of bursts we observed a strong 'steady' source at a level of ~ 8 per cent of the Crab (Lewin *et al.* 1978a; Lewin 1978a). I alerted Rick Dower at MIT and Richard Griffiths at SAO who worked with the HEAO-1 A3 data. Subsequently, the HEAO A3 group determined multiple accurate positions of which one was selected (Griffiths *et al.* 1978); see fig. 18.8. The HEAO A3 position lies inside the 1976 SAS-3 error box (~ 5 arcmin radius) which was determined from burst data only (Lewin *et al.* 1976b), and there is little doubt that the source of steady emission in March 1978 and that of bursts in October 1976 is the same (bursts from MXB 1659−29 were also observed during the world-wide burst watch in the summer of 1977; Lewin 1977c). With the accurate position from HEAO

A3, the optical identification (with an ~ 18 mag blue star) was made immediately (Grindlay 1978a).

This was the second time that it was found that burst activity is not present when a source is in a 'high' state. This correlation was first observed for MXB 1820−30, located in the globular cluster NGC 6624 (Grindlay et al. 1976; Clark et al. 1976; Clark et al. 1977).

A unique series of 22 Type I bursts was observed from this source by Clark and his co-workers (Clark et al. 1977). During their observations, the 'steady' flux of the source increased continuously, the burst intervals

Fig. 18.8. Location of MXB 1659−29 at a distance of about 20 arcmin from the globular cluster NGC 6266. The 1976 position (Lewin et al. 1976b) was determined from bursts only (there was no observable steady flux at that time). The 1978 positions (Lewin et al. 1978a; Lewin 1978a; Griffiths et al. 1978) are for the steady source (there were no bursts at the time). The source has been identified with a faint blue star (Grindlay 1978a). This figure was prepared by Oren Grad.

MXB 1659−29

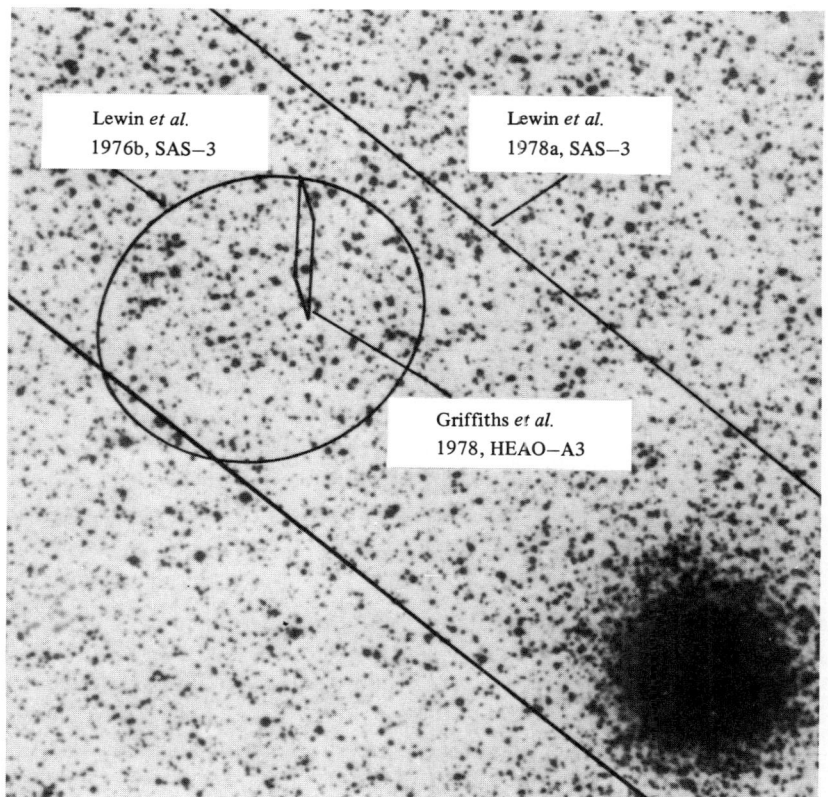

become shorter, and the 'tails' in the bursts become less pronounced (see fig. 18.9). When the burst activity finally came to a halt, the 'steady' flux was still increasing.

Clark and Li have extensively studied the X-ray source in the globular cluster NGC 1851. No bursts were ever observed when the steady emission was low (which is the case most of the time). However, they did observe

Fig. 18.9. Series of X-ray bursts from the source located in the globular cluster NGC 6624 (upper curves, 2–6 keV; lower curves, 6–11 keV). During the SAS-3 observations the intervals between the bursts gradually decreased from ~ 3.4 hours to ~ 2.2 hours. At the same time, the steady emission increased by at least a factor of five. Notice that the burst 'tails' become shorter as time goes on. The differences in amplitude can be attributed entirely to the random modulations produced by the collimator (from Clark *et al.* 1977).

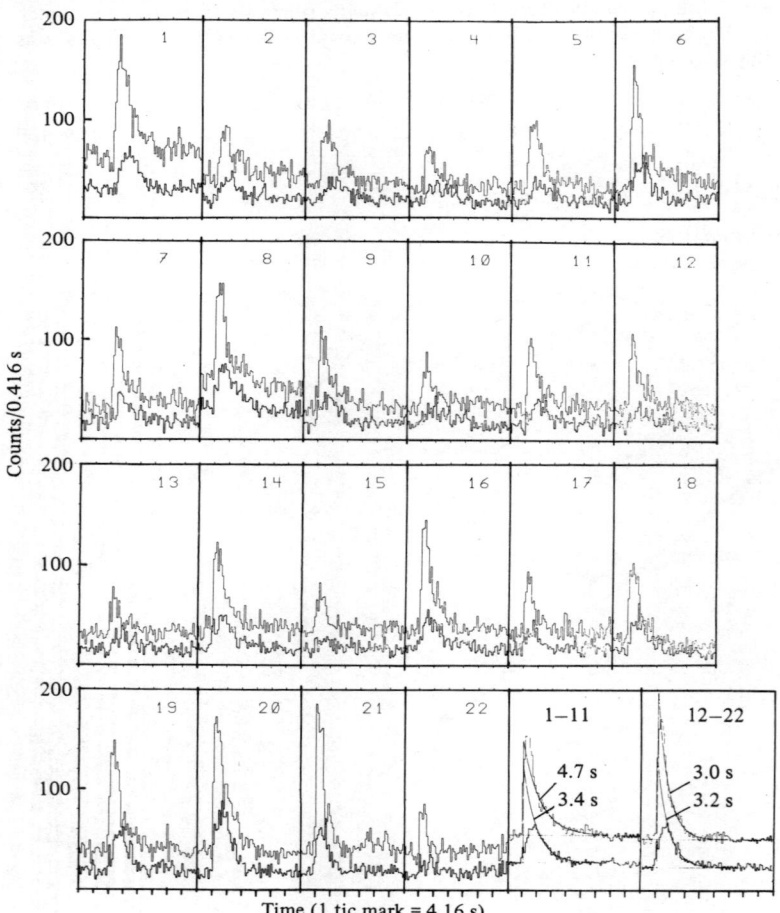

bursts when the steady source was much higher than normal (Clark & Li 1977). Thus, based on very limited statistics, I tentatively conclude:
- Type I bursts are not *observed* if the 'steady' emission is either very high or very low.

The 'absence' of Type I bursts for very low steady emission (so far only observed for the source in NGC 1851) could be the result of a selection effect. There is an obvious observational bias against the detection of bursts at very low frequencies. Thus, if the burst frequencies decrease with decreasing steady emission (as expected from the thermonuclear flash models and from other models), one would not expect to detect bursts from sources with very low luminosities.

18.4.5 Spectra and sizes of Type I burst sources

Fig. 18.10 shows five bursts from different burst sources. In all five cases, the softening of the spectrum during the burst decay is clearly visible in the burst tails. This distinct spectral softening is much less pronounced in the Type II bursts from the Rapid Burster (Marshall *et al.* 1979; Hoffman *et al.* 1979). It has been suggested (Alcock & Hatchett 1978) that the observed lengths, energy dependence, and variability of the burst tails can

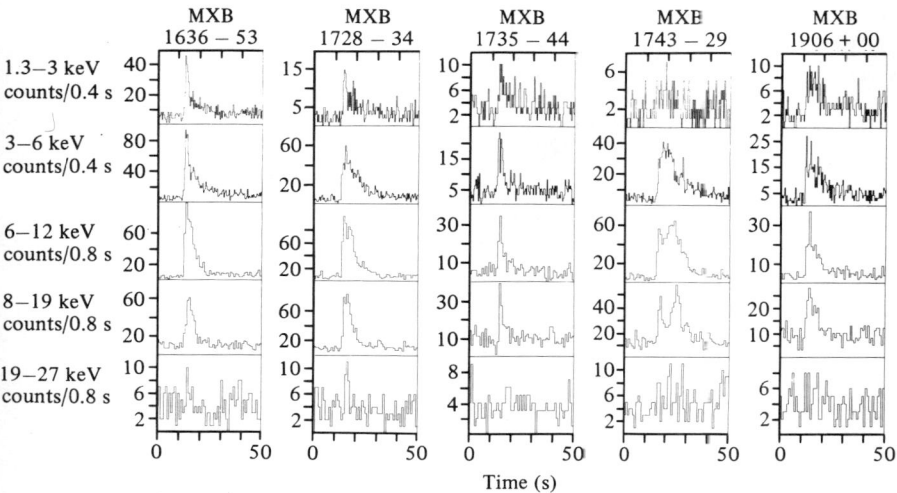

Fig. 18.10. Profiles (five energy channels) of Type I bursts from five different sources (SAS-3 data). Note that in all cases the gradual decay (tail) persists longer at lower energies than at higher energies. The burst profiles are reasonably distinctive for each particular source (from Lewin & Joss 1977).

be accounted for by a population of large (~ 3 μm) interstellar grains. This is perhaps true for some sources but certainly not for the majority. The simple reason is that the tails occasionally vary quite a bit from burst to burst. In fact, we have observed tails from several sources which varied, on a time scale of hours, from a few seconds to over one hundred seconds. Thus, a large part of the tails must be intrinsic to the bursts. Other reasons why the grain model does not work are given by van Paradijs & Lewin (1978).

Jean Swank and co-workers first showed that the best spectral fit to a Type I burst is that of a black body (Swank et al. 1977); see fig. 18.11. Hoffman, Lewin & Doty (1977a, b) showed that the same is true for bursts from MXB 1728−34 and MXB 1636−53. Both groups found peak temperatures of 25×10^6–30×10^6 K (reached at burst maximum) and subsequent cooling. For spherical symmetry and at an assumed source distance of 10 kpc, the calculated radii are comparable to that of a neutron star: Swank et al. (1977) observed a decrease in size from ~ 100 km to

Fig. 18.11. Incident spectra (three time intervals) of a very long Type I burst (I designated this source XB 1724−30 in table 18.3). The solid curves show the best black-body fits. The observations were made with OSO-8 (from Swank et al. 1977).

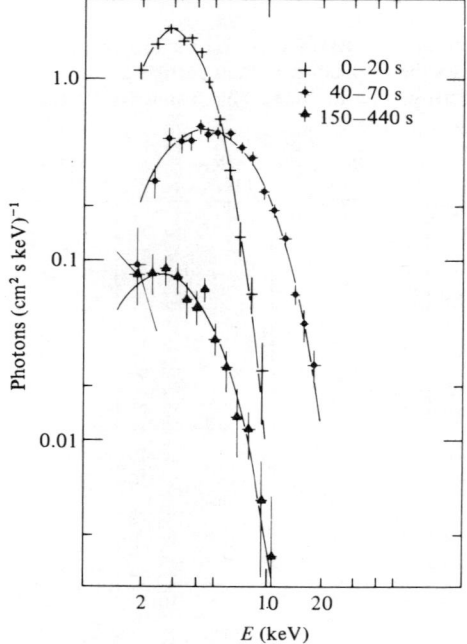

~ 15 km, while Hoffman et al. (1977a, b) found sizes of 10–15 km which remained constant throughout the bursts.

Jan van Paradijs (1978b) carried this idea a bit further. He made an *ad hoc* assumption that the Type I bursts are standard candles at burst maximum. Thus, $F_{max} \times 4\pi d^2 = L_0$. Here, F_{max} is the maximum observed flux, d is the distance to the source and L_0 is the luminosity during burst maximum (the latter was assumed to be a constant for all Type I bursts). Van Paradijs analysed SAS-3 data from ten burst sources and assumed black body radiation for all of them. For spherical symmetry, the radius R of an object is:

$$R(t) = T^{-2}(t) \left[\frac{F(t)}{F_{max}} \right]^{\frac{1}{2}} \left[\frac{L_0}{\sigma 4\pi} \right]^{\frac{1}{2}}.$$

Here, $T(t)$ and $F(t)$ are the observed temperature and X-ray flux at time t. Van Paradijs (1978b) demonstrated that under the above assumptions, the values for $R(t)$ are independent of time and are about the same for all ten burst sources. This, of course, greatly strengthens his assumption that the Type I bursts are standard candles at burst maximum. If L_0 is chosen to be $\sim 1.8 \times 10^{38}$ erg s^{-1} (the Eddington limit for a 1.4 M$_\odot$ object), the radii for all ten burst sources are $\sim$ 7 km! *The conclusion that burst sources are neutron stars (or perhaps black holes of a few solar masses) becomes almost inescapable.*

Fig. 18.12. Discovery of Type I and Type II bursts from the Rapid Burster. The Type I bursts ('special') occur independently of the sequence of the rapidly repetitive Type II bursts (numbered independently) (from Hoffman et al. 1978b).

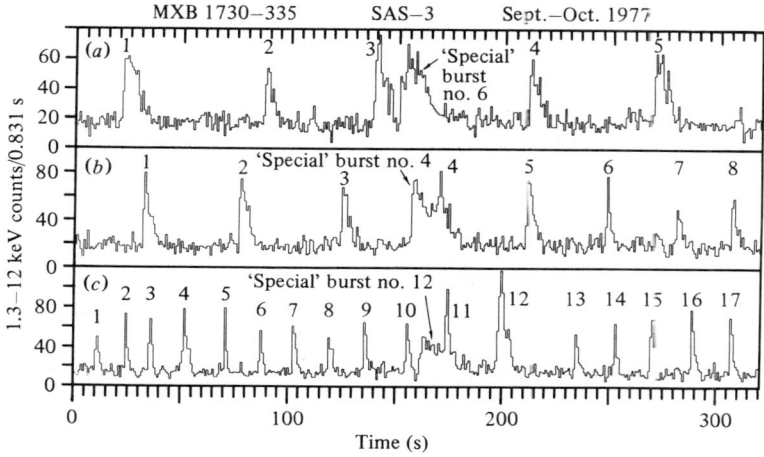

18.4.6 Type I bursts from the Rapid Burster

The Rapid Burster is the only object that is known to emit both Type I and Type II bursts (Hoffman *et al.* 1978b). It was this discovery that led to the burst classifications (figs. 18.12 and 18.13). It was shown recently (Marshall *et al.* 1979) that the Type I bursts from the Rapid Burster (like those of other sources) are also best fit by black-body spectra (with decreasing temperatures) and that the radius of the object is 8 ± 2 km (assuming spherical symmetry and a source distance of 10 kpc). Thus, both types of bursts from this object indicate sizes comparable to that of a neutron star (or a black hole of a few solar masses).

If Type I bursts are the result of thermonuclear flashes on the surface of a neutron star, and Type II bursts are chopped up 'steady' emission, one expects (assuming isotropic emission) the time-averaged ratio of energy

Fig. 18.13. Spectral information for the data shown in fig. 18.12.(*a*). The Type I burst spectrum softens during burst decay; this softening is much less pronounced in the rapid (Type II) bursts (from Hoffman *et al.* 1978b).

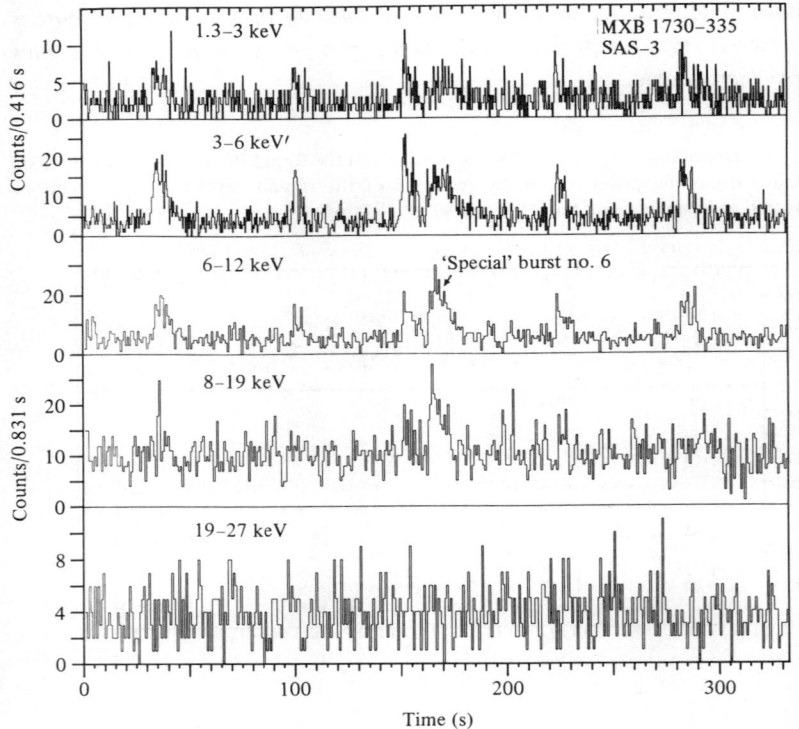

emitted in Type II and Type I bursts to be $\geq 10^2$ (see section 18.5). The observed value is ~ 130 (Hoffman et al. 1978b).

18.4.7 'Steady' X-ray emission from burst sources

There are presently ten Type I burst sources which are known (with a very high degree of certainty) to be associated with a source of 'steady' (though variable) X-ray emission (Lewin 1977a; Lewin & Joss 1977; Lewin 1977d). They are listed in table 18.2. Several more have been suggested (table 18.3), but in view of possible source confusion (large error boxes for the burst sources in combination with high source density), their association is still somewhat uncertain (in some cases very uncertain). I have added the Rapid Burster to table 18.2 since it is fairly certain that the Type II bursts from this object 'replace' the 'steady' emission as observed from many (not all) Type I burst sources (see section 18.3.1).

Several large error boxes for burst sources contain more than one known 'steady' source. There are also some (at least two) that do not contain any known source of 'steady' emission.

18.4.8 How steady is 'steady'?

The galactic bulge X-ray sources, of which class the burst sources are members, are often referred to as 'mildly' variable. This may be true for some, but it certainly is not true for all. For example, the source 4U 1820−30 (in NGC 6624) is highly variable, even on a time scale of ten minutes (Canizares & Neighbours 1975). Long-term variability for this source, for MXB 1659−29, and for the Rapid Burster (counting the Type II bursts as steady emission), is in excess of a factor of 10. In fact both MXB 1659−29 and the Rapid Burster could be called 'transients' depending on whose criterion one applies. A1742−289, near the galactic centre, is a transient (Eyles et al. 1975). Triple-peaked bursts at intervals of ~ 35 hours have been observed from MXB 1743−29 (fig. 18.10). The transient and burst source may well be the same (Lewin et al. 1976c); see also table 18.3.

Fabbiano & Branduardi (1979) argued that as many as seven Type I burst sources are associated with transient sources. However, one has to be very careful in accepting their associations since their selection criteria were very loose and generous. Several of their proposed associations of burst sources with sources of 'steady' emission are uncertain and some are almost certainly incorrect. (Surprisingly, they do not list the Rapid Burster as a transient.)

Rapid variability in MXB 1659−29 has been reported; several intensity changes by factors of four were observed on time scales of tens of seconds (Lewin et al. 1978a; Lewin 1978a). In fig. 18.14 I show five such changes. The low states of short duration ($\leqslant 15$ min) are probably due to absorption. A careful comparison of the data has shown that in a transition from 'high' to 'low', the 1.3–3 keV counting rate decreases first, then the 3–6 keV and finally the 6–12 keV counting rate. When a transition occurs from 'low' to 'high', the reverse is the case.

Fig. 18.14. SAS-3 observations of the 'steady' component of MXB 1659−29. The data were taken in March 1978 when the 'steady' emission was first observed (Lewin et al. 1978a; Lewin 1978a). Notice at least five rapid changes (in ~ 40 s) in the 3–6 keV data. There are periods of low emission, two of about 15 min. Others were observed (not shown here) which did not last that long. The recurrence of these low states was not periodic. These low states are probably due to absorption (see text). The source was occulted by the Earth during the gaps in the data. This figure was prepared by Lynn Cominsky and Bill Ossmann.

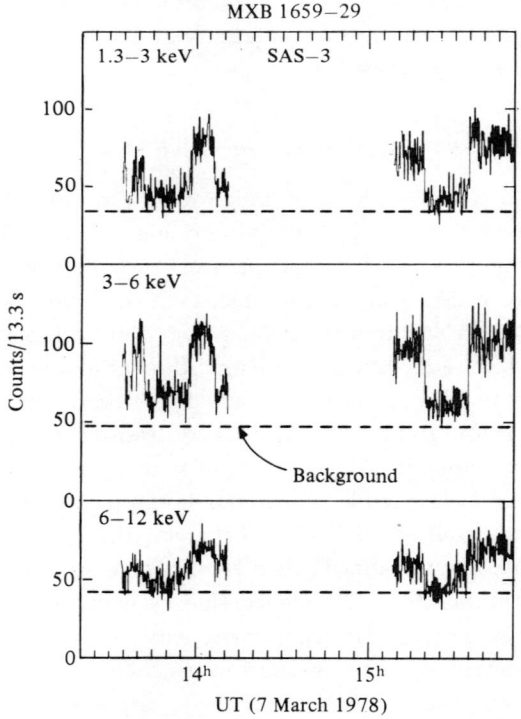

18.4.9 Search for periodicities and eclipses

More than 15 burst sources have been studied extensively and for long periods (1–3 weeks) by the Horizontal Tube detectors of SAS-3 (Lewin et al. 1976d; Buff et al. 1977). The associated 'steady' emission of the strongest sources (see table 18.2) has been searched for periodicities and the presence of eclipses. None have been found so far. In table 18.4 I summarize some of the results as provided by Lynn Cominsky, Herman Marshall and Fuk Li.

The absence of pulsations, periodicities and eclipses in the 'steady' emission makes one wonder about proposed binary-neutron star models for burst sources. Yet I do not believe on the basis of the presently available information that we have compelling evidence against such models.

- The absence of pulsations may simply mean that the magnetic fields are weak, which may be related to their old age. It has been suggested that magnetic fields of neutron stars decay in time (Gunn & Ostriker 1970; Chanmugan & Gabriel 1971; Ewart, Guyer & Greenstein 1975; Flowers & Itoh 1976; Flowers & Ruderman 1977).
- The absence of eclipses may be due to the geometry of the binary systems. Ed van den Heuvel has suggested Roche Lobe overflow systems with a very low-mass nuclear-burning companion. This idea was worked out in some detail by Joss & Rappaport (1979). In such systems the probability of an eclipse is low. M. Milgrom (preprint) suggested a

Table 18.4. *Upper limits to periodicities in the 'steady' components of Type I burst sources*

Source	Upper limits (90% confidence) to pulsed fraction[a] in %, for periods in the ranges listed		
	0.002–2 s	1.6–66 s	66–1000 s
1636−53	—	1.9	3.5
1659−29	—	3.5	9.5
1728−34	—	3.6	9.2
1735−44	8	1.9	13.0
1820−30	1.5	1.5	1.5[b]
1837+05	—	1.1	6.
1916−05	—	1.3	15

[a] Fraction of total flux.
[b] Period range 66–200 s.

geometry that explains in a simple way why eclipses may not have been seen from the strongest (best studied) galactic bulge sources. His model predicts that eclipses are expected, however, from the faint (not well studied) sources. I have proposed SAS-3 observations to search for eclipses from 4U 1624−49, 1702−42, 1705−32, 1708−23, 1746−37, 1755−33 and 1822−00. These are all relatively faint (less that 80 Uhuru counts) galactic bulge sources, yet they are strong enough to be studied with the SAS−3 Y-axis detectors (Lewin et al. 1976d; Buff et al. 1977). Milgrom's model also predicts that the 'steady' X-ray emission is strongly anisotropic. Thus values for α lower than $\sim 10^2$ could be expected (see section 18.5).

Another possibility is perhaps that some of these objects have wide eccentric orbits as observed for the transient pulsar 4U 0115+63 (Rappaport et al. 1978).

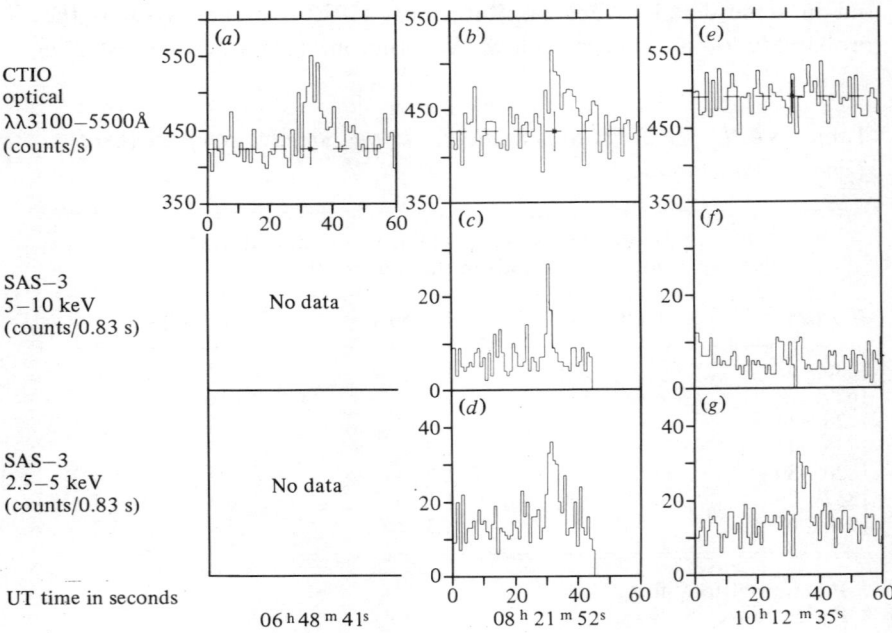

Fig. 18.15. Optical and X-ray bursts from the X-ray burst source MXB 1735−44. Optical data for 3 bursts are shown in (a) burst 1, (b) burst 2, and (c) burst 3, at the indicated times on UT 2 June 1978. No SAS-3 data were recorded for optical burst 1, but the X-ray data are shown for bursts 2 and 3. The average optical count rate on the source (which was ~ 170 counts/s above sky background) derived before the burst onset is shown in the optical burst plots as the dashed lines with $\pm 1\,\sigma$ error bars. The pre-burst average X-ray count rates (including the total background of ~ 4 counts/0.83 s) in the (2.5–5, 5–10) keV energy ranges were (13.5, 5.6) and (12.2, 6.2) counts/0.83 s for bursts 2 and 3, respectively (from Grindlay et al. 1978).

- It is perhaps possible that the reason for the absence of eclipses is simply that the compact objects are single, but this seems unlikely.

18.4.10 Radio/optical/X-ray bursts

A world-wide coordinated burst watch was organized in 1977 by the SAS-3 group (Lewin 1977c). Forty-four observatories from 14 different countries participated. A total of 120 bursts were observed by SAS-3 in 35 days from ten different Type I burst sources. No simultaneous positive detection was made in either the radio, optical, or infrared. Upper limits were reported for MXB 1837+05 and MXB 1916−05 (Thomas, Duldig, Haynes & Murdin 1978; Johnson *et al.* 1978; M. P. Ulmer *et al.*, in preparation; Abramenko *et al.* 1978; Takagishi *et al.* 1978; Bernacca *et al.* 1979).

On 2 June 1978, an optical burst was observed simultaneously with an X-ray burst (Grindlay *et al.* 1978) from MXB 1735−44 (McClintock's star, $m_V \sim 17.5$); see fig. 18.15. The fast rise and the absence of delay between the X-ray and optical bursts indicate that the optical region must be within ~ 1 light second of the X-ray emitting region. This result, in combination with the absence of optical variability of this star (other than bursts) makes it likely that the optical burst comes from an accretion disc heated by the X-ray burst rather than from a possible companion.

Comparing the steady flux with the flux at burst maximum, the X-ray

Fig. 18.16. Comparison of the spectra of two blue stars identified with the X-ray sources 4U 1735−44 and Sco X-1 (from McClintock *et al.* 1978).

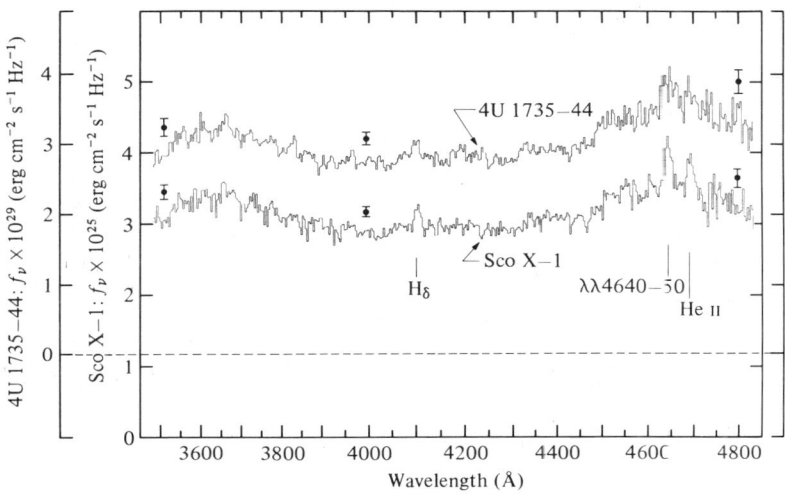

Table 18.5. *Results from simultaneous optical and SAS–3 X-ray burst observations*

Burst source	$\left(\dfrac{E_{opt}}{E_X}\right)^a$ burst	$\left(\dfrac{E_{opt}}{E_X}\right)$ steady	Number of bursts	Telescope aperture (m)	Method	References
MXB 1837+05	$< 10^{-4}$	$\sim 4 \times 10^{-5}$	2	0.5	Photography $\sim$ B	Abramenko et al. (1978) Crimean Observatory
	$< 5 \times 10^{-5}$	$\sim 4 \times 10^{-5}$	2	1.2	Photometry no filter	Bernacca et al. (1979) Asiago Observatory
	$< 3 \times 10^{-5}$	$\sim 4 \times 10^{-5}$	2	1.0	Photometry U B	Walker; Bernacca et al. (1979) South African Astrophysical Observatory
	$< 6 \times 10^{-5}$	$\sim 4 \times 10^{-5}$	2	1.3	Photometry V	Backman & Canizares; Bernacca et al. (1979) McGraw Hill Observatory
MXB 1916−05 (no optical identification)	$< 2 \times 10^{-3}$		2	0.6	Photometry U B V	Takagishi et al. (1978) Miyazaki University, Japan
MXB 1735−44	$\sim 2 \times 10^{-5}$	$\sim 1.3 \times 10^{-4}$	1	1.5	Photometry 3100–5500 Å	Grindlay et al. (1978) Cerro Tololo Inter-American Observatory

[a] Observed ratio of energy densities during the first 10 s of the burst (not corrected for interstellar extinction). Upper limits are 3σ. Optical energies in the range 4000–6000 Å.

signal increased by a factor 6.3 ± 1 and the optical signal by a factor 1.5 ± 0.3. Calculations of the effects of X-ray heating predict that the factor by which the optical flux increases (at burst maximum) should equal that of the X-ray increase raised to the power β: the predicted value of β is ~ 0.25 for a hot black body. Our observations give a value $\beta = 0.22 \pm 0.1$ which is consistent with the simple X-ray heating of a black body.

On the basis of our simultaneous observations we cannot conclude whether MXB 1735−44 is in a binary system. However, since the steady source spectrum of the star shows the characteristic features expected from X-ray heating (fig. 18.16), in combination with our observed value for β, it seems possible that most (perhaps all) of the optical emission from the star (thus not only the optical burst but also the steady optical emission) comes from the inner part of an accretion disc.

In table 18.5 I summarize the results of the ratios of energy density in the optical and X-ray bursts and in the steady fluxes. We will be pursuing

Table 18.6. *Burst summary*

	Type I	Type II[b] (Rapid Burster)
Rise time	$\leqslant 1$ s smooth	$\leqslant 1$ s 50 ms structure
Decay time	3–30 s	1–30 s
Tails	10–10^3 s	< 10 s
Best-fit spectra	Black body	Black body
T_{max}	$\sim 30 \times 10^6$ K cooling	$\sim 20 \times 10^6$ K No cooling
L_{max}^a	$\sim 10^{38}$ erg s^{-1}	$\sim 10^{38}$ erg s^{-1}
Energy in burst[a]	$\sim 10^{39}$ erg	10^{38}–10^{40} erg
Radius[a] (black body)	~ 10 km Remains constant during burst decay	~ 15 km Decreases during burst decay
Mechanism	Nuclear flash on surface of neutron star?	Instability in accretion flow, probably onto a neutron star (or black hole of solar mass)
Stellar object	Probably neutron star (or black hole of solar mass)	Probably neutron star (or black hole of solar mass)
Single or binary	Probably binary	Probably binary

[a] Assumed distance of 10 kpc.
[b] Our knowledge on Type II bursts from sources other than the Rapid Burster is very limited (see text).

the simultaneous optical observations in 1979 and a large amount of dark time has been reserved on SAS-3 to make such observations possible (Lewin 1978b).

To conclude the observational part of this chapter, I summarize in table 18.6 the salient features of Type I and Type II X-ray bursts.

18.5 Models

Various models have been proposed for X-ray bursts. Most of them fall into two broad categories: (i) instabilities in the accretion of matter onto a compact object; and (ii) thermonuclear flashes in matter accreted onto the surface of a neutron star. These two very different types of models have been the subject of many heated discussions over the past two years. After the detection by Hoffman *et al.* (1978b) of both Type I and Type II bursts from the same object (the Rapid Burster), it has become clear that perhaps both types of models may apply: (i) accretion instabilities for Type II bursts; and (ii) thermonuclear flashes for Type I bursts. Work on accretion instabilities (Lamb, Fabian, Pringle & Lamb 1977; Wheeler 1977; Grindlay 1978b; Liang 1977a, b; Baan 1977; Joss & Rappaport 1977) was reviewed recently (Lewin & Joss 1977). Thus I will discuss here only recent very promising work on thermonuclear flash models.

The idea that bursts might be the result of thermonuclear flashes in matter accreted onto a neutron star was put forward by Maraschi & Cavaliere (1977) and by Woosley & Taam (1976). Subsequent work on this idea was done by Joss (1977), Lamb & Lamb (1978), and R. E. Taam & R. E. Picklum (preprint). Only recently did Joss (1978) succeed in constructing a detailed numerical model for thermonuclear flashes. This model assumes a non-rotating, non-magnetized spherically accreting neutron star of mass 1.41 $M_\odot$, radius 6.57 km, core temperatures in the range $(2.5–5.7) \times 10^8$ K, and accretion rates in the range $(0.3–3) \times 10^{17}$ g s^{-1}. His model calculations reproduce the observed: (i) burst luminosities, (ii) total burst energies, (iii) rise times, (iv) decay time scales, (v) recurrence times, (vi) black-body radiation, (vii) maximum temperature of $(25–30) \times 10^6$ K, (viii) spectral evolution, and (ix) ceasing of burst activity at high steady emission. Fig. 18.17 shows some of Joss's results. As promising as these results are, the model is not without problems, as discussed below.

First, from the model one expects the burst intervals to decrease rapidly with increasing steady emission. Clark and co-workers (Clark *et al.* 1977) noticed that during a 3-day observation of MXB 1820−30 the burst intervals decreased by only ∼ 50 per cent while the steady flux increased by ~ 500 per cent at which time the burst activity came to a halt. In the case of MXB 1837+05 (Li *et al.* 1977) and MXB 1735−44 (W. H. G. Lewin, L. Cominsky & J. van Paradijs, unpublished SAS-3 results), there is no relation at all between the burst intervals and the steady flux. The burst intervals at times are erratic (they vary from ~ 1 hour to ~ 50 hours) and simultaneous changes in the steady flux are less than a factor of two.

Second, the model predicts that the ratio (α) of the energy emitted in steady emission to that in bursts is $\geq 10^2$ (see also Joss 1977 and Lamb & Lamb 1978). The observed value of $\alpha < 25$ for MXB 1659−29 is very low (Lewin & Joss 1977). Of course, if the steady emission is not isotropic, α can be smaller than ~ 10^2 and the observed low value of α for MXB 1659−29 does not necessarily impose a problem.

Finally, multiple peaks as observed from various sources (most strikingly from MXB 1743−29; see fig. 18. 10) are not accounted for by Joss's model calculations. The above problems (the first, I believe, is the most serious one) can perhaps be overcome if one removes the spherical symmetry (e.g. disc accretion rather than spherical accretion, as assumed in the model).

18.6 What are X-ray burst sources?

This question goes to the heart of one of the most intriguing unsolved problems in X-ray astronomy. The right answer would probably tell us what the origin is of most of the galactic bulge sources (including X-ray sources in globular clusters).

Fig. 18.17. Thermonuclear flashes on the surface of a neutron star. Temporal evolution of Type I X-ray bursts as calculated by Joss (1978) (parameters shown at the top). The dashed line indicates the level of persistent emission due to accretion. (*a*) The burst occurred after ~ 15.2 hours of accumulation of accreted material. (*b*) The burst recurred after ~ 1.4 hours of accumulation (higher accretion rate and higher neutron star core temperature). (*c*) No bursts are observed any time (still higher accretion rate and core temperature; see text). This figure by courtesy of P. C. Joss.

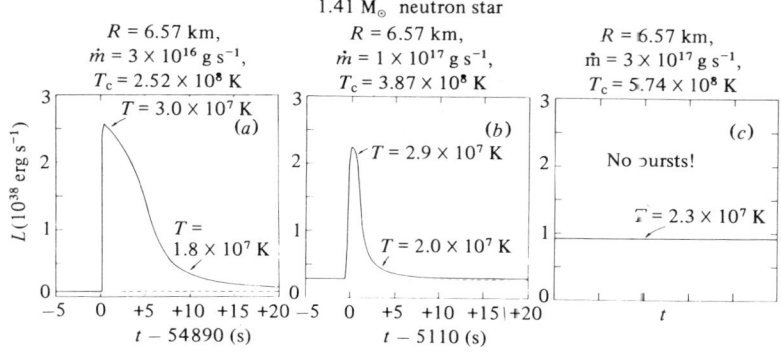

Though we do not have conclusive data in hand, the evidence in favour of neutron stars (or perhaps black holes of a few solar masses) is growing, as I have tried to show. The absence of pulsations is not an embarrassment; it may simply indicate that these objects have no strong magnetic fields. Maybe they were born that way or the field decayed due to old age. It is also possible that the axis of rotation and that of the magnetic dipole gradually coaligned. It remains to be seen whether the burst sources and galactic bulge sources are in binary systems but that seems very likely. The absence of observed eclipses would argue against this. Yet the geometry of the binary systems may be such that the chance of observing such eclipses is small (Joss & Rappaport 1979; M. Milgrom, preprint). The observed strong variability in the 'steady' emission of burst sources and globular cluster X-ray sources, but particularly the recurrent transient behaviour of the Rapid Burster, favours binary systems. The available information is clearly not yet conclusive. The best way to find out about the binary nature of these intriguing objects may well be to make detailed optical studies of the identified stars. Large telescopes and lots of observing time (dark time!) are needed.

I thank Roberta Caruso, Lynn Cominsky, Garrett Jernigan, Paul Joss, Fuk Li, S. Rappaport, Bruno Rossi, Jan van Paradijs and Lorraine Wright for their useful comments and contributions to this manuscript.

Note added in proof. Since this article was written in August 1978, several important contributions have been made. None, however, has changed the basic ideas described in my article that the burst sources and globular cluster X-ray sources are probably neutron stars with a low-mass companion. The thermonuclear flash model (helium flash on the surface of the neutron star) still is the most plausible explanation for Type I bursts even though several observed features are unexplained by this model to date (see article on MXB1735−44 by Lewin, Van Paradijs, Cominsky & Holzner to be published in *Mon. Not. Roy. astron. Soc.*, 1980). There is now convincing evidence that the optical emission from X-ray burst sources is predominantly due to reprocessing (X-ray heating) in the disc surrounding the compact object and that the companion star, which is almost certainly present, is of low mass (see Canizares, C. R., McClintock, J. E. & Grindlay, J. E., *Astrophys. J.* in press, 1980). Optical bursts observed simultaneously with X-ray bursts have been observed from two more sources: MXB 1837+05 (Hackwell, J. A. *et al. IAU Circ.* no. 3331, 1979; *Astrophys. J. Lett.* **233** (1979) L115) and MXB 1636−53 (Pedersen, A. *et al. IAU Circ.* no. 3399, 1979). The optical bursts always arrive a few seconds after the X-ray bursts (see also McClintock, J. E. *et al. Nature*, **279** (1979) 47); this delay is interpreted as a difference in travel time (the accretion disc has a radius of order of 1 lightsecond). Very bright infrared bursts have been detected from the Rapid Burster (Kulkarni, P. V. *et al. Nature*, **280** (1979) 819 and A. W. Jones *et al.* preprint). It is unclear at present whether they are related to the X-ray bursts but that seems likely. Very long Type II bursts (several hundred seconds) were detected from the Rapid Burster (Marshall, F. J. *IAU Circ.* no. 3336, 1979; M. Inoue *et al.* preprint; Basinska, E. *et al. Astrophys. J.* in press, 1980). Important contributions were recently made by the Einstein Observatory (J. Grindlay and co-workers). The globular clusters 47 Tuc and Terzan 2 are

X-ray sources, the latter is a burst source. From the scatter of the positions of the X-ray sources relative to the centers of the cores in globular clusters (see Bahcall, J. N. & Wolf, R. *Astrophys. J.* **209** (1976) 214; Jernigan & Clark 1979; Jernigan, p. 356, this volume), Grindlay and co-workers determined that the mass of the X-ray systems lies between ~ 1 and 5 $M_\odot$ (2σ level of confidence). Even though this result was expected, it is an independent confirmation of the low-mass character of these systems and as such an important result. For a more recent review on X-ray globular clusters and burst sources see W. H. G. Lewin & G. W. Clark, Proceedings of the NATO Advanced Study Institute in Cape Sounion, Greece, June 1979 and X-Ray Symposium in Tokyo, Japan, August 1979.

References

Abramenko, A. N., Gershberg, R. E., Pavlenko, E. P., Prokof'eva, V. V., Lewin, W. H. G., van Paradijs, J. A., Hoffman, J. A. & Li, F. K. (1978). *Mon. Not. Roy. astron. Soc.* **184**, 27P.
Alcock, C. & Hatchett, S. (1978). *Astrophys. J.* **222**, 456.
Apparao, K. M. V., Bradt, H. V., Dower, R. G., Doxsey, R. E., Jernigan, J. G. & Li, F. (1978). *Nature*, **271**, 225.
Baan, W. A. (1977). *Astrophys. J.* **214**, 245.
Bahcall, J. N. & Ostriker, J. P. (1975). *Nature*, **256**, 23.
Bernacca, P. L., Bianchini, A., Walker, A., Backman, D., Canizares, C. R., van Paradijs, J., Hoffman, J. A., Doty, J., Marshall, H., Wheaton, W., Jernigan, J. G. & Lewin, W. H. G. (1979). *Mon. Not. Roy. astron. Soc.* **186**, 287.
Brecher, K., Morrison, P. & Sadun, A. (1977). *Astrophys. J. Lett.* **217**, L139.
Buff, J., Jernigan, G., Laufer, B., Bradt, H., Clark, G. W., Lewin, W. H. G., Matilsky, T., Mayer, W. & Primini, F. (1977). *Astrophys. J.* **212**, 768.
Canizares, C. R. & Neighbours, J. E. (1975). *Astrophys. J. Lett.* **199**, L97.
Canizares, C. R. & Oda, M. (1977). *Astrophys. J. Lett.* **214**, L119.
Chanmugam, G. & Gabriel, M. (1971). *Astron. Astrophys.* **11**, 268.
Clark, G. W. (1975). *Astrophys. J. Lett.* **199**, L143.
Clark, G. W., Jernigan, J. G., Bradt, H., Canizares, C., Lewin, W. H. G., Li, F. K., Mayer, W., McClintock, J. & Schnopper, H. (1976). *Astrophys. J. Lett.* **207**, L105.
Clark, G. W. & Li, F. K. (1977). *IAU Circ.* no. 3092.
Clark, G. W., Li, F. K., Canizares, C., Hayakawa, S., Jernigan, J. & Lewin, W. H. G. (1977). *Mon. Not. Roy. astron. Soc.* **179**, 651.
Cruddace, R. G., Fritz, G., Shulman, S., Friedman, H., McKee, J. & Johnson, M. (1978). *Astrophys. J. Lett.* **222**, L95.
Davidsen, A. (1975). *IAU Circ.* no. 2824.
Davidsen, A., Malina, R., Smith, H., Spinrad, H., Margon, B., Mason, K., Hawkins, F. & Sanford, P. (1974). *Astrophys. J. Lett.* **193**, L25.
Doty, J. (1976). *IAU Circ.* no. 2922.
Doxsey, R. E., Apparao, K. M. V., Bradt, H. V., Dower, R. G. & Jernigan, J. G. (1977). *Nature*, **269**, 112.
Doxsey, R., Bradt, H., Gursky, H., Johnston, M., Schwartz, D. A. & Schwartz, J. (1978). *Astrophys. J. Lett.* **221**, L53.
Epstein, A., Delvaille, J., Helmken, H., Murray, S., Schnopper, H. W., Doxsey, R. & Primini, F. (1977). *Astrophys. J.* **216**, 103.
Ewart, G. M., Guyer, R. A. & Greenstein, G. (1975). *Astrophys. J.* **202**, 238.
Eyles, C. J., Skinner, G. K., Willmore, A. P. & Rosenberg, F. D. (1975). *Nature*, **257**, 291.
Fabbiano, G., Bradt, H. V., Doxsey, R. E., Gursky, H., Schwartz, D. A. & Schwartz, J. (1978). *Astrophys. J.* **221**, L49.

Fabbiano, G. & Branduardi, G. (1979). *Astrophys. J.* **227**, 294.
Fabian, A. C., Pringle, J. E. & Rees, M. J. (1975). *Mon. Not. Roy. astron. Soc.* **172**, 15P.
Flowers, E. & Itoh, N. (1976). *Astrophys. J.* **206**, 218.
Flowers, E. & Ruderman, M. (1977). *Astrophys. J.* **215**, 302.
Forman, W. & Jones, C. (1976). *Astrophys. J. Lett.* **207**, L177.
Forman, W. *et al.* (1979). *Astrophys. J. Suppl.* **38**, no. 4, 357.
Griffiths, R., Johnston, M., Bradt, H., Doxsey, R., Gursky, H., Schwartz, D. & Schwartz, J. (1978). *IAU Circ.* no. 3190.
Grindlay, J. E. (1978a). *IAU Circ.* no. 3229.
Grindlay, J. E. (1978b). *Astrophys. J.* **221**, 234.
Grindlay, J. E. (1978c). *Astrophys. J. Lett.* **224**, L107.
Grindlay, J. E. & Gursky, H. (1976). *Astrophys. J. Lett.* **209**, L61.
Grindlay, J. E. & Gursky, H. (1977). *Astrophys. J. Lett.* **218**, L117.
Grindlay, J. E., Gursky, H., Schnopper, H., Parsignault, D. R., Heise, J., Brinkman, A. C. & Schrijver, J. (1976). *Astrophys. J. Lett.* **205**, L127.
Grindlay, J. E. & Liller, W. (1978). *Astrophys. J. Lett.* **220**, L127.
Grindlay, J. E., McClintock, J. E., Canizares, C. R., van Paradijs, J., Cominsky, L., Li, F. K. & Lewin, W. H. G. (1978). *Nature*, **274**, 567.
Gunn, J. E. & Ostriker, J. P. (1970). *Astrophys. J.* **160**, 979.
Hills, J. G. (1975). *Astron. J.* **80**, 1075.
Hoffman, J. A., Lewin, W. H. G. & Doty, J. (1977a). *Mon. Not. Roy. astron. Soc.* **179**, 57P.
Hoffman, J. A., Lewin, W. H. G. & Doty, J. (1977b). *Astrophys. J. Lett.* **217**, L23.
Hoffman, J., Lewin, W., Marshall, H., Primini, F., Wheaton, W. & Cominsky, L. (1978a). *IAU Circ.* no. 3190.
Hoffman, J. A., Marshall, H. L. & Lewin, W. H. G. (1978b). *Nature*, **271**, 630.
Hoffman, J. A., Wheaton, W. A., Primini, F. A. *et al.* (1979). *Nature*, (in press).
Jernigan, J. G., Apparao, K. M. V., Bradt, H. V., Doxsey, R. E. & McClintock, J. E. (1977). *Nature*, **270**, 321.
Jernigan, J. F. & Clark, G. W. (1979). *Astrophys. J. Lett.* **231**, L125.
Jernigan, J. G., McClintock, J. E., Marshall, H., Chartres, M., Hoffman, J. A. & Lewin, W. H. G. (1978). *IAU Circ.* no. 3204.
Johnson, H. M., Catura, R. C., Lamb, P. A., White, N. E., Sanford, P. W., Hoffman, J. A., Lewin, W. H. G. & Jernigan, J. G. (1978). *Astrophys. J.* **222**, 664.
Jones, C. (1977). *Astrophys. J.* **214**, 856.
Joss, P. C. (1977). *Nature*, **270**, 310.
Joss, P. C. (1978). *Astrophys. J. Lett.* **225**, L123.
Joss, P. C. & Rappaport, S. (1977). *Nature*, **265**, 222.
Joss, P. C. & Rappaport, S. (1979). *Astron. Astrophys.* **71**, 217.
Katz, J. I. (1975). *Nature*, **253**, 698.
Kelley, R. & Bradt, H. (1978). *IAU Circ.* no. 3165.
Lamb, D. Q. & Lamb, F. K. (1978). *Astrophys. J.* **220**, 291.
Lamb, F. K., Fabian, A. C., Pringle, J. E. & Lamb, D. Q. (1977). *Astrophys. J.* **217**, 197.
Lewin, W. H. G. (1976). *IAU Circ.* no. 2922.
Lewin, W. H. G. (1977a). In *Proceedings of the 8th Texas Symposium on Relativistic Astrophysics*, ed. M. Papagiannis, *Ann. N.Y. Acad. Sci.* **302**, 210.
Lewin, W. H. G. (1977b). *Amer. Scientist*, **65**, no. 5, 605.
Lewin, W. H. G. (1977c). *IAU Circ.* no. 3078.
Lewin, W. H. G. (1977d). *Mon. Not. roy. astron. Soc.* **179**, 43.
Lewin, W. H. G. (1978a). *IAU Circ.* nos. 3190, 3193.
Lewin, W. H. G. (1978b). *IAU Circ.* no. 3252.

Lewin, W. H. G., Doty, J., Clark, G. W., Rappaport, S. A., Bradt, H. V. D., Doxsey, R., Hearn, D. R., Hoffman, J. A., Jernigan, J. G., Li, F. K., Mayer, W., McClintock, J., Primini, F. & Richardson, J. (1976a). *Astrophys. J. Lett.* **207**, L95.
Lewin, W. H. G., Hoffman, J. A. & Doty, J. (1976b). *IAU Circ.* no. 2994.
Lewin, W. H. G., Hoffman, J. A., Doty, J., Hearn, D. R., Clark, G. W., Jernigan, J. G., Li, F. K., McClintock, J. E. & Richardson, J. (1976c). *Mon. Not. Roy astron. Soc.* **177**, 838.
Lewin, W. H. G., Hoffman, J. A., Doty, J., Clark, G. W., Swank, J. H., Becker, R. H., Pravdo, S. H. & Serlemitsos, P. J. (1977a). *Nature*, **267**, 28.
Lewin, W. H. G., Hoffman, J. A. Doty, J., Li, F. K. & McClintock, J. E. (1977b). *IAU Circ.* no. 3075.
Lewin, W. H. G., Hoffman, J., Marshall, H., Primini, F., Wheaton, W., Cominsky, L., Jernigan, G. & Ossman, W. (1978a). *IAU Circ.* no. 3190.
Lewin, W. H. G. & Joss, P. C. (1977). *Nature*, **270**, 211.
Lewin, W. H. G., Li, F. K., Hoffman, J. A., Doty, J., Buff, J., Clark, G. W. & Rappaport, S. (1976d). *Mon. Not. Roy. astron. Soc.* **177**, 93P.
Lewin, W. H. G., Marshall, H. & Cominsky, L. (1978b). *IAU Circ.* no. 3211.
Li, F. K. & Clark, G. W. (1977). *IAU Circ.* no. 3095.
Li, F. K., Lewin, W. H. G., Clark, G. W., Doty, J., Hoffman, J. A. & Rappaport, S. A. (1977). *Mon. Not. Roy astron. Soc.* **179**, 21P.
Liang, E. P. T. (1977a). *Astrophys. J. Lett.* **211**, L67.
Liang, E. P. T. (1977b). *Astrophys. J.* **218**, 243.
McClintock, J. E. (1977). *IAU Circ.* no. 3088.
McClintock, J. E., Canizares, C. R. & Backman, D. E. (1978). *Astrophys. J. Lett.* **223**, L75.
McClintock, J. E., Canizares, C. R., Bradt, H. V., Doxsey, R. E., Jernigan, J. G. & Hiltner, W. A. (1977a). *Nature*, **270**, 320.
McClintock, J. E., Rappaport, S. A., Nugent, J. J. & Li, F. K. (1977b). *Astrophys. J. Lett.* **216**, L15.
Maraschi, L. & Cavaliere, A. (1977). *Highlights of Astronomy*, vol. 4, part 1, ed. E. A. Müller, p. 127. Dordrecht: D. Reidel.
Margon, B., Kwitter, K. B. & Parkes, G. E. (1978). *IAU Circ.* no. 3246.
Markert, T. H., Canizares, C. R., Clark, G. W., Hearn, D. R., Li, F. K., Sprott, G. F. & Winkler, P. F. (1977). *Astrophys. J.* **218**, 801.
Marshall, H. & Lewin, W. H. G. (1978). *IAU Circ.* no. 3208.
Marshall, H., Li, F. K. & Rappaport, S. (1977). *IAU Circ.* no. 3134.
Marshall, H. L., Ulmer, M. P., Hoffman, J. A., Doty, J. & Lewin, W H. G. (1979). *Astrophys. J.* **227**, 555.
Mason, K. O., Bell-Burnell, S. J. & White, N. E. (1976). *Nature*, **262**, 474.
Murdin, P., Penston, M. J., Penston, M. V., Glass, I. S., Sanford, P. W., Hawkins, F. J., Mason, K. O. & Willmore, A. P. (1974). *Mon. Not. Roy. astron. Soc.* **169**, 25.
Ostriker, J. P. (1977). In *Proceedings of the 8th Texas Symposium on Relativistic Astrophysics*, ed. M. Papagiannis, *Ann. N.Y. Acad. Sci.* **302**, 229.
Proctor, R. J., Skinner, G. K. & Willmore, A. P. (1978). *Mon. Not. Roy. astron. Soc.* **185**, 745.
Rappaport, S., Clark, G. W., Cominsky, L., Joss, P. C. & Li, F. K. (1978). *Astrophys. J. Lett.* **224**, L1.
Silk, J. & Arons, J. (1975). *Astrophys. J. Lett.* **200**, L131.
Swank, J. H., Becker, R. H., Boldt, E. A., Holt, S. S., Pravdo, S. H. & Serlemitsos, P. J. (1977). *Astrophys. J. Lett.* **212**, L73.
Swank, J. H., Becker, R. H., Boldt, E. A., Holt, S. S. & Serlemitsos, P. J. (1978). *Mon. Not. Roy. astron. Soc.* **182**, 349.

Swank, J. H., Becker, R. H., Pravdo, S. H., Saba, J. R. & Serlemitsos, P. J. (1976). *IAU Circ.* no. 3010.
Takagishi, K., Nagareda, K., Matsuoka, M. *et al.* (1978). *ISAS Res. Note*, ISAS RN60, CSR-P-78-31.
Tananbaum, H., Chaisson, L. J., Forman, W., Jones, C. & Matilsky, T. A. (1976). *Astrophys. J. Lett.* **209**, L125.
Thomas, R. M., Duldig, M. L., Haynes, R. F. & Murdin, P. (1978). *Mon. Not. Roy. astron. Soc.* **185**, 29P.
Thorstensen, J., Charles, P. & Bowyer, S. (1978). *IAU Circ.* no. 3253.
Ulmer, M. P., Lewin, W. H. G., Hoffman, J. A., Doty, J. & Marshall, H. (1977). *Astrophys. J. Lett.* **214**, L11.
van den Heuvel, E. P. J. (1977). In *Proceedings of the 8th Texas Symposium on Relativistic Astrophysics*, ed. M. Papagiannis, *Ann. N.Y. Acad. Sci.* **302**, 14.
van Paradijs, J. (1978a). *IAU Circ* no. 3197.
van Paradijs, J. (1978b). *Nature*, **274**, 650.
van Paradijs, J. & Lewin, W. H. G. (1978). *Nature*, **276**, 249.
Wheeler, J. C. (1977). *Astrophys. J.* **214**, 560.
White, N. E., Mason, K. O., Carpenter, G. F. & Skinner, G. K. (1978). *Mon. Not. Roy. astron. Soc.* **184**, 1P.
Woosley, S. E. & Taam, R. E. (1976). *Nature*, **263**, 101.

19
X-rays from globular clusters: steady emission

J. G. JERNIGAN

19.1 Introduction

Seven X-ray sources are definitely associated with globular clusters, as is known from position determinations. Three of these were discovered as part of the Uhuru survey: 3U 1820−30 in NGC 6624, 3U 1746−37 in NGC 6441, and 3U 2131+11 in NGC 7078 (M15). Subsequently two more, MX 0513−40 in NGC 1851 and MX 1746−20 in NGC 6440, were revealed through observations with the OSO-7 satellite. A sixth association, A 1850−08 in NGC 6712, was suggested by observations from ANE/5 and subsequently confirmed with Uhuru, ANS and SAS-3 data. The seventh association identifies the 'Rapid Burster' MXB 1730−335 with Liller 1, a highly-reddened cluster discovered within the X-ray error box. In fig. 19.1 I have plotted the celestial sphere in galactic coordinates showing the

Fig. 19.1. A plot in galactic coordinates of all globular clusters listed in Harris (1976). The clusters which emit X-rays are indicated.

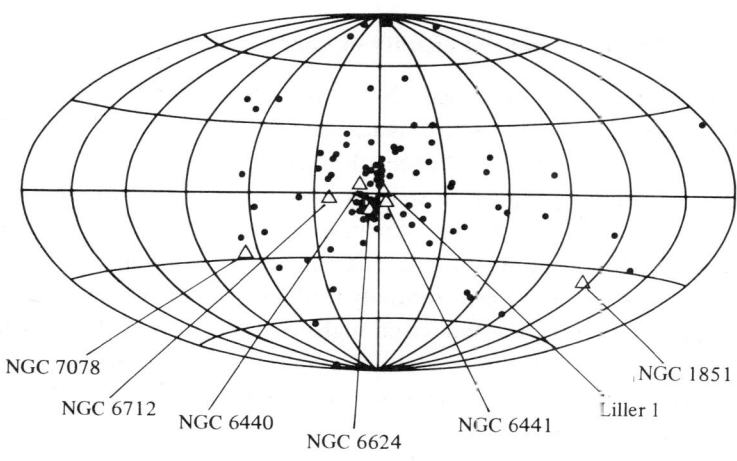

locations of all known galactic globular clusters. The seven X-ray emitters are indicated.

Of the seven X-ray sources associated with globular clusters, certainly three and probably five have been shown to emit slowly recurrent X-ray bursts of Type I (see Lewin, p. 315, this volume). All of these sources have been shown also to be steady emitters of X-rays, with the sole exception of MXB 1730−335. In one case, 3U 1820−30, the first source whose burst pattern was discovered to exhibit a quasi-periodic nature, the quenching of bursts has been correlated with an increase in the steady emission. The similarity of the bursts from these sources and the clear evidence for some physical association between bursts and steady emission in 3U 1820−30 greatly increases the likelihood that a single model may account for both the steady and the burst emission from all the sources. I will exclude the X-ray burst source MXB 1730−335 (the 'Rapid Burster') from the subsequent discussion since it does not have a steady counterpart. The properties of this unique source are reviewed by Lewin (p. 322, this volume).

The steady (i.e. continuous) emission from these sources has exhibited a highly variable character on time scales ranging from ~ 10 minutes to months. However, no evidence for pulsations, eclipses, or binary periods has been obtained. The fastest variation other than bursts was a ~ 10-minute flare of 3U 1820−30, reported by Canizares and based on OSO-7 data. However, 3U 1820−30 has demonstrated episodes of steady emission at a constant level over periods of weeks. The other 'steady' emitters are typically much weaker than 3U 1820−30 and are therefore less well studied; however, there is evidence for variations by factors from 2 to 10.

19.2 The dynamical identity of globular cluster sources

The discovery of high luminosity X-ray sources in globular clusters has raised several interesting questions concerning the relationship of these objects to the dynamical evolution of clusters. These sources are more frequent per star in globular clusters by two orders of magnitude than all other galactic X-ray sources per star in the galaxy. Table 19.1 contains a summary of the properties of these X-ray source and cluster pairs. As indicated, the sources are preferentially found in clusters with large escape velocities, small relaxation times and high central concentrations, properties indicative of an advanced state of dynamical evolution and possible core collapse. It must be remembered though that NGC 6712 is an exception to this correlation.

Table 19.1. *X-ray and optical data on X-ray sources in globular clusters*

Globular cluster X-ray source designations	Core radius r_c (arcsec)	Tidal radius r_t (arcmin)	Concentration $\rho = \dfrac{r_t}{r_c}$	Escape velocity v_e (km s^{-1})	Relaxation time t_r (yr)	Optical X-ray difference (arcsec)	Typical X-ray flux density (μJy)[b]
NGC 1851 MX 0513−40[a]	8	22	165	∼ 30	0.24×10^8	15 (9.3)	4
NGC 6440 MX 1746−20	⩽ 10	5	⩾ 30	⩽ 48	⩽ 0.3	—	10
NGC 6441 3U 1746−37[a]	9	8	53	71	1.8	8 (14.0)	13
NGC 6624 3U 1820−30[a]	8	13	98	28	0.3	9 (9.3)	∼ 200
NGC 6712 A 1850−08[a]	49	⩾ 8	> 9.7	—	3.4	18 (14.0)	4
NGC 7078 (M15) 3U 2131+11	10	15	90	37	1.2	14 (9.3)	6
Average for all globular clusters	∼ 50	∼ 20	∼ 24	∼ 23	∼ 6	—	—

[a] Emitters of Type I X-ray bursts.
[b] 1.0 μJy is equivalent to 2.2×10^{-11} erg^{-1} cm^{-2} in the energy range 2–11 kev. ($I_{\mathrm{CRAB}} = 1060\ \mu\mathrm{Jy}$.)

It is generally assumed that these and all other high-luminosity X-ray stars derive their energy from accretion of matter onto compact objects, and that in most or all cases the matter for accretion is supplied by a nuclear-burning companion in a close binary system. Within a globular cluster the available channels for the evolution of a primordial binary into a presently active binary X-ray source are narrow or non-existent (see Heggie, p. 281, this volume). Thus to explain the presence of any such sources in globular clusters and, even more, their high relative occurrence rate and strong preference for centrally condensed clusters, most speculations about the nature of the globular cluster sources have been based on the assumption that the conditions which are peculiar to the condensed cores result in the formation of X-ray binaries through capture of field stars by neutron stars or black holes (Clark 1975; Fabian, Pringle & Rees 1975; Hills 1975), or give rise to more or less massive black holes which accrete the debris from surrounding stars (Bahcall & Ostriker 1975; Grindlay & Gursky 1976).

Fig. 19.2. X-ray error circles (90 per cent confidence) superimposed on optical photographs of globular clusters. The photograph of NGC 1851 is from the ESO quick survey. All other photographs are taken from red plates of the Palomar Sky Survey (National Geographic Society). All the photographs are displayed with the same scale, and all circles have diameters of 40 or 60 arcsec. For each globular cluster the 2S designation for the error circle is indicated. North is up and east is to the left. Note that in all five cases the tidal radius lies outside the boundary of the photograph (see table 19.1) (from Jernigan & Clark 1979).

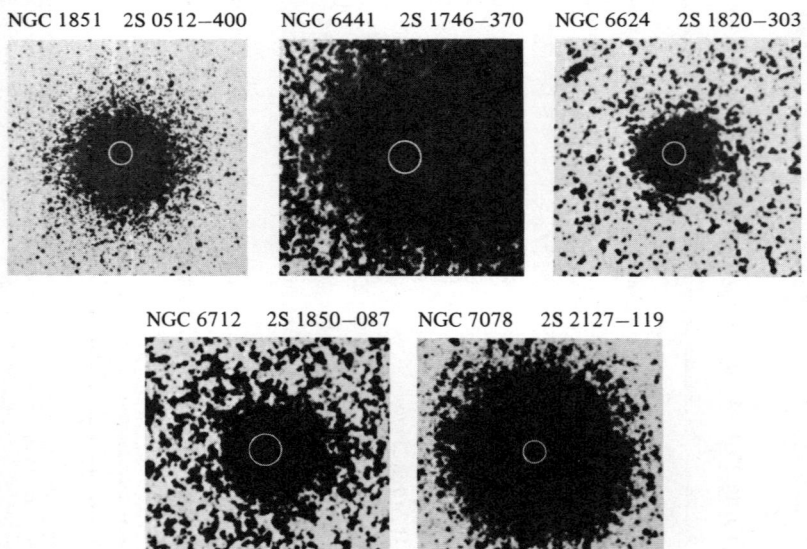

X-rays: steady emission

Five of these X-ray globular clusters have been observed with the rotating modulation collimators (RMC) on SAS-3. These instruments have apertures of 2.2 and 4.5 arcmin FWHM and can locate a typical galactic source to an accuracy of 20–30 arcsec in an observation of $\leq 10^5$ s. The 90 per cent confidence error circles of five of these sources are superimposed on photographs of the clusters in fig. 19.2. The tidal and core radii of the clusters are presented in table 19.1 along with the separation of the centre of the cluster from the centre of the X-ray error circle. The expected one sigma separations based on the accuracy of the position determinations are given in parentheses. In all cases the centre of the cluster is well within the 90 per cent confidence error circle.

In fig. 19.3 (a)–(e) the integrated radial projected profiles of the visible stars are shown along with the likelihood functions for the radial position of the X-ray source within the cluster. The likelihood function has fallen to a value of less than 0.1 at the projected radius which contains less than one-half of the visible stars in each case. These results clearly indicate that the X-ray sources are members of the cluster cores.

Fig. 19.3. (a)–(e) The likelihood functions for the projected radial displacement of each source from its cluster centre, and the normalized integrated distributions of stars. The maximum likelihood is normalized to unity in each case. The integrated distributions increase to 1 as the projected radii increase to the tidal radii. (f) Probability that an X-ray source of mass 2, 5 or 10 times the mass of an average cluster member will lie within a given projected radius for the case of NGC 7078 (from Jernigan & Clark 1979).

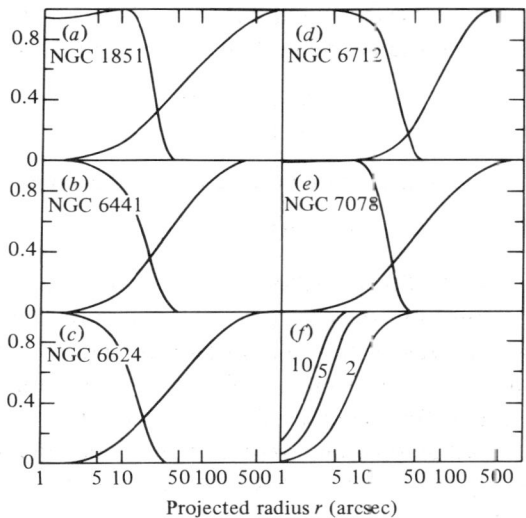

19.3 Masses of the globular cluster X-ray sources

In all theories of globular cluster X-ray sources the hypothetical objects are substantially more massive than the visible stars and are therefore expected to lie close to the cluster centres. Since the expectation distance of an X-ray source from the centre of its cluster must be a function of its mass, measurements of the radial distances of several X-ray sources can in principle provide a statistical measure of the mean mass of the sources.

A simple model can be adopted to permit the estimation of the expectation distance of an X-ray source as a function of its mass. Assume that the density profile of a globular cluster is described by the approximation to an isothermal sphere in three dimensions,

$$f(r) \propto \left[1 + \left(\frac{r}{r_c}\right)^2\right]^{-\frac{3}{2}} \quad r < r_t,$$

where r_c is the core radius and r_t is the tidal radius. Assuming that the cluster and the X-ray source are dynamically relaxed, one can show that the three-dimensional probability distribution for the position of an X-ray source of mass m_X is proportional to $[f(r)]^q$, where

$$q = \frac{m_X}{m_*}$$

and m_* is the mass of a typical cluster member. Assuming that $r_t \gg r_c$, which is approximately true for globular clusters, then $\langle r \rangle$, the expectation of the projected radial displacement of the X-ray source, is given by

$$\langle r \rangle \approx \frac{2r_c}{\pi^{\frac{1}{2}}} \frac{\Gamma(\frac{3}{2}q - \frac{3}{2})}{\Gamma(\frac{3}{2}q - 1)}.$$

If one considers only large q then this equation becomes

$$\langle r \rangle \approx 0.9 r_c q^{-\frac{1}{2}}.$$

Fig. 19.4 shows a graph of this function. The graph indicates that an X-ray source which is confined to the core ($\langle r \rangle \leqslant r_c$) will have a mass which is greater than twice the typical mass of a cluster member. Fig. 19.3(f) shows the integrated probability distribution of the projected radius r if $q = 2$, 5 or 10 (as indicated) for the case of NGC 7078 (M15); compare figs. 19.3(f) and (e). Based on this simple equilibrium model, the SAS-3 RMC measurements imply that the X-ray sources are at least twice as massive as a typical cluster member.

Fig. 19.4. The normalized projected radius, $\langle r \rangle/r_c$, as a function of the normalized mass of the X-ray source, $q = m_X/m_\star$. For large q the asymptotic limit of $\langle r \rangle$ is proportional to $q^{-\frac{1}{2}}$ (from Jernigan & Clark 1979).

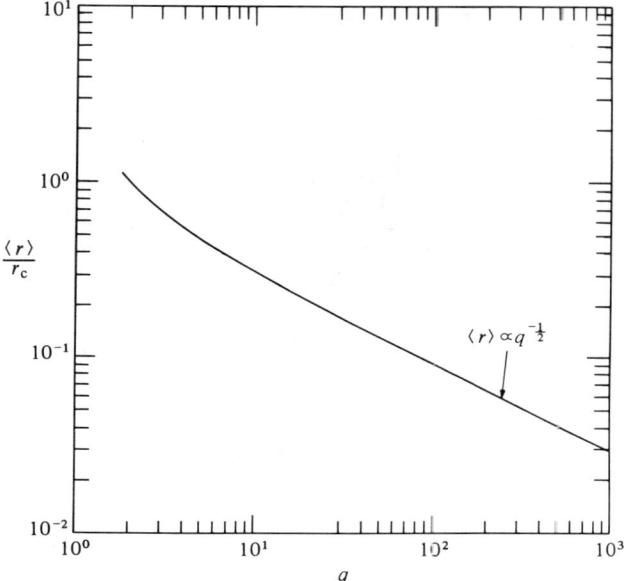

Fig. 19.5. A plot of the positions of X-ray sources in galactic coordinates in the region near the galactic centre. The bulge is arbitrarily defined as the region within the circle of radius 30°.

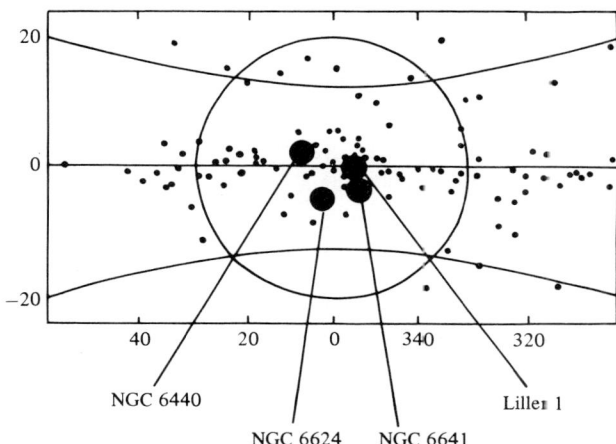

19.4 Cluster sources versus galactic bulge sources

As a class of objects, the globular cluster X-ray emitters have been linked with the galactic bulge X-ray sources since both groups are associated with the Population II stellar component of the Galaxy. The X-ray burst sources which are not in globular clusters are predominantly located in the galactic bulge, and a significant fraction of these burst sources are identified with the steady emitters in the bulge. Fig. 19.5 is a plot of the X-ray sources in the galactic bulge (which is arbitrarily defined as the region within 30° of the galactic centre). This region contains ~ 80 X-ray sources of which 4 are in globular clusters.

Fig. 19.6 shows a graph of the number of sources $N(S)$ brighter than flux density S as a function of S. The solid line is the curve derived for all galactic X-ray sources (those at $|b| \leqslant 20°$), whereas the points and dashed line are derived from the ~ 80 sources within 30° of the galactic centre. (One transient X-ray source and Sco X-1 have been arbitrarily omitted from the sample.) The brightest 14 bulge sources are listed in table 19.2 along with their flux densities. As a group, these sources do not exhibit any pulsations or periodicities. They correspond to the bright end of the lg N versus lg S curve (fig. 19.6) which has a slope of -1.5, while the sources for which $S < \sim 200$ have a slope of ~ -0.4. The shapes of the

Fig. 19.6. A plot of the number of sources $N(S)$ with flux density greater than S as a function of S. The solid line is the curve for all galactic X-ray sources. The dashed line is the curve for sources within 30° of the galactic centre.

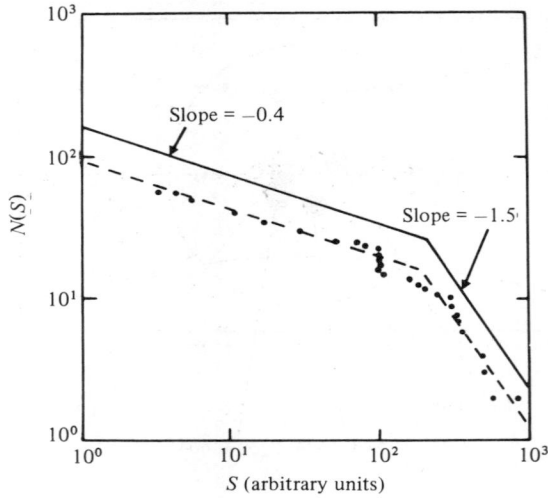

lg N versus lg S curves for the bulge sources and for all galactic sources are approximately the same. At the high flux density limit, the similar shape is understandable since the curve is dominated by the 14 bright sources in each case. However, it is somewhat surprising that the low luminosity limits are similar. Moreover, it should be noted that three of the weaker sources within the 14 brightest (namely 4U 1636−53, 4U 1820−30 and 4U 1735−44) are steady counterparts of X-ray bursters.

The point at which the slope of the lg N versus lg S curve changes corresponds to the flux density of the brightest burst sources in the bulge. This implies either that the most luminous bulge emitters are a separate class of objects or that their enhanced emission is physically correlated with the lack of X-ray bursts.

19.5 Future prospects

Currently there are ~ 34 identified optical counterparts of galactic X-ray sources. There are ~ 27 unidentified counterparts for sources which have positions determined to an accuracy of better than 1 arcmin. Of these, all but three are located within 40° of the galactic centre. The HEAO-B satellite will locate these objects to an accuracy of ~ 1 arcsec.

The HEAO-B measurement of the X-ray source in NGC 6712 will be of particular importance because of the large core radius (~ 50 arcsec) in this cluster. Based on the simple model presented earlier and summarized in fig. 19.4, an X-ray source located within 1 arcsec of the centre of NGC 6712 would have an indicated mass of $\sim 10^3$ $M_\odot$, though a model allowing for a non-equilibrium core collapse could indicate a lower mass.

Since the nature of the bulge sources, globular cluster sources, and burst

Table 19.2. *The brightest fourteen galactic bulge sources*

Source	X-ray flux density (μJy)	Source	X-ray flux density (μJy)
GX 5−1	860	GX 13+1	330
GX 349+2	580	GX 9+9	300
GX 17+2	490	4U 1755−33	300
GX 9+1	480	4U 1636−53	240
GX 3+1	400	4U 1730−22	200
4U 1630−47	370	4U 1820−30	180
GX 340+0	340	4U 1735−44	160

sources are now linked by observations, it would be a significant advance to determine the as yet unknown structure of one source which is a member of any of these three groups. Perhaps the resolution of the nature of these objects is the most important outstanding problem in galactic X-ray astronomy.

References

Bahcall, J. N. & Ostriker, J. P. (1975). *Nature*, **256**, 23.
Clark, G. W. (1975). *Astrophys. J. Lett.* **199**, L143.
Fabian, A. C., Pringle, J. E. & Rees, M. J. (1975). *Mon. Not. Roy. astron. Soc.* **172**, 15P.
Grindlay, J. E. & Gursky, H. (1976). *Astrophys. J. Lett.* **205**, L131.
Harris, W. E. (1976). *Astron. J.* **81**, 1095.
Hills, J. G. (1975). *Astron. J.* **80**, 1075.
Jernigan, J. G. & Clark, G. W. (1979). *Astrophys. J. Lett.* **231**, L125.

20
Summary of contemporary research on globular clusters

K. C. FREEMAN

20.1 Introduction

Globular clusters are now a vigorous field after many quiet years. Why? Here are four reasons.

(i) There is a great interest in galaxy formation and evolution, and for this work globular clusters are vital. Many researchers are now discussing cluster properties in the context of galaxy formation and chemical evolution; see for instance Searle (1977) and van den Bergh (p. 175, this volume).

(ii) The chemical inhomogeneity of globular clusters, discovered only in the last few years, gives us new opportunities for studying (*a*) the chemical evolution of individual stellar systems (for those clusters with inhomogeneities going back to the time of cluster formation), and (*b*) mixing processes in the late stages of stellar evolution; see for instance Kraft (p. 87, this volume) and Norris (p. 113, this volume).

(iii) The discovery of X-rays from globular clusters and the advances in core collapse theory (although it is not so clear now that these two items are connected); see for instance Lewin (p. 315, this volume) and Heggie (p. 281, this volume).

(iv) The great increase in telescope aperture and detector capabilities that have made many impossible things possible.

I will try now to pick out some high spots and low spots in present-day studies and to identify some topics that are particularly confusing.

20.2 Dynamics of globular clusters

20.2.1 The quest for a good dynamical description of a static globular cluster

This is the essential starting point for a realistic understanding of globular cluster dynamics. Here are a few problem areas:

(i) *Anisotropy*. Most of the well-used globular cluster models assume an isotropic lowered Maxwellian distribution function (King, 1966). If this turns out to be unrealistic, then the problems of making unique models become much more difficult (see King, p. 263, this volume). The problem is observational, and so Gunn & Griffin's (1979; and see Gunn, p. 271, this volume) radial velocity work and Cudworth's (1976, 1978) proper motion studies are most important. The data so far reported point towards an anisotropic distribution function, but the interpretation of the observations is not entirely straightforward.

(ii) *The tidal radius.* Two interesting aspects of the nature of the tidal radius for globular clusters have come up. First, is the conventional assumption correct, that the cluster's tidal radius is set at the closest approach to the galactic centre? Maybe not, inasmuch as the orbital period in the cluster for stars near the tidal radius is of the same order as the cluster's orbital period in the galaxy. So, if tidal radii are to be used as probes of cluster orbits or of the galactic mass distribution, then what is the appropriate prescription for relating the tidal radius to the cluster's orbit? Secondly, Keenan (1978; see also Keenan & Innanen 1975) has pointed out how the tidal radius is not spherical because of the restricted three-body nature of the cluster and galaxy potential field. This asymmetry of the potential in the outer parts of the cluster means that stars near the tidal radius will feel a torque. However, for the distribution function to be anisotropic in the usual sense, the stars reaching the outer parts of the cluster must have even lower angular momentum than they would have in an isotropic system. But how can they continue to have low angular momenta in the presence of this 'tidal torque'? It would probably be worth some effort to identify cluster stars near the tidal radius of a rich cluster like ω Cen (there will only be a few such stars): their kinematics and chemistry could be quite interesting.

(iii) *Stellar content.* Thermal equilibrium models of globular clusters require a significant fraction of dark stars, including some more massive than the present giants. This causes some problems. Da Costa's (1977) 47 Tuc models require far more white dwarfs than any reasonable extrapolation of the observed mass function would allow. Gunn's models for M3 (p. 276, this volume) invoke 1.2 $M_\odot$ neutron stars, but again others argue from the high velocities of pulsars that these neutron stars may be difficult to contain in the cluster.

(iv) *Why are globular clusters spherical?* Evaporation of angular momentum has been proposed, and this is consistent with the relatively high ellipticities and long relaxation times of ω Cen and the dE galaxies. But on the other hand most of the very young blue globular clusters in the LMC are fairly spherical and they are much younger than their relaxation times.

20.2.2 The dynamical evolution of globular clusters

(i) *Internal effects.* Spitzer & Shull (1975) have clearly shown how the core-halo structure develops. The inner parts contract and the outer parts expand to absorb the energy, and by $\approx 15T_{rh}$ the core collapse is well advanced (where T_{rh} is a reference relaxation time for the cluster). A significant population of initially hard binaries can at least delay this collapse; see for instance Heggie (p. 289, this volume). What actually happens at core collapse? Opinion seems to be that probably no real singularity results. It may be that the core's binding energy is transferred to a central hard binary which finally gets ejected from the cluster: then the whole core collapse process could begin again. Some people argue that the halo could become so distended that the cluster would then be easy prey for disruption by tidal shocks (this is consistent with the presence of low density systems at large galactic radii only). Others conjecture that nothing interesting happens at core collapse: the characteristic length scale for most of the cluster population is not significantly affected by the core collapse. Results reported by Heggie (p. 292, this volume) support this view. The production of a massive central object through core collapse is no longer so attractive as an explanation for cluster X-ray emission: the radii of Type I bursters appear to be about 10 km which corresponds to neutron stars (Lewin, p. 335, this volume).

(ii) *External effects.* Tidal shocks can be very effective at heating and destroying clusters, particularly those crossing the galactic plane within a few kpc of the galactic centre, while dynamical friction in the galactic disc is effective at destroying massive disc globular clusters, if these ever existed; see Fall (p. 309, this volume).

20.2.3 Binaries in globular clusters

There are probably a few binaries in globular clusters: several novae and U Gem stars are known to be members. The blue stragglers in some clusters may also be binaries; and the anomalous cepheids in NGC 5466 and the dE galaxies are interpreted as resulting from mass transfer in binaries. But Gunn & Griffin's (1979; and see Gunn, p. 272, this volume) velocities show no binaries in the 0.3–7 AU range, from a sample of about 110 giants. Furthermore Trimble (1977, 1978) has found no evidence for eclipsing binaries on the main sequence of M55. Although hard and soft binaries may be ejected or destroyed in cluster cores, it would be surprising if initial binaries did not survive in the cluster envelopes. The situation remains confusing.

20.2.4 Dwarf spheroidal galaxies and globular clusters

What is the difference? (Zinn, p. 191, this volume). Why do we see the predominance of low surface brightness systems (the Palomar clusters and the dwarf spheroidals) far out in the galaxy, where the higher surface brightness clusters could survive just as well? There seems to be a fairly abrupt change in the surface brightness distribution, from the normal clusters to the low surface brightness objects. Is this due to some dynamical process? For example, maybe many low surface brightness systems formed in the inner parts of the Galaxy and were destroyed by the tidal field, or perhaps they were prevented from forming at all by the tidal field of the Galaxy. Do the low surface brightness systems perhaps have a very flat mass function, so there are few giants per unit mass? A direct estimate of the mass for one of these systems would be very valuable, though probably impossible.

20.3 The horizontal branch and the second parameter

Horizontal branch (HB) morphology is sensitive to several chemical and physical parameters, like Z, Z_{CNO}, Y, age, mass-loss rate and initial rotation, *inter alia* (Castellani, p. 68, this volume). This is good because it sets some constraints on these parameters, and bad because they are difficult to disentangle. It seems clear that the intermediate and low abundance clusters show a spread in HB morphology at a given abundance: what is the dominant second parameter? Some important new results have appeared, but confusion remains.

Zinn (1978, and p. 194, this volume) and Schommer (1978) showed the striking result that the second parameter is most active in the outer parts of the Galaxy. This could imply that the second parameter is age: metal-weak clusters in the inner parts of the Galaxy were all formed very early, while the metal-weak clusters further out have an age spread. Also, for these outer clusters it may be that the CNO production and the heavy element production did not go together. Cohen, Frogel & Persson (1978; and see Persson & Frogel, p. 149, this volume) have studied CO in giants of M13 and M3. These clusters have similar [Fe/H] but M3 shows stronger

CO. Again, for the pair M5 and M10, M5 has stronger CO. The strong CO clusters have RHB and BHB stars, the others have BHB stars only. However, studies by others of distant systems with anomalous HB stars confuse the problem, and it is still not clear what the second parameter is.

The origin of the Oosterhoff dichotomy has come up on several occasions. Castellani proposes (see p. 83, this volume) that it comes from a helium-abundance difference: $Y \approx 0.22$ for the metal-weak Oosterhoff II systems, and $Y \approx 0.27$ for Oosterhoff I.

Understanding mass loss still seems very important for understanding the HB. $H\alpha$ emission consistent with the required mass loss rate is seen in many globular cluster giants and field Population II giants (Mallia 1978; Cohen 1976), but it is not completely certain that the $H\alpha$ emission really is associated with mass loss. No gas has yet been detected in a globular cluster: the limits are significantly below 1 $M_\odot$ (Hesser & Shawl 1977; Knapp, Rose & Kerr 1973). Globular cluster wind theory predicts unobservable amounts of gas for the smaller clusters, but in the larger ones the 21 cm predictions are getting into the observable range.

20.4 Chemical abundances

20.4.1 The large-scale distribution in the Galaxy

Searle (1977, 1978) has reviewed the distribution of [Fe/H] for globular clusters with distance from the galactic centre. For $R \gtrsim 5$ kpc the mean abundance level is about -0.5. There is a steep abundance gradient between 5 and 10 kpc and then beyond 10 kpc the mean abundance is roughly constant with radius; clusters in this outer region have abundances between about -1 and -2, with the peak at -1.4. For M31 the distribution is qualitatively similar, but the gradient is not so steep and the peak in the distribution for the outer clusters is at about -1.0.

How were these galactic abundance gradients set up? Searle has argued against the slow diffusive enrichment picture because the galactic distribution of clusters is so spherical. He proposes that the enrichment occurs in fragments: the amount of enriched matter lost from the fragments determines the mean abundance in the outer parts of the galaxy. This picture successfully predicts the width of the [Fe/H] distribution for the outer globular clusters; it also predicts that high mass loss from the fragments, which should lead to a small bulge-to-disc ratio for the galaxy, produces a low mean [Fe/H] in the outer parts. This is consistent with the comparison of M31 and the Galaxy: M31 has a larger bulge-to-disc ratio and a higher mean abundance in the outer parts. In the inner parts of the galaxy, the enriched material cannot escape so easily, and a relatively high level of [Fe/H] is established fairly rapidly.

The dependence of the HB second parameter on galactocentric distance should be mentioned again. This fits easily into Searle's picture if the second parameter is identified as age: metal-weak clusters can form over a longer time interval in the outer parts of the galaxy than in the inner parts.

20.4.2 Individual globular clusters

The galactic radial abundance story, as outlined by the globular clusters, is replayed in miniature by ω Cen and 47 Tuc (Freeman, p. 105, this volume). ω Cen shows clear radial changes in Ca and CN; for 47 Tuc, so far only the CN gradient has been found. In the outer parts of ω Cen the abundance histogram is qualitatively like the histogram for the globular cluster abundance distribution in the outer parts of the galaxy. It is not yet clear how these radial gradients were set up in the clusters (or in the galaxy!). In the cluster case it is no small effect, because the gradients themselves involve at least 1000 $M_\odot$ of metals alone. Understanding the cluster gradients is now just as urgent a problem as the galaxy gradients but is probably more tractable, because many more observational tests are possible for the clusters.

ω Cen is inhomogeneous in CNO and heavy elements, and this produces the wide giant branch and the great CN and CO variations. From Searle & Zinn's (1978) and from Cohen's (1978) work it is seen that no other clusters show comparable heavy element variations. 47 Tuc has CN inhomogeneities, which still allow the tight giant branch: the histograms of CN strengths for 47 Tuc and for the smaller cluster NGC 6752 appear to be bimodal. It seems clear now (by acclamation!) that the heavy element inhomogeneities in ω Cen were established at the time of cluster formation: this is probably true also for the CN because it correlates so well with Ca (Norris, p. 113, this volume). Kraft's results for M92, where large NH and CH variations occur without C–N anticorrelation (Kraft, p. 95, this volume; Carbon et al. 1979), may indicate that the CN spread results from middle-mass supernovae early in the cluster's life.

20.4.3 Carbon depletion

Dickens (1978) and Kraft (p. 97, this volume) have reported increasing C/Fe depletion with later evolutionary stage in cluster giants, beginning near $M_V = -0.7$. Sweigart & Mengel (1979) explain this by meridional circulation, which convects matter from the surface down towards the hydrogen-burning shell where the carbon is burnt. Because the profiles of C and N with radius in the star are different, a C–N anticorrelation need not be produced. To drive the meridional circulation, a main sequence angular velocity of about 10^{-4} s^{-1} is needed. The onset of the circulation is predicted to occur near $M_V = -0.7$, in good agreement with the observed onset of carbon depletion.

20.5 X-rays from globular clusters

Lewin (p. 335, this volume) has discussed the important result that the radius of the emitting object in Type I bursters is about 10 km: the favoured X-ray mechanism is then thermonuclear helium flashing of accreted material on a neutron star. Ostriker (1977, 1978) has proposed that X-ray sources occur in clusters of high escape velocity because these have most chance of containing a few neutron stars. These neutron stars diffuse to the cluster core where they capture main sequence companions. The orbits circularize, the stars spiral together, and accretion then proceeds. On the other hand, Fabian (1978) has argued that neutron stars in clusters have a fair chance of finding enough red giant mass loss ejecta to produce X-rays without the need for binary formation.

Many colleagues gave me advice on the style and content of this summary, and I am very grateful to them.

References

Carbon, D. F., Langer, G. E., Butler, D., Kraft, R. P., Trefzger, Ch. F., Suntzeff, N., Kemper, E. & Nocar, J. (1979). *Astron. J.* (in press).
Cohen, J. G. (1976). *Astrophys. J. Lett.* **203**, L127.
Cohen, J. G. (1978). *Astrophys. J.* **223**, 487.
Cohen, J. G., Frogel, J. A. & Persson, S. E. (1978). *Astrophys. J.* **222**, 165.
Cudworth, K. M. (1976). *Astron. J.* **81**, 975.
Cudworth, K. M. (1978). Results reported at NATO Advanced Study Institute on Globular Clusters, Cambridge.
Da Costa, G. S. (1977). Thesis, Australian National University.
Dickens, R. J. (1978). Results reported at NATO Advanced Study Institute on Globular Clusters, Cambridge.

Fabian, A. C. (1978). Results reported at NATO Advanced Study Institute on Globular Clusters, Cambridge.
Gunn, J. E. & Griffin, R. F. (1979). *Astron. J.* **84**, 752.
Hesser, J. E. & Shawl, S. J. (1977). *Astrophys. J. Lett.* **217**, L143.
Keenan, D. W. (1978). Results reported at NATO Advanced Study Institute on Globular Clusters, Cambridge.
Keenan, D. W. & Innanen, K. A. (1975). *Astron. J.* **80**, 290.
King, I. R. (1966). *Astron. J.* **71**, 64.
Knapp, G. R., Rose, W. K. & Kerr, F. J. (1973). *Astrophys. J.* **186**, 831.
Mallia, E. A. (1978). Results reported at NATO Advanced Study Institute on Globular Clusters, Cambridge.
Ostriker, J. P. (1977). In *Proceedings of the 8th Texas Symposium on Relativistic Astrophysics*, ed. M. Papagiannis, *Ann. N.Y. Acad. Sci.* **302**, 229.
Ostriker, J. P. (1978). Results reported at NATO Advanced Study Institute on Globular Clusters, Cambridge.
Schommer, R. (1978). Results reported at NATO Advanced Study Institute on Globular Clusters, Cambridge.
Searle, L. (1977). In *The Evolution of Galaxies and Stellar Populations*, ed. B. M. Tinsley & R. B. Larson, p. 219. New Haven: Yale University Observatory.
Searle, L. (1978). Results reported at NATO Advanced Study Institute on Globular Clusters, Cambridge.
Searle, L. & Zinn, R. (1978). *Astrophys. J.* **225**, 357.
Spitzer, L. Jr & Shull, J. M. (1975). *Astrophys. J.* **200**, 339.
Sweigart, A. V. & Mengel, J. G. (1979). *Astrophys. J.* **229**, 624.
Trimble, V. (1977). *Mon. Not. Roy. astron. Soc.* **178**, 335.
Trimble, V. (1978). Results reported at NATO Advanced Study Institute on Globular Clusters, Cambridge.
Zinn, R. (1978). *Astrophys. J.* **225**, 790.

GLOSSARY

This list is not intended to be exhaustive. The terms are often given in full, but in chapters where they occur frequently are abbreviated as follows.

AGB – asymptotic giant branch
AHB – above horizontal branch
BC – bolometric correction
BHB – blue horizontal branch
$(B-V)_{0,g}$ – colour of the giant branch at the level of the horizontal branch
CM diagram – colour–magnitude diagram
dE – dwarf elliptical
dex – abbreviation for decimal exponent (e.g. 2 dex = 10^2)
Dsph – dwarf spheroidal
DSS – double shell source
[Fe/H] – the logarithm of the ratio of abundances of the elements shown, in units of the same ratio for the sun
FGB – first giant branch
FWHM – full width at half maximum
GB – giant branch
HB – horizontal branch
HR – Hertzsprung–Russell
LMC – Large Magellanic Cloud
MS – main sequence
NS – neutron star
RG – red giant
RGB – red giant branch
RGT – red giant tip
RHB – red horizontal branch
r.m.s. – root mean square
SGB – subgiant branch
SMC – Small Magellanic Cloud
TO – turnoff point
UHB – upper horizontal branch
ZAHB – zero-age horizontal branch
ZAMS – zero-age main sequence

AUTHOR INDEX

Italics indicate the initial page of an article in this volume

Aaronson, M. 148, 157, 159, 160, 161, 162, 163, 165, 166, 167, 168, 170, 173
Aarseth, S. J. 9, 17, 281, 292, 295, 296, 297, 298
Ables, H. D. 206, 211, 228, 246
Abramenko, A. N. 341, 342, 347
Agekian, T. A. 288, 295, 298
Alcaino, G. 5, 17, 21, 31, 61, 93, 100, 255, 268
Alcock, C. 333, 347
Aller, L. H. 6, 17, 173
Alter, G. 21, 61, 223, 246
Andrews, P. J. 15, 17
Anosova, Zh. P. 285, 288, 295, 298
Apparao, K. M. V. 319, 329, 347, 348
Arai, K. 140, 142
Arnett, W. D. 142, 172, 173
Arons, J. 320, 349
Arp, H. C. 3, 4, 5, 7, 13, 14, 17, 21, 24, 31, 32, 34, 35, 36, 48, 49, 54, 59, 61, 62
Audouze, J. 14, 17
Aurière, M. 253, 269

Baade, W. 3, 4, 13, 17, 22, 23, 61, 194, 208, 211
Baan, W. A. 344, 347
Backman, D. E. 329, 342, 347, 349
Bahcall, J. N. 10, 17, 320, 347, 354, 360
Bahcall, N. A. 262, 268
Bailey, S. 11, 14
Baker, N. H. 35, 63, 82, 86, 201, 212
Balász, B. 21, 61
Baldwin, J. R. 147, 148, 157, 161, 173
Balkowski, C. 246, 247
Barnes, J. V. 32, 61
Basinska, E. 346
Basinska-Grezesik, E. 17, 85, 189
Baum, W. A. 217, 231, 246
Becker, R. H. 349, 350

Becklin, E. E. 187, 189
Bell, R. A. 7, 17, 87, 88, 91, 97, 100, 113, 117, 119, 120, 121, 122, 150, 157, 179, 188, 201, 207, 208, 209, 211
Bell-Burnell, S. J. 325, 349
Berendzen, R. 13, 17
Bernacca, P. L. 341, 342, 347
Bernard, A. 186, 188
Bessell, M. S. 47, 61, 106, 109, 111, 113, 114, 116, 117, 120, 122, 123, 150, 154, 156, 157, 207, 208, 209, 211
Beyer, W. H. 185, 188
Bianchini, A. 347
Bigay, J. H. 186, 188
Blaauw, A. 4, 17, 19, 61, 141
Blanco, B. M. 141
Blanco, V. M. 141
Boesgaard, A. M. 136, 141
Böhm-Vitense, E. 275, 279
Boksenberg, A. 189
Boldt, E. A. 349
Bond, H. E. 126, 141, 156, 157
Bottinelli, L. 245, 246
Bouvier, P. 8, 17, 18, 19, 85, 100, 142, 157, 212
Bowers, P. F. 205, 212
Bowyer, S. 329, 350
Bradt, H. 327, 347, 348, 349
Branduardi, G. 337, 348
Brecher, K. 323, 347
Brinkman, A. C. 348
Brodie, J. P. 6
Bruzual, G. 183, 189
Buff, J. 327, 339, 340, 347, 349
Burr, E. J. 104, 105, 110
Burstein, D. 31, 34, 61, 186, 188
Burton, W. B. 100, 189
Butler, D. 7, 17, 43, 47, 61, 88, 91, 95, 99, 100, 105, 106, 110, 121, 122, 127, 141, 160, 161, 162, 173, 179, 188, 195, 201, 211, 367

Cacciari, C. 108, 198, 199, 201, 211
Caloi, V. 81, 85
Cameron, A. G. W. 130, 142
Canizares, C. R. 262, 268, 327, 329, 337, 342, 346, 347, 348, 349, 352
Cannon, R. D. 90, 100, 104, 110, 150, 157, 192, 193, 211
Canterna, R. 43, 47, 61, 164, 173, 195, 196, 207, 208, 211
Caputo, F. 71, 77, 78, 79, 80, 81, 83, 85, 201, 211
Carbon, D. F. 87, 95, 96, 97, 100, 126, 138, 141, 366, 367
Carpenter, G. F. 325, 350
Caruso, R. 346
Castellani, V. v, vii, 7, 11, 12, 17, 48, 50, 51, 61, 65, 77, 78, 80, 81, 85, 94, 100, 132, 150, 157, 200, 201, 202, 211, 364, 365
Catchpole, R. M. 126, 141
Catura, R. C. 348
Cavaliere, A. 344, 349
Cayrel de Strobel, G. 18, 19
Chaisson, L. J. 350
Chanmugan, G. 339, 347
Charles, P. A. 268, 329, 350
Chartres, M. 348
Chevalier, R. A. 302, 307, 310, 313
Chiosi, C. 71, 72, 74, 76
Christy, R. F. 11, 17, 35, 39, 61, 204, 211
Chun, M. S. 105, 109, 110
Ciardullo, R. B. 147, 157, 168, 171, 173
Ciatti, F. 66, 67, 85
Clark, G. W. 10, 17, 281, 298, 315, 317, 319, 320, 329, 331, 332, 333, 344, 347, 348, 349, 354, 355, 357, 360
Cohen, J. G. v, vii, 6, 9, 17, 87, 88, 91, 92, 94, 98, 99, 100, 109, 110, 140, 143, 144, 145, 146, 147, 148, 149, 153, 157, 159, 161, 168, 173, 196, 201, 211, 225, 364, 365, 366, 367
Cominsky, L. 319, 325, 330, 338, 339, 345, 346, 348, 349
Contreras, C. 31, 61
Copeland, H. 275, 279
Cottrell, P. L. 119, 122
Coutts, C. 199, 211
Cowley, A. P. 186, 196, 207, 208, 211
Crawford, D. L. 32, 61
Cruddace, R. G. 319, 347
Cruz-González, C. 295, 298
Cudworth, K. 9, 17, 96, 100, 258, 268, 362, 367

Da Costa, G. S. 10, 19, 104, 105, 109, 172, 173, 265, 266, 267, 269, 272, 273, 275, 277, 278, 279, 362, 367
Dahn, C. C. 204, 212

Dalgarno, A. 306, 307
D'Antona, F. A. 85
Danziger, I. J. 7, 17, 127, 142, 186, 188, 206, 211
Davidsen, A. 319, 329, 347
Davis Philip, A. G. 5, 17, 20, 21, 28, 36, 42, 43, 50, 51, 54, 61, 100, 110, 111, 123, 188, 211, 212, 269
Davoust, E. 293, 298
Dawe, J. A. 246
de Kort, J. 14, 17
Delplace, A. M. 18, 19
Delvaille, J. 347
Demarque, P. 12, 17, 30, 49, 61, 84, 86, 133, 135, 138, 142, 147, 157, 168, 171, 173, 204, 205, 211, 275
Demers, S. 193, 194, 196, 199, 201, 207, 211, 212
Deupree, R. G. 206, 211
Deutsch, A. J. 44, 61
de Vaucouleurs, A. 231, 246
de Vaucouleurs, G. 58, 61, 182, 188, 206, 211, 214, 216, 221, 222, 223, 224, 231, 232, 233, 235, 241, 245, 246
Dicke, R. H. 176, 177, 189, 305, 307, 309, 313
Dickens, R. J. 7, 17, 39, 43, 47, 50, 51, 61, 87, 88, 100, 105, 110, 113, 117, 121, 122, 127, 141, 150, 154, 157, 179, 188, 199, 200, 201, 207, 209, 211, 246, 367
Dixon, M. E. 227, 228, 246
Dopita, M. A. 206, 211
Doroshkevich, A. G. 177
Doty, J. 329, 330, 334, 347, 348, 349, 350
Dower, R. G. 347
Doxsey, R. E. 315, 317, 319, 347, 348, 349
Duldig, M. L. 341, 350
Duncan, B. J. 268

Edmondson, F. K. 14, 17
Edwards, A. C. 137, 139, 142
Eggen, O. J. 31, 46, 61, 150, 157, 180, 189, 306, 307
Eigenson, A. M. 51, 61
Elitzur, M. 141
Epps, E. 7, 17, 88, 100, 105, 110, 121, 122, 127, 141, 179, 188, 199, 201, 211
Epstein, A. 327, 347
Eriksson, K. 119, 120, 122
Evans, D. S. 20
Evans, T. Lloyd (see Lloyd Evans, T.)
Ewart, G. M. 339, 347
Eyles, C. J. 319, 337, 347

Fabbiano, G. 319, 337, 347, 348
Faber, S. M. 46, 62, 162, 173, 209, 211
Fabian, A. C. 320, 344, 348, 354, 360, 367, 368

Author index

Fall, S. M. vi, vii, 8, 129, 142, 217, 245, 247, 290, 295, 298, 301, 302, 307, *309*, 310, 311, 312, 313, 363
Faulkner, D. A. 9, 17
Faulkner, D. J. 6, 17
Faulkner, J. 68, 85, 92, 100
Faÿ, T. D. 262, 268
Feast, W. M. 32, 62, 120, 122, 126, 141, 142, 143, 144, 146, 150, 153, 157
Fernie, J. D. 13, 17, 18, 19, 20, 38, 62, 63, 86, 212
Fisher, J. R. 245, 247
Flowers, E. 339, 348
Flügge, S. 19
Forman, W. 315, 317, 319, 321, 329, 348, 350
Freeman, K. C. v, vi, vii, 2, 4, 7, 8, 9, 15, 17, 18, 58, 62, 87, 88, 91, 100, *103*, 104, 105, 110, 113, 122, 127, 142, 150, 154, 155, 157, 207, 211, 273, 275, 277, 278, 279, 303, 304, 307, *361*, 366
Friedman, H. 347
Fritz, G. 347
Frogel, J. A. v, vii, 6, 7, 11, 15, 16, 87, 94, 100, 140, *143*, 146, 147, 148, 149, 153, 157, *159*, 160, 161, 162, 163, 166, 167, 168, 173, 196, 211, 225, 233, 245, 364, 367

Gabriel, M. 339, 347
Gallagher, J. 141
Gascoigne, S. C. B. 15, 18, 104, 105, 110, 186, 189, 196, 211
Geisler, J. E. 49, 61
Gershberg, R. E. 347
Giannone, P. 48, 61, 77, 85
Gingold, R. A. 108, 110, 134, 142, 168, 173, 202, 211
Glass, I. S. 143, 144, 146, 153, 157, 349
Goad, J. W. 173, 189
Gott, J. R. 305, 307
Gougenheim, L. 245, 246
Grad, O. 331
Graham, J. A. 11, 18, 204, 211
Grasdalen, G. L. 162, 173
Greenstein, G. 339, 347
Griffin, R. F. 271, 272, 275, 279, 362, 363, 368
Griffiths, R. 317, 330, 331, 348
Grindlay, J. E. 10, 18, 268, 315, 319, 320, 321, 326, 327, 329, 331, 340, 341, 342, 344, 346, 347, 348, 354, 360
Gross, P. G. 65, 72, 74, 76, 80, 81, 86, 204, 205, 211
Gunn, J. E. vi, vii, 8, 16, 258, *271*, 272, 275, 279, *301*, 304, 305, 307, 339, 348, 362, 363, 368
Gurevich, L. E. 282, 298

Gursky, H. 319, 326, 347, 348, 354, 360
Gustafsson, B. 7, 18, 87, 100, 119, 120, 122, 208, 209, 211
Guyer, R. A. 339, 347

Hackwell, J. A. 346
Hanes, D. A. v, vi, vii, ix, xi, *1*, 6, 15, 16, 58, 60, 62, 166, 173, 177, 183, 189, *213*, 214, 216, 223, 224 225, 226, 227, 228, 229, 230, 231, 232 233, 234, 235, 237, 238, 240, 241, 242 244, 245, 246, 247, 268, 302, 307
Harding, G. A. 47, 62, 88, 100
Härm, R. 57, 63, 125, 133, 134, 137, 142
Harris, H. C. 164, 173
Harris, W. E. 2, 4, 5, 13, 14, 15, 16, 18, 21, 28, 29, 30, 31, 34, 35, 36, 37, 44, 58, 59, 60, 61, 62, 94, 100, 154, 157, 170, 173, 176, 181, 182, 185, 189, 195, 196, 201, 211, 212, 223, 228, 231, 234, 236, 241, 244, 245, 246, 247, 260, 268, 351, 360
Hart, M. H. 290, 293
Hart, R. 13, 17
Hartwick, F. D. A. 4, 18, 28, 30, 35, 42, 43, 49, 50, 61, 62, 87, 94, 100, 113, 122, 150, 157, 176, 179, 186, 189, 192, 193, 195, 196, 208, 211, 223, 247
Harvel, C. A. 262, 268
Hatchett, S. 333, 347
Hauck, B. 17, 20
Hawarden, T. G. 192, 206, 211
Hawkins, F. 347, 349
Hawley, S. A. 92, 100
Hayakawa, S. 347
Hayes, D. S. 17, 20, 100, 111, 123, 188, 211, 212
Hayli, A. 18, 20, 269, 295, 298, 307, 314
Haynes, R. F. 341, 350
Hearn, D. R. 349
Heck, A. 216, 247
Hedemann, E. 5, 18, 51, 62, 269
Heggie, D. C. vi, vii, 8, 9, 18, 272, *281*, 282, 283, 284, 285, 286, 287, 288, 290, 292, 293, 294, 295, 298, 354, 361, 363
Heise, J. 348
Hejlesen, P. M. 73, 74
Helmken, H. 347
Hemenway, M. K. 216, 247
Hénon, M. H. 9, 18, 264, 268, 279, 293, 295, 296, 298
Henry, R. C. 6, 20, 29, 45, 63
Hesser, J. E. 9, 13, 18, 35, 62, 87, 88, 89, 100, 107, 110, 113, 120, 122, 186, 189, 195, 211, 365, 368
Hills, J. G. 272, 279, 281, 286, 287, 288, 292, 294, 296, 298, 320, 348, 354, 360
Hiltner, W. A. 268, 349
Hirshfeld, A. W. 30, 61, 204, 205, 211

Hodge, P. W. 16, 18, 177, 178, 187, 189, 192, 193, 194, 196, 197, 203, 204, 206, 207, 211, 212, 222, 224, 247
Hodge, S. M. 5, 18, 51, 62, 269
Hodson, S. W. 206, 211
Hoffman, J. A. 321, 323, 324, 325, 327, 330, 333, 334, 335, 336, 337, 344, 347, 348, 349, 350
Hogg, H. B. *see* Sawyer(-Hogg), H. B.
Holt, S. S. 349
Holzner, R. 346
Hoover, R. B. 268
House, F. C. 264, 269
Hoyle, F. 68, 85
Huchra, J. 80, 85
Hyland, A. R. 119

Iben, I., Jr v, vii, 11, 12, 18, 24, 35, 47, 50, 62, 63, 68, 70, 76, 78, 80, 85, 88, 104, 110, 121, 122, *125*, 127, 132, 134, 136, 140, 142, 172, 194, 202, 212
Ihle, A. xi
Illingworth, G. 8, 18, 58, 62, 187, 188, 189, 266, 267, 269
Illingworth, W. 58, 62, 266, 267, 269, 292, 298
Innanen, K. A. 196, 212, 264, 269, 362, 368
Inoue, M. 346
Itoh, N. 339, 348

Jaschek, C. O. R. 15, 18
Jefferys, W. H. 264, 269
Jensen, J. O. 275, 279
Jernigan, J. G., Jr. vi, vii, 10, 262, 315, 317, 319, 325, 326, 346, 347, 348, 349, *351*, 354, 355, 357, 360
Johnson, H. L. 144, 147, 157, 163, 173, 227, 247
Johnson, H. M. 341, 347, 348
Johnson, S. L. 29, 62
Johnston, M. 347, 348
Jones, A. W. 346
Jones, C. 315, 320, 329, 348, 350
Jørgensen, H. E. 275, 279
Joss, P. C. 330, 333, 337, 339, 344, 345, 346, 348, 349

Kaminisi, K. 140, 142
Katem, B. 90, 101
Katz, J. I. 320, 348
Keenan, D. W. 264, 269, 362, 368
Keenan, P. C. 17, 20
Keisuke, K. 140, 142
Kelley, R. 327, 329, 348
Kemper, E. 95, 100, 367
Kerr, F. J. 9, 17, 18, 205, 212, 365, 368
Kienle, H. 13, 18

Kimura, H. 25, 30, 63
King, I. R. vi, vii, ix, 5, 8, 9, 10, 18, 19, 51, 54, 56, 57, 62, 170, 173, 192, 212, *249*, 250, 255, 257, 258, 260, 261, 262, 263, 266, 267, 268, 269, 273, 274, 278, 279, 289, 290, 292, 298, 299, 362, 368
Kinman, T. D. 13, 18, 32, 42, 44, 61, 62
Kireeva, N. N. 30, 33, 62
Knapp, G. 9, 18, 205, 212, 365, 368
Kogon, C. S. 201, 212
Kormendy, J. 183, 189
Kowal, C. T. 223, 247
Kozai, Y. 299
Kraft, R. P. v, vii, ix, 2, 6, 7, 11, 19, 47, *87*, 90, 98, 100, 104, 113, 126, 138, 141, 150, 183, 189, 194, 196, 209, 361, 366, 367
Kristian, J. 17, 20, 247
Kron, G. E. 3, 19, 32, 44, 62, 213, 218, 219, 222, 241, 247
Kukarkin, B. V. 5, 6, 19, 21, 28, 30, 31, 33, 37, 42, 43, 45, 46, 49, 51, 62, 195, 212
Kulkarni, P. V. 346
Kunkel, W. E. 193, 194, 196, 207, 211, 212
Kurilienė, G. 35, 63
Kwitter, K. B. 329, 349

Lamb, D. Q. 330, 344, 345, 348
Lamb, F. K. 330, 344, 345, 348
Lamb, P. A. 348
Lambert, D. L. 95, 100, 135, 136, 142
Langer, E. 95, 100, 126, 141, 367
Langford, W. R. 31, 62
Laques, P. 253, 269
Larson, R. B. 17, 19, 111, 123, 179, 189, 211, 247, 311, 313, 314, 368
Larsson-Leander, G. 12, 19
Lasker, B. M. 262, 268
Laufer, B. 347
Layzer, D. 310, 313
Lecar, M. 9, 17, 298
Ledoux, P. 18
Lee, S. W. 78, 85
Leroy, J. L. 253, 269
Levin, B. Yu. 282, 298
Lewin, W. H. G. vi, vii, 10, 19, 251, 262, 281, *315*, 319, 321, 322, 324, 325, 326, 327, 328, 329, 330, 331, 333, 334, 337, 338, 339, 340, 341, 344, 345, 346, 347, 348, 349, 350, 352, 361, 363, 367
Li, F. K. 315, 319, 327, 329, 330, 332, 333, 339, 345, 346, 347, 348, 349
Liang, E. P. T. 344, 349
Lightman, A. P. 8, 10, 19, 181, 182, 189, 290, 292, 295, 297, 298, 302, 307
Liller, W. 10, 19, 268, 319, 329, 348

Lloyd Evans, T. 15, 17, 105, 111, 113, 118, 122, 123, 126, 142, 147, 150, 157
Longair, M. S. 247
Lynden-Bell, D. 180, 189, 193, 212, 306, 307
Lynds, C. R. 189

McCarthy, M. F. 141
McClintock, J. E. 268, 327, 329, 341, 346, 347, 348, 349
McClure, R. D. 12, 17, 45, 62, 87, 94, 100, 113, 122, 141, 150, 157, 195, 196, 208, 211
McCray, R. A. 306, 307
McDonald, L. H. 31, 34, 61, 186, 188
McKee, J. 347
McNamara, D. H. 29, 31, 62
McVittie, G. C. 18, 247
Madore, B. F. v, vii, ix, xi, 4, 5, 6, 11, 13, 14, *21*, 59, 62, 268
Maeder, A. 18, 19, 85, 100, 142, 157, 212
Malina, R. 347
Malkan, M. 159, 173
Mallia, E. A. 87, 88, 92, 98, 99, 100, 108, 113, 123, 365, 368
Manduca, A. 91, 100
Maraschi, L. 344, 349
Margon, B. 329, 347, 349
Markert, T. H. 320, 349
Marshall, H. 319, 321, 322, 324, 325, 326, 333, 336, 339, 346, 348, 349, 350
Martin, P. G. 236, 241, 247
Martin, W. Ch. 103, 105, 111
Martins, D. H. 262, 268
Mason, K. O. 325, 347, 349, 350
Mathewson, D. S. 193, 212
Matilsky, T. 347, 350
Matsuoka, M. 350
Matthews, K. 157, 159, 173
Mavridis, L. N. 17, 18
Mayall, N. U. 3, 6, 7, 13, 19, 32, 44, 61, 62, 213, 218, 219, 222, 241, 247
Mayer, W. 347, 349
Mayor, M. 17, 85, 189
Mazzitelli, I. 71, 72, 74
Mengel, J. G. 97, 101, 133, 135, 138, 142, 205, 212, 367, 368
Merrill, P. W. 136, 142
Michie, R. W. 8, 19, 196, 212
Middlehurst, B. M. 173
Mihalas, D. 31, 62
Milgrom, M. 339, 340, 346
Miller, J. S. 92, 100
Miller, R. H. 283, 298
Mironov, A. V. 43, 51, 62, 195, 212
Mitalas, R. 140, 142
Monaghan, J. J. 284, 285, 299
Moretti, M. 73, 85

Morgan, W. W. 3, 7, 14, 19, 32, 42, 44, 62
Morrison, P. 323, 347
Morton, D. C. 187, 189
Mould, J. R. 159, 168, 173
Mouschovias, T. 141
Mufson, S. L. 268
Müller, A. B. 17, 18, 20, 349
Murdin, P. 319, 341, 349, 350
Murray, S. 347

Nagareda, K. 350
Naranan, S. 329
Neff, J S. 156, 157
Neighbours, J. E. 337, 347
Neugebauer, G. 187, 189
Newell, E. B. 10, 19, 228, 246, 254, 255, 262, 264, 265, 266 269
Nishida, M. 68, 85
Nocar, J. 100, 367
Noels, A. 18
Nordlund, Å. 119, 120, 122
Norris, J. v, vii, 2, 10, 11, 19, 47, 61, 62, 87, 88, 89, 97, 100 106, 107, 109, 111, *113*, 116, 117, 119, 120, 122, 123, 127, 150, 154, 155, 156, 157, 202, 204, 205, 207, 208, 209, 211, 212, 265, 266, 269, 361
Nugent, J. J. 327, 349

O'Connell, R. W. 168, 173, 189
Oda, M. 327, 347
Odenwald, S. F. 181, 182, 189
Oke, J. B. 6, 19, 187, 189, 228, 233, 247
Olszewski, E. W. 194, 212
O'Neil, E. J. 228, 245, 254, 255, 256, 262, 264, 265, 269
Oort, J. H. 14, 19, 53, 62, 103, 111, 180, 181, 182, 189, 257, 269
Oosterhoff, P. Th. 38, 39, 62
Osborn, W. 144, 157
Ossmar, W. 319, 338, 349
Ostriker, J. P. 10, 17, 19, 187, 189, 302, 304, 307, 310, 313, 314, 320, 329, 339, 347, 348, 349, 354, 360, 367, 368

Paczyński, B. 137, 138, 142
Pagel, B. 7, 19, 108
Papagiannis, M. 19, 348, 349, 350, 368
Parker, P. D. 140, 142
Parkes, G. E. 329, 349
Parsignault, D. R. 348
Pavlenko, E. P. 347
Payne-Gaposchkin, C. 4, 17, 59, 62
Pedersen, A. 346
Peebles, P. J. E. 176, 177, 189, 245, 247, 305, 307, 309, 313
Peimbert, M. 6, 9, 19, 20
Penston, M. J. 349

Penston, M. V. 349
Persson, S. E. v, vii, 6, 7, 11, 87, 94, 100, 140, *143*, 147, 148, 157, *159*, 160, 161, 162, 163, 166, 167, 168, 173, 196, 211, 225, 364, 367
Peterson, C. J. 5, 19, 54, 56, 57, 62, 170, 173, 192, 212, 250, 260, 262, 269, 289, 292, 299
Peterson, R. C. 92, 99, 100
Petrie, P. L. 182, 188, 231, 234, 244, 247
Philip, A. G. D. *see* Davis Philip, A. G.
Picklum, R. E. 344
Pilachowski, C. 91, 143, 147, 149, 157, 161, 173, 196, 212
Plaut, L. 7, 11, 19, 181, 182, 189
Poveda, A. 295, 298
Pravdo, S. H. 349, 350
Press, W. H. 181, 182, 189, 291, 299
Preston, G. W. 11, 19, 47, 62, 88, 99, 100
Primini, F. 347, 348, 349
Pringle, J. E. 320, 344, 348, 354, 360
Pritchet, C. 162, 173
Proctor, R. J. 319
Prokof'eva, V. V. 347

Racine, R. 2, 4, 5, 6, 13, 14, 15, 16, 18, 19, 21, 28, 29, 30, 31, 32, 33, 34, 44, 46, 58, 62, 184, 187, 189, 196, 212, 213, 218, 219, 220, 223, 224, 225, 226, 227, 228, 233, 234, 240, 241, 245, 246, 247
Rappaport, S. 319, 327, 339, 340, 344, 346, 348, 349
Read, M. 199, 211
Rees, M. J. ix, 129, 142, 217, 245, 247, 298, 302, 305, 307, 310, 311, 312, 313, 314, 320, 348, 354, 360
Reiz, A. 17
Renzini, A. 7, 9, 12, 19, 48, 61, 68, 76, 77, 85, 90, 94, 100, 132, 133, 135, 142, 150, 157, 194, 198, 199, 201, 205, 211, 212
Retterer, J. M. 290, 298, 299
Rice, W. 173, 189
Richardson, J. 349
Richstone, D. 188, 189
Rodgers, A. W. 7, 17, 18, 87, 88, 91, 100, 104, 105, 107, 110, 113, 122, 127, 142, 150, 154, 157, 207, 211
Rood, R. T. 24, 63, 65, 80, 85, 97, 101, 140, 142, 146, 147, 157, 202, 212
Rose, W. K. 365, 368
Rosenberg, F. D. 347
Rosino, L. 11, 19, 37, 38, 39, 63, 66, 67, 85
Rossi, B. 346
Różyczka, M. 138, 142
Ruderman, M. 339, 348
Ruprecht, J. 21, 61, 223, 246

Saar, E. M. 177
Saba, J. R. 350
Sadun, A. 323, 347
Sandage, A. R. 4, 17, 20, 30, 31, 35, 36, 48, 49, 63, 90, 101, 180, 187, 189, 194, 212, 213, 214, 216, 218, 219, 221, 245, 247, 255, 256, 257, 258, 269, 275, 279, 304, 306, 307
Sandage, M. 17, 20, 247
Sanders, R. H. 137, 142
Sanford, P. W. 268, 347, 348, 349
Sargent, W. L. W. 14, 18, 28, 61, 62, 176, 188, 189, 192, 193, 196, 211, 223, 234, 241, 247
Savage, A. 216, 247
Sawyer (-Hogg), H. B. 1, 3, 5, 9, 11, 19, 21, 29, 35, 37, 63, 258
Scalo, J. M. 127, 142
Schmidt, M. 17, 19, 61, 141, 304, 307
Schmidt-Kaler, Th. 35, 63
Schnopper, H. 347, 348
Schommer, R. A. 47, 61, 94, 101, 194, 195, 196, 207, 208, 211, 212, 364, 368
Schramm, D. N. 142
Schrijver, J. 348
Schwartz, D. A. 347, 348
Schwartz, J. 347, 348
Schwarz, M. P. 193, 212
Schwarzschild, M. 68, 85, 126, 133, 134, 137, 142
Schweizer, F. 164, 173
Searle, L. 6, 11, 14, 19, 88, 101, 106, 109, 111, 115, 116, 123, 146, 157, 160, 161, 162, 163, 164, 166, 169, 173, 179, 187, 189, 195, 197, 203, 205, 212, 228, 233, 247, 304, 307, 313, 361, 365, 366, 368
Seeley, D. 13, 17
Serlemitsos, P. J. 349, 350
Shakeshaft, J. R. 20, 189, 246
Shandarin, S. F. 177
Shapiro, S. L. 8, 10, 19, 292, 297, 298, 302, 307
Shapley, H. xi, 3, 4, 11, 13, 19, 36, 54, 57, 63, 258, 269
Shara, M. 234, 240, 241, 246, 247
Sharov, A. S. 58, 63
Shawl, S. J. 9, 18, 236, 241, 247, 365, 368
Shortridge, K. 189
Shull, J. M. 363, 368
Shulman, S. 347
Silk, J. 320, 349
Simoda, M. 25, 30, 50, 63, 76, 85
Skinner, G. K. 319, 325, 347, 349, 350
Smith, H. 347
Smith, K. 135
Smith, L. L. 48, 63
Smith, M. G. 231, 234, 244, 246, 247
Sneden, C. 95, 99, 100, 101

Spencer Young, P. 42, 49, 63
Spinrad, H. 6, 7, 14, 20, 164, 173, 347
Spite, F. 98, 99, 101
Spite, M. 98, 99, 101
Spitzer, L. 9, 20, 57, 63, 187, 189, 279, 288, 290, 294, 297, 299, 302, 307, 310, 311, 313, 314, 363, 368
Sprott, G. F. 349
Spulak, R. 141, 142
Stellingwerf, R. F. 81, 82, 85, 201, 212
Stetson, P. 30, 63, 194
Stobie, R. S. 35, 63, 79, 85, 90, 100, 104, 110, 150, 157
Straižys, V. 35, 63
Strand, K. Aa. 63, 247
Strittmatter, P. A. 175, 189
Strom, K. M. 24, 63, 173, 183, 189, 202, 212
Strom, S. E. 24, 63, 159, 162, 173, 183, 189, 202, 212
Strömgren, B. 46, 63
Struck-Marcell, C. 164, 173
Struve, O. 4, 20
Sturch, C. 31, 63
Sūdižius, J. 35, 63
Sugimoto, D. 137, 139, 142
Suntzeff, N. 90, 95, 100, 189, 367
Sussi, M. G. 66, 67, 85
Swank, J. H. 315, 319, 329, 334, 349, 350
Sweigart, A. V. 12, 20, 65, 72, 74, 76, 80, 81, 84, 86, 97, 101, 133, 134, 135, 138, 142, 205, 212, 367, 368
Swope, H. H. 194, 207, 208, 211
Szkody, P. 275, 279

Taam, R. E. 344, 350
Takagishi, K. 341, 342, 350
Talbot, R. J. 172, 173
Tammann, G. A. 213, 214, 245, 247
Tananbaum, H. 319, 350
Tanikawa, K. 30, 63
Terzian, Y. 19
Teukolsky, S. A. 291, 299
Thackeray, A. D. 3, 15, 20, 32, 62, 120, 122
Thomas, H.-C. 137, 138, 142
Thomas, R. M. 341, 350
Thorstensen, J. 329, 350
Thuan, T. X. 187, 189
Tifft, W. G. 178, 189, 196, 204, 212
Tinsley, B. M. 4, 14, 17, 19, 20, 111, 123, 163, 164, 169, 173, 179, 189, 211, 247, 311, 313, 314, 368
Tomkin, J. 135, 136, 142
Toomre, A. 177, 189
Tornambè, A. 77, 78, 81, 85, 94, 100, 150, 157, 201, 211
Trefzger, Ch. F. 90, 95, 100, 183, 189, 367

Tremaine, S. D. 9, 20, 137, 138, 142, 187, 189, 310, 313, 314
Trimble, V. 363, 363
Tritton, S. B. 192, 211
Truran, J. W. 122, 127, 130, 136, 141, 142
Tsesevich, V. 11, 20
Tully, R. B. 245, 247

Ugarte, P. 186, 189
Ulmer, M. P. 325, 341, 349, 350
Ulrich, R. K. 137, 142

Valtonen, M. J. 196, 212, 287, 288, 299
van Agt, S. L. T. J. 11, 16, 20, 39, 63, 177, 189, 194, 197, 198, 212
van Albada, T. S. 35, 63, 82, 86, 201, 212, 295, 299, 302, 307
Vander Berg, D. A. 9, 20
van den Bergh, S. vi, vii, 4, 6, 9, 13, 15, 16, 20, 28, 29, 30, 31, 32, 34, 36, 45, 46, 49, 62, 63, 128, 159, 164, 165, 166, 170, 173, 175, 176, 178, 183, 184, 185, 187, 188, 189, 193, 194, 206, 212, 214, 221, 223, 232, 233, 234, 236, 241, 245, 247, 305, 361
van der Heuvel, E. F. J. 320, 339, 350
van Herk, G. 103, 111, 216, 247, 257, 269
van Paradijs, J. 319, 329, 330, 334, 335, 345, 346, 347, 348, 350
Vanysek, V. 223, 246
Veeder, G. J. 275, 279
Vetešník, M. 169, 173, 223, 247
von Hoerner, S. 13, 20, 295, 299
von Weizsäcker, C. F. 180, 189
von Zeipel, H. 5
Vrba, F. J. 173, 189

Wakamatsu, K. I. 233, 245, 247
Walker, A. 347
Walker, N. 342
Wallerstein, G. 48, 63, 91
Wamsteker, W. 262, 268
Warner, B. 127, 142
Webbink, R. 141
Webster, B. L. 206, 211
Weedman, D. W. 246, 247
Wehlau, A. 199, 201, 212
Weistrop, D. 225, 247
Wesselink, A. J. 32, 62
Westerlund, B. E. 18, 19, 246, 247
Wheaton, W. 347, 343, 349
Wheeler, J. C. 344, 350
White, N. E. 268, 325, 326, 348, 349, 350
White, R. E. 5, 18, 20, 21, 51, 62, 63, 251, 269
White, S. D. M. 311, 314
Whitford, A. E. 45, 63
Whitney, C. A. 4, 13, 20

Wielen, R. 9, 20
Wildey, R. 49, 63, 194, 212
Williams, R. E. 175, 189
Willmore, A. P. 319, 347, 349
Wilson, C. 256, 258
Winkler, P. F. 349
Wisniewski, W. 268
Wolf, R. 347
Woltjer, L. 5, 14, 20, 21, 30, 44, 58, 61, 63, 216, 247
Wood, P. R. 77, 85, 134, 142
Woolley, R. v. d. R. 150, 154, 157, 216, 247
Woosley, S. E. 344, 350

Wright, F. W. 197, 203, 204, 212
Wright, L. 346
Wyatt, S. 141

Yoshinaga, K. 140, 142
Young, P. J. 189

Zebergs, V. 4, 20
Zinn, R. vi, vii, 7, 11, 14, 16, 19, 20, 39, 47, 62, 63, 87, 88, 95, 97, 100, 101, 104, 109, 111, 120, 123, 146, 157, 160, 161, 162, 173, 177, 178, *191*, 195, 197, 202, 203, 204, 205, 208, 209, 212, 303, 304, 307, 364, 366, 368

Object index

Parentheses around page numbers indicate object mentioned in figure only.

GLOBULAR CLUSTERS
Our own Galaxy
ω Gen (NGC 5139) 4, 7, 26, (38), 40, 52, 88, 89, 90, 91, 103, 104, 105, 106, 107, 108, 109, 113–22, 127, 132, 143, 144, 145, 150, (151), 152, 153, 154, 155, 156, 157, 179, 191, 192, 201, 207, 208, 209, 259, 362, 366
47 Tuc (NGC 104) 26, (37), 40, (50), 52, 58, 78, 88, 89, 94, 104, 105, 107, 108, 109, 110, 120, 143, 145, 147, 186, 251, (266), 267, 346, 362, 366
M2 (NGC 7089) 28, 42, (50), 54, (200), 201
M3 (NGC 5272) (25), 26, 30, 31, (37), (38), 40, 52, 77, 84, 88, 91, 92, (93), 94, (98), 99, 143, 146, (147), (148), 149, 150, 151, 153, 154, (155), 156, (200), 201, 208, 253, 255, 257, 258, 271, 272, 275, 276, 304, 362, 364
M4 (NGC 6121) 26, 40, 52, 78, 84
M5 (NGC 5904) 26, 30, 38, 40, (50), 52, (69), 84, 120, 143, 149, 150, 209, 365
M9 (NGC 6333) 26, 40, 52
M10 (NGC 6254) 26, 40, 52, 143, 149, 150, 365
M12 (NGC 6218) 26, 40, 52
M13 (NGC 6205) 4, (25), 26, 30, 31, 40, 47, (50), 52, 88, 91, 92, (93), 94, (98), 99, 143, 146, (147), (148), 149, 150, (151), (153), 154, (155), 156, 209, 214, (215), 249, (250), 259, 364
M14 (NGC 6402) 28, 42, 54, 84, (200), 201
M15 (NGC 7078) 9, 10, 28, 31, 42, 47, 54, 90, 91, 188, 207, 253, (254), 255, 256, 262, 263, (264), 265, (266), 267, 315, 351, 353, 354, (355), 356
M19 (NGC 6273) 26, 40, 52
M22 (NGC 6656) xi, 28, 42, (50), 54, 92, 108, 143

M28 (NGC 6626) 28, 42, 54
M30 (NGC 7099) 28, 38, 42, 54
M53 (NGC 5024) 25, 40, 52, 259, (260)
M54 (NGC 6715) 28, 42, 54, 58
M55 (NGC 6809) 28, 42, 54, 364
M56 (NGC 6779) 28, 42, 54
M62 (NGC 6266) 26, 40, 52, 58, (331)
M68 (NGC 4590) 26, 40, 52, (200), 201
M69 (NGC 6637) 28, 42, 54, 168
M70 (NGC 6681) 28, 42, 54
M71 (NGC 6838) 28, 42, 54, 143, 146, (147), 149, (151), 152, (153), (154), 155, 156, 168
M72 (NGC 6981) 28, (38), 42, 54, 201
M75 (NGC 6864) 28, 42, 54, 58
M79 (NGC 1904) 26, 40, 52
M80 (NGC 6093) 26, 40, 52, 58
M92 (NGC 6341) (25), 26, 30, 31, 40, (45), 47, (50), 52, 66, 90, 91, 95, 96, 97, 98, 99, 106, 126, 138, 143, (146), (147), (148), 149, (151), 152, 153, 154, 155, 156, 188, 207, 208, 209, 210, 366
M107 (NGC 6171) 26, (38), 40, (50), 52, 143
NGC 104 *see* 47 Tuc
NGC 288 26, 40, 52
NGC 362 26, 40, (50), 52, 58, 251
NGC 1261 26, 40, 52
NGC 1851 26, 40, 52, 58, 262, 315, 316, (328), 329, 332, 333, 351, 353, 354, (355)
NGC 1904 *see* M79
NGC 2298 26, 40, 52
NGC 2419 26, 40, 52, (195), 196
NGC 2808 26, 40, 52, 58, 109
NGC 3201 26, 40, 52, 78, 84
NGC 4147 26, 40, 52
NGC 4372 26, 40, 52
NGC 4590 *see* M68
NGC 4833 26, 40, 52
NGC 5024 *see* M53

379

NGC 5053 26, 40, 52, 259, 260, (261)
NGC 5139 see ω Cen
NGC 5272 see M3
NGC 5286 26, 40, 52
NGC 5466 26, 40, 52, (203), 204, 205, 363
NGC 5634 26, 40, 52
NGC 5694 26, 40, 52
NGC 5824 26, 40, 52, 262
NGC 5897 26, 40, 52
NGC 5904 see M5
NGC 5927 26, 40, 52
NGC 5946 26, 40, 52
NGC 5986 26, 40, 52
NGC 6093 see M80
NGC 6101 26, 40, 52
NGC 6121 see M4
NGC 6139 28, 42, 40, 52
NGC 6144 26, 40, 52
NGC 6171 see M107
NGC 6205 see M13
NGC 6218 see M12
NGC 6229 26, 40, 52, (195)
NGC 6235 26, 40, 52
NGC 6254 see M10
NGC 6256 see TRZ 12
NGC 6266 see M62
NGC 6273 see M19
NGC 6284 26, 40, 52
NGC 6287 26, 40, 52
NGC 6293 26, 40, 52
NGC 6304 26, 40, 52
NGC 6316 26, 40, 52
NGC 6325 26, 40, 52
NGC 6333 see M9
NGC 6341 see M92
NGC 6342 26, 40, 52
NGC 6352 26, 40, 52
NGC 6355 26, 40, 52
NGC 6356 26, 40, (45), (50), 52
NGC 6362 26, 40, 52
NGC 6366 26, 40, 52
NGC 6380 see Ton 1
NGC 6388 28, 42, 54, 58, 179
NGC 6397 28, 42, 47, 54, 92
NGC 6401 28, 42, 54
NGC 6402 see M14
NGC 6426 28, 42, 54
NGC 6440 28, 42, 54, 262, 315, 320, 351, 353, (357)
NGC 6441 28, 42, 54, 58, 262, 315, 318, 329, 351, 353, 354, (355), (357)
NGC 6453 28, 42, 54
NGC 6496 28, 42, 54
NGC 6517 28, 42, 54
NGC 6522 13, 28, 42, 54
NGC 6528 28, 42, 54
NGC 6535 28, 42, 54
NGC 6539 28, 42, 54

NGC 6541 28, 42, 54
NGC 6544 28, 42, 54
NGC 6553 28, 42, 54
NGC 6558 28, 42, 54
NGC 6569 28, 42, 54
NGC 6584 28, 42, 54
NGC 6624 28, 42, 54, 262, 315, 316, (328), 329, 331, (332), 337, 351, 353, 354, (355), (357)
NGC 6626 see M28
NGC 6637 see M69
NGC 6638 28, 42, 54
NGC 6642 28, 42, 54
NGC 6652 28, 42, 54
NGC 6656 see M22
NGC 6681 see M70
NGC 6712 28, 42, (50), 54, 262, 315, 318, 329, 351, 352, 353, 354, (355), 359
NGC 6715 see M54
NGC 6717 see Pal 9
NGC 6723 28, 42, 54
NGC 6749 28, 42, 54
NGC 6752 28, 42, 54, 92, 366
NGC 6760 28, 42, 54
NGC 6779 see M56
NGC 6809 see M55
NGC 6838 see M71
NGC 6864 see M75
NGC 6934 28, 42, 54, 84
NGC 6981 see M72
NGC 7006 28, (38), 42, 54, 94, 194, (195)
NGC 7078 see M15
NGC 7089 see M2
NGC 7099 see M30
NGC 7492 28, 42, 54
IC 1276 see Pal 7
IC 4499 26, 40, 52
Pal 1 26, 40, 52
Pal 2 26, 40, 52, 262
Pal 3 26, 40, 52, (195), 196
Pal 4 26, 40, 52, (195), 196
Pal 5 26, 40, 52
Pal 6 28, 42, 54
Pal 7 (IC 1276) 28, 42, 54
Pal 8 28, 42, 54
Pal 9 (NGC 6717) 28, 42, 54
Pal 10 28, 42, 54
Pal 11 28, 42, 54
Pal 12 28, 42, 54, 94, 196
Pal 13 28, 42, 54, 66, (67)
Pal 14 (1608+15) 26, 40, 52
Pal 15 26, 40, 52
TRZ 1 26, 40, 52
TRZ 2 26, 40, 52, 315, 318, 320, 321, 329, 346
TRZ 4 26, 40, 52
TRZ 5 28, 42, 54
TRZ 6 28, 42, 54

Object index

TRZ 7 28, 42, 54
TRZ 9 28, 42, 54
TRZ 11 28, 42, 54
TRZ 12 (NGC 6256) 26, 40, 52
Ton 1 (NGC 6380) 26, 40, 52
Ton 2 28, 42, 54
Arp 2 (1925−30) 28, 42, 54
AM-01 26, 40, 52
Eri (0423−21) 26, 40, 52
HP 1 26, 40, 52
Liller 1 (1730−33) 26, 40, 52, 262, 315, 317, 322, 329, 351, (357)
0423−21 see Eri
1608+15 see Pal 14
1730−33 see Liller 1
1925−30 see Arp 2

Large Magellanic Cloud
NGC 1466 196
NGC 1841 196
NGC 1978 178
NGC 2257 196
Reticulum system 196

Small Magellanic Cloud
Kron 3 315, 321
NGC 121 178, 196

NGC 205
NGC 205: III 187

M31
M11 221

OPEN CLUSTERS
Hyades 272
M47 (NGC 2422) (50)
M67 (NGC 2682) (146), (147), (148), 149

INDIVIDUAL STARS
HD 122563 95
HD 126778 272
G55 (155)
G78 (155)
RGO 40 88
RGO 65 88
ROA 40 (151), 152, (154)
ROA 43 (151), (152)
ROA 55 (151), 152, 154
ROA 70 (151), 152, 154, (155)
ROA 84 (151), (152), (154), (155)
ROA 139 118, (151), (152), (154), (155)
ROA 150 (152), (154), (155)
ROA 159 118
ROA 162 118, (119), (154), (155), 156
ROA 171 (151), (152), (154), (155)
ROA 213 118, (119)
ROA 219 (151), (152), (154), (155)
ROA 253 116, 117, (154), (155)
ROA 270 (151), (152), (154), (155)
ROA 279 (151), (152), (154), (155)
ROA 287 (151), (152), (154), (155)
ROA 371 (151), (152), (154), (155), 156
ROA 394 (151), (152), (154), (155)
ROA 480 (154), (155)
ROA 513 118, (119)
ROA 537 (151), (152), (154), (155)
ROA 577 (151), (152), (154), (155)
von Zeipel 318 272
von Zeipel 764 273
von Zeipel 803 272
von Zeipel 911 273

X-RAY SOURCES
Cygnus X-1 327, 328
LMC X-4 327, 328
Rapid Burster see MXB 1730−335
Sco X-1 (341)
Ser X-1 see MXB 1837+05
A 1742−289 318, 337
A 1742−294 318
A 1850−08 318, 351, 353
A 1905+00 316
2A 0512−399 316
GCX 328
GCX-1 318
GX 3+1 359
GX 5−1 359
GX 9+1 359
GX 9+9 359
GX 13+1 359
GX 17+2 359
GX 301−2 327, 328
GX 304−1 327, 328
GX 340+0 359
GX 349+2 359
H 1658−298 317
H 1730−333 317
MX 0513−40 351, 353
MX 1608−52 318, 329
MX 1746−20 351, 353
MXB 0512−40 316, (328), 329
MXB 1636−53 316, (328), 329, (333), 334, 339, 346
MXB 1659−29 317, (328), 329, 330, (331), 337, 338, 339, 345
MXB 1706−43 318
MXB 1728−34 316, (321), (328), 330, (333), 334, 339
MXB 1730−335 (Rapid Burster) 317, (321), 322, 323, (324), 325, 326, 327, (328), 329, 333, (335), 336, 337, 344, 346, 351, 352
MXB 1735−44 316, (328), 329, 330, (333), 339, (340), 341, 342, 343, 345, 346
MXB 1742−29 318

MXB 1743−29 318, (333), 337, 345
MXB 1746−37 318, 329
MXB 1820−30 316, (328), 329, 331, 339, 344
MXB 1837+05 (Ser X-1) 316, (328), 329, 330, 339, 341, 342, 345, 346
MXB 1850−08 318, 329
MXB 1906+00 316, (328), (333)
MXB 1916−05 316, (328), 339, 341, 342
2S 0512−400 316, 354
2S 1608−523 318
2S 1636−536 316
2S 1702−429 318
2S 1705−440 318
2S 1728−337 316
2S 1735−444 316
2S 1742−294 318
2S 1746−370 318, 354
2S 1820−303 316, 354
2S 1837+049 316
2S 1850−087 318, 354
2S 1905+000 316
2S 1916−053 316
2S 2127−119 354
3U 1746−37 351, 353
3U 1820−30 351, 352, 353
3U 2131+11 351, 353
4U 0115+63 340
4U 0513−40 316
4U 0614+09 318
4U 1608−52 318, 329
4U 1624−49 340
4U 1630−47 359
4U 1636−53 316, 359
4U 1702−42 318, 340
4U 1704−30 317
4U 1705−32 340
4U 1705−44 318
4U 1708−23 340
4U 1722−30 318
4U 1728−33 316
4U 1730−22 359
4U 1735−44 316, (341), 359
4U 1746−37 318, 340
4U 1755−33 340, 359
4U 1820−30 316, 337, 359
4U 1822−00 340
4U 1837+04 316
4U 1850−08 318
4U 1857+01 316
4U 1915−05 316
XB 06??+?? 318
XB 1608−52 318
XB 1724−30 318, 329, (334)

LOCAL GROUP GALAXIES
M31 (NGC 224) 3, 6, 46, 107, 159–172, 176, 178, 183, 184, 185, 186, 187, 188, 193, 218, 219, 221, 223, 224, 225, 233, 234, 236, 240, 241, 242, 243, 365, 366
M32 (NGC 221) 186, 187, 188, 191
M33 (NGC 598) 187, 188
Large Magellanic Cloud 3, 4, 6, 8, 11, 15, 126, 128, 141, 167, 177, 178, 179, 186, 191, 192, 193, 195, 196, 204, 205, 206, 219, 258, 301, 303, 307, 363
Small Magellanic Cloud 3, 4, 6, 8, 11, 15, 126, 128, 141, 177, 178, 179, 186, 191, 192, 193, 195, 196, (203), 204, 205, 219, 251, 258, 301, 303, 307
Carina 191, 192
Draco 30, 39, 191, 192, 194, 195, 196, 197, 198, 199, 200, 201, 202, 205, 208, 209, 210
Fornax 177, 186, 191, 192, 194, 206, 209, 224
Leo I 191, 197, 204, 207
Leo II 191, 192, 194, 195, 196, 197, 198, 199, 200, 201, 202, 207
Sculptor 177, 191, 194, (195), 196, 197, 198, 199, (200), 202, 207, 209, 224
Ursa Minor 39, 191, 192, 194, (195), 196, 197, 198, 199, (200), 201, 202, 203, 207, 208, 209
NGC 147 191, 192, 224
NGC 185 191, 224
NGC 205 187, 188, 191, 224
NGC 221 see M32
NGC 224 see M31
NGC 598 see M33
NGC 6822 192

OTHER GALAXIES
M49 (NGC 4472) 182, 216, (229), 230, (232), 233, (241), 244
M87 (NGC 4486) 6, 182, 187, 214, (215), 217, 218, 219, (220), 221, 223, (224), 225, (226), (227), 228, (229), 230, 232, 233, 234, 235, (241), 244, 246
NGC 3115 183
NGC 4216 (229), 230, 231
NGC 4374 (232), 233, (241)
NGC 4406 (232), 233, (241)
NGC 4472 see M49
NGC 4486 see M87
NGC 4486B (167)
NGC 4526 (241)
NGC 4569 (229), 230, 232
NGC 4596 (241)
NGC 4649 (229), 230, (232), 233, 235, (241)
NGC 4762 183

Object index

CLUSTERS OF GALAXIES
Local Group 193, 213, 217, 218, 219, 221, 222, 224, 225, 228, 233, 234, 242, 243, 246
Virgo cluster 162, 213, 214, 215, 216, 217, 218, 219, 221, 223, 224, 225, 228, 229, 232, 233, 234, 235, (236), 237, 238, 239, 240, 241, 242, 243, 244, 245, 246
Coma cluster 162

OTHER OBJECTS
NGC 1976 (M42) Orion nebula 206
Magellanic Stream 93

SUBJECT INDEX

a, b, c type pulsators *see* RR Lyrae stars
above horizontal branch (AHB) stars 24
absolute magnitudes of globular clusters
   in the Galaxy 26, 28, 170–2, 178, 238
   in M31 170–2, 178
   correlated with colour 170–2, 178–9, 225
abundances (*see also* abundance anomalies; abundance determination methods)
   in globular clusters 2, 6, 27, 29, 87–100, 103–10, 113–22, 125–41, 143–57, 159–72, 183–7, 233–4, 365–7
   data tabulations 6, 27, 29, 41, 43
   dependence upon galactocentric position 7, 14, 60, 128–31, 176, 178–9, 182–3, 365–6
   effects on CMD morphology 22–5, 48–51, 70, 92–4, 140, 146–50, 153–5, 364–5
   second-parameter effect *see separate entry*
   change during stellar evolution 125–7, 132–4
   in extragalactic globular clusters 159–60, 163–72, 183–7, 206, 228, 233–4, 365–6
   frequency distribution of metallicities 165–6, 183–7, 233–4
   in Dsph galaxies 206–10
   in galaxies 2, 162–3, 165–9, 187–8, 194–7, 206–10
abundance anomalies (*see also* abundances; abundance determination methods; peculiar stars)
   in globular clusters (*see also* Omega Cen) 2, 6, 7, 47, 87–100, 103–10, 113–22, 125–41, 149–57, 196–7, 303, 361, 366–7
   abundance gradients in clusters 88–9, 103–10, 366
   globular clusters vs. halo stars 2, 95–9
   in dwarf galaxies 194–7, 206–10, 303, 364
   in CNO elements *see* CNO
   in α-process elements *see* α-process
   in s-process elements *see* s-process
   in iron-peak elements 91, 127
   second-parameter effect *see separate entry*
   Zinn effect 7, 47, 95–8, 366
   proposed explanations: close binary stars 134–7; cluster self-polllution 88, 109–10, 128, 303, 366; external pollution 88, 127–32; helium flash effects 137–41; mixing 47, 87–8, 95–8, 113–22, 133, 156, 209, 367; primordial differences 88–9, 98, 104, 113–22, 127–8, 156, 209, 366
abundance determination methods
   reviews 2, 6, 7, 10
   CMD morphology *see* colour–magnitude diagram
   integrated photometry 32–3, 45–6, 184–5, 233–4
   integrated spectra (line strengths) 44, 46, 165–6, 183–7, 234
   integrated spectrophotometry 45, 163–4
   integrated infrared photometry 161–3, 165–9
   stellar photometry 46–7 (and *see* colour–magnitude diagram)
   spectra of single stars 47, 87–100, 113–22
   stellar infrared photometry 143–57
   CO molecular band strengths 94, 147–57, 160–3, 167–9, 364–5
   pulsation theory 79–85
   model stellar atmospheres 96, 118–21
accretion *see* X-ray burst sources; Binary stars

Subject index

ages of clusters
  effect on CMD morphology 12, 25, 78, 94
  measurements, ranges 12, 78
  effect on integrated colours 164
  as second parameter see second-parameter effect
α-process elements (see also abundance anomalies)
  abundances in clusters 91, 98–9, 113–22
  abundances in halo stars 98–9
  Ca inhomogeneities 88, 91, 104–6, 113–22, 127, 366
anisotropy of cluster velocity distributions 257–8, 263–4, 268, 273–9, 362
anomalous cepheids
  in globular clusters 204
  in Dsph galaxies 203–6, 210
  in the SMC 203–4
  models 204–5, 363
asymptotic giant branch (AGB) stars 24, 68, 77, 81, 133–4

$(B-V)_{TO}$ metallicity indicator 41, 43, 50
$(B-V)_{0,g}$ metallicity indicator 41, 43, 48
binary stars
  in cluster dynamics: review 281–99, 363–4; formation 281, 283, 286–91, 293, 295, 297; disruption 283, 285–90, 293; hard binaries 282, 284–94, 296–7, 364; soft binaries 282–3, 286, 288–90, 293, 295, 364; encounters 205, 282–9, 292, 294, 297; core collapse 293–4, 296–7; Aarseth binary 296; initial binaries 272, 293, 297
  as X-ray sources see X-ray burst sources
  as anomalous cepheids through mass transfer 205, 363
  as blue stragglers 363
  mass transfer and abundance anomalies 128, 134–6, 363
  observations 38, 272, 363–4
black holes (see also X-ray burst sources)
  luminosity distributions in clusters 10, 262–8
blanketing corrections 31
blue stragglers 363
bolometric corrections for red giants 144–5

C (carbon) stars see peculiar stars
C-parameter (metallicity indicator) 40, 42, 49, 50
calcium see α-process
cepheids (see also W Vir; anomalous cepheids) 37–8
CH stars see peculiar stars
chemical composition see abundances

classical aging effect 92, 98–9
CNO
  and CMD morphology 73, 75–8, 90, 125, 132, 364
  burning, in post-MS stars 75
  abundance ratios 91–2, 95–9, 125, 367
  CN abundance variations 82–92, 95–8, 103, 106–7, 113–22, 126–7, 149–53, 366
  CO feature strength variations (see also CO) 149–57
  mixing 97–8, 113–22, 367
  primordial enhancements 121–2, 141
  atmospheric cooling effects of CO 120
  as second parameter see second-parameter effect
colour–magnitude diagram (CMD)
  importance 21–22, 31, 68
  components (see also separate entries) 23–30
  morphology 4, 23–30, 48–51, 69
  dependence of morphology on: age 12, 68; CNO abundances 76, 90; heavy element abundances 10, 15, 48–51, 68, 76, 90; helium abundance 49–50, 69–70, 78
  tabulation of morphological indicators 40–3
  second parameter see second-parameter effect
  for integrated cluster light 170–2, 218–19, 225–6
  in Dsph galaxies 194–7, 207
  for extragalactic globular clusters 169–172, 218–9, 225–6
colours
  of clusters in the Galaxy 27, 29, 34, 227
  of extragalactic clusters 218–19, 225–8
  dependence upon mass function 171–2
  dependence upon composition 171–2
  gradients within clusters 103–5, 109
  correlation with luminosity 170–2, 225–6
colour–colour diagrams
  in reddening determinations 31
  of extragalactic globular clusters 227–8
CO molecular absorption feature
  importance 160–1
  photometry, stellar 94, 147–57, 364–5
  photometry, integrated light 159–61, 167–8
  feature strength variations 149–57
  atmospheric cooling 120
  second-parameter effect see separate entry
  in galaxies 167
completeness of cluster sample 14, 26, 28, 59, 61, 240

composition *see* abundances
concentration 52, 54, 57, 260, 268
concentration class (Shapley) 52, 54, 57
convection *see* mixing
core collapse *see* dynamics
core of galaxy 180
core mass (stellar) and HB morphology 69, 75, 77, 94, 108
core radius (*see also* structure of globular clusters) 52, 54, 56, 260–2

data compilations for galactic globular clusters
  general description 21–63
  absolute magnitudes 26, 28
  abundances 6, 27, 29, 41, 43
  central density (spatial) 52–5
  central surface brightness 52–5
  CMD morphology 40–3
  concentration 52, 54
  concentration class 52, 54
  core radii 52, 54
  distances 13, 27, 29
  distance moduli 26, 28
  escape velocities 53, 55
  helium abundances 40, 42
  HB morphology 41, 43
  identifications 14, 26, 28
  integrated colours 27, 29, 34
  kinematics 5, 13, 26, 28
  limiting (tidal) radii 52, 54
  line strengths 6, 27, 29
  masses 58
  mass functions 8
  mass-to-light ratios 8, 58
  metallicity-indicative parameters 40–3
  metallicities 6, 27, 29, 41, 43
  Morgan metallicity class 40, 42
  Oosterhoff group classifications 27, 29
  photometry 5, 27, 29
  positions 5, 14, 26, 28
  reddenings 5, 27, 29, 31
  relaxation times 53, 55
  resolution 26, 28
  richness 41, 43
  spectral types, integrated 5, 6, 40, 42
  spectrophotometry 6, 27, 29
  star counts 5
  structure 5
  turnoff colour 41, 43
  variable stars 5, 27, 29, 37, 84
  velocity dispersions 53, 55, 58
  velocities 5, 13, 26, 28
  X-ray sources, burst 315–19
  X-ray sources, steady 353
density profiles 249–68
Deutsch class (metallicity indicator) 44
disc population of clusters 44
disruption of clusters
  general 8, 175, 177, 217, 309–13, 363
  cluster shocking 311–13
  core collapse 182
  disc shocking 217, 310
  dynamical friction 310–13
  evaporative 57, 310
  by supernovae 303
  tidal 56, 129, 175–7, 263, 302, 304, 363
  time scales 310–11
distances of globulars
  tabulations 26–9
  methods: luminosity of brightest stars 36, 147; luminosity of HB 35; luminosity of RR Lyrae stars 35, 39; main sequence fits 35; richness index 37; visual resolution 36
double shell source (DSS) stars (*see also* horizontal branch) 24, 68, 126, 133
dust *see* interstellar matter
dwarf spheroidal galaxies
  reviewed 191–210, 364
  abundances 109, 194, 206–10, 304
  colour–magnitude diagrams 194–7, 207
  compared with globulars 177, 179, 191–3, 210, 364
  compared with galaxies 193, 209
  Oosterhoff dichotomy 39, 198–201, 210
  as precursors of galaxies 177–9, 192, 197, 210, 364
  second-parameter effect 194–7, 210
  variable stars 11, 16, 39, 177–8, 197–206, 210
dynamical friction
  and galactic nuclei 9, 187–8, 278
  cluster disruption 310, 312–13, 363
dynamics of globular clusters
  review 8, 51–8, 271–9
  anisotropy 257–8, 273–9, 361–2
  binary stars *see* binary stars
  core collapse 181–2, 292–4, 296–7, 302, 359, 363
  models 273–9, 361–2
  observations 271–3, 361
  simulations 9, 294–7
  star counts 249, 257
ellipticities 109, 150, 178, 251, 258–9, 273, 362
escape
  of stars 53, 55, 57–8, 281, 292, 294–6, 363
  and pollution 131
evaporation *see* disruption
evolution, stellar *see* stellar evolution
evolution, galactic
  importance of globulars 2, 4, 15, 58–9, 110

Subject index 387

evolution (cont.)
  chemical evolution 7, 14, 21, 58, 79, 110, 365
extragalactic distance measurements 213–46
extragalactic globular clusters
  reviews 15, 16, 213–46
  as distance indicators 213–46
  in M31: abundances 46, 107, 159, 163–73, 178–9, 183–7, 233–4, 365; photometry 159, 166, 169–72, 178, 184–5, 241
  in M87: photometry 218–19, 223, 225–9
  in Virgo cluster: photometry 213–17, 229–46
  in Magellanic Clouds: comparison with galactic clusters 3, 4, 15–16, 126, 128, 141, 177–9, 186, 196, 258; young clusters 4, 15, 196, 301–3, 362–3; formation 301, 307
  in other galaxies: composition 186–7, 206

first giant branch (FGB) see red giant branch
formation of clusters 15–16, 44, 59, 109, 129, 176–9, 301–7, 309
formation of galaxy 16, 175–80, 310, 361
formation of stars 130, 172, 179, 272, 303

galactic centre – distance, via globular clusters 13, 219
Galaxy, our own (see also galactic structure)
  formation and evolution 175–88, 192, 306–7, 361
  importance of globulars 175, 306–7
  relationship with Dsph galaxies 192–3, 197
  mass, via globular clusters 14
galaxies (see also extragalactic globular clusters)
  abundances 110, 162, 183, 209, 361
  elliptical, compared with Dsph galaxies 209
  elliptical, compared with globulars 103, 110, 165–9, 180–1
  nuclei 187–8
galactic structure
  importance of globulars 2–4, 13, 44, 58–9, 181–3, 362
  importance of RR Lyrae stars 7, 11, 13, 181–2
  distribution of clusters 44, 59–61, 180–1, 217, 230–1, 244, 302–6
  distribution as function of metallicities 7, 14, 44, 58–60, 128–30, 176–8, 182–3, 305–6

halo of galaxy 61, 180–1, 302, 305–6, 312–13
gas within globular clusters see interstellar matter
giant branch (GB) see red giant branch
globular clusters
  general reviews 1, 2
  historical reviews 1, 3
  importance of studying 2, 175, 213–14, 366
  topical reviews 2, 3

halo of galaxy see galactic structure
halo population of globulars 44
halo stars
  abundances, compared with globular cluster stars 2, 95, 98–9
hard binaries see binary stars
helium
  abundance effects on CMD morphology 7, 49–50, 69–71, 78, 80, 94, 121
  abundance anomalies 121, 137–41
  Oosterhoff dichotomy 78, 83–5
  pulsation theory 7, 81–5
  as second parameter see second-parameter effect
  in stellar evolution 23–4, 65–8, 73, 76–7, 108, 132, 137–41
  in Dsph galaxies 206
horizontal branch (HB) (see also RR Lyrae stars)
  morphology: definitions 23–4, 50–1, 80–1, 195–6; effect of helium 7, 50, 69, 78, 83; effect of age 50, 94, 364; tabulations 41, 43; effect of heavy metal abundances 48, 90, 364; effect of CNO abundances 77, 90, 94, 364
  stellar evolutionary understanding 23–4, 65, 68, 70, 76–81, 94, 107–9, 133–4, 205, 365
  second-parameter effect see separate entry
  Oosterhoff dichotomy 200–1
  in Dsph galaxies 194–5, 200–1, 210
HR diagrams (see also colour–magnitude diagrams) 146, 153
Hubble constant 218, 221, 223, 245–6
Hubble diagram 216
Hubble law in elliptical galaxies 180–1
hydrogen burning 23, 25, 30, 65, 68, 73, 132
hysteresis phenomenon (and Oosterhoff effect) 201–2

infrared photometry
  integrated light of clusters 159–63, 165–72

infrared photometry (*cont.*)
  integrated light of galaxies 162, 165–9
  stellar 143–57
inhomogeneities *see* abundance anomalies
initial mass function *see* mass function
intergalactic globular clusters 59, 177
interstellar absorption, dependence on cluster spectral type 32–3
interstellar matter in clusters (*see also* mass loss)
  dust 9
  expectation, models 9
  searches, evidence 9, 92, 108–9, 365
  gas, neutral 9
  gas, ionized 9, 92, 108–9, 365
  novae 38
  planetary nebulae 9, 38, 133
interstellar matter in Dsph galaxies 205
isochrones 12, 75

Jeans mass (in cluster formation) 305–6

kinematics (*see also* velocities)
  of galactic globular clusters 5, 13–14, 26, 28, 58, 61, 196

limiting (tidal) radius
  data tabulation 52, 54, 56
  physical interpretation 56, 259–62, 268, 362
  cluster orbits 263–4, 362
line-strength parameter
  galactic globulars 6, 27, 29, 165, 183, 185–6
  M31 globulars 6, 165, 183, 185–6, 234
luminosity functions
  of stars in globular clusters 25, 30, 256–8, 304
  radial dependence 104–5, 107–9
  of globular cluster systems 15, 213–14, 216–25, 229–46, 302

Magellanic Clouds (*see also* extragalactic globular clusters)
  variable stars 11, 178
  galaxy formation 177
magnetic fields and abundance anomalies 138–9
main sequence (MS)
  stellar evolutionary understanding 25, 65, 68, 70–2
  sensitivity to abundance 35–6, 71, 73
  photometry and fitting 31, 35
mass of galaxy (via globular clusters) 14
mass functions of cluster populations 311–12
mass functions of stars within clusters
  importance 8, 171–2, 267, 301, 362

galactic 267, 272, 274–7, 301, 362
extragalactic 8, 16
mass loss (*see also* interstellar matter)
  and stellar evolution 9, 12, 23, 70, 76–7, 80, 108–9, 133, 267, 365
  evidence for 9, 92, 108–9, 365
  from cluster as a whole 131, 301, 303–4, 365
mass segregation and cluster dynamics 8, 294, 297, 304, 367
mass transfer *see* binary stars; X-ray burst sources
masses of globulars 58, 265–8
mass-to-light ratios
  dynamical models 265–8, 275–7
  cluster formation 304–6, 313
  in galaxy haloes 313
mergers (in galaxy formation) 176–9
metallicities *see* abundances
mixing
  in stellar evolution 70–1, 77, 84, 87–8, 361, 367
  and abundance anomalies 47, 87–8, 95, 97–8, 137–9, 367
mixing length, effect in models 72–6, 121–2, 147
Morgan metallicity class 40, 42, 44

N-body calculations 9, 295
neutron stars
  cluster models 266–7, 362
  X-ray sources 320, 322–3, 326, 335–6, 339, 344–6, 354
new globular clusters 59
nuclei of galaxies *see* galaxies

observations, need for 77–8, 99, 209
Omega Cen (ω Cen)
  abundance inhomogeneities 7, 88, 103–10, 113–22, 127, 150–7, 179, 366
  abundance gradients 88, 103–10, 366
  the giant branch 90, 104, 106, 113–22, 150–7, 366
  stellar populations within 103–10
Oosterhoff dichotomy
  definitions 27, 29, 39, 200–2
  abundance effects 39, 78, 83, 200–2, 365
  in Dsph galaxies 39, 198–201, 210
open clusters, compared with globulars 59
orbits of globular clusters
  estimates 13, 263–4
  systemic rotation 13–14, 61, 183, 259
origin of globular clusters *see* formation of clusters

peculiar stars (*see also* Zinn effect)
  Ba stars 126–7, 135, 137, 141, 156
  C stars 126–8, 141

peculiar stars (cont.)
  CH stars 47, 88, 127–8, 137
  CN-strong stars 47
  weak G-band stars 47, 94–7
  S stars 136–7
  technetium stars 136
photometry, stellar 31–2, 46–7
photometry, integrated (see also infrared photometry; Q-parameter; $\mathscr{R}$-parameter)
  galactic globulars 27–9, 32–4, 45–7, 184–5, 224
  extragalactic globulars 15, 213–46
photometry, surface see surface photometry
populations, stellar (see also Omega Cen) 4, 22, 103–10, 305, 366
population synthesis, importance of globular clusters 6, 169
positions of galactic globular clusters 5, 26, 28, 59, 60
primordial abundance variations see abundance anomalies
proper motion of cluster stars 9, 96, 99, 362
protoclusters (see also formation of clusters) 176, 304
pulsation theory (see also RR Lyrae stars) 7, 11, 35, 39, 79–85

Q-parameter (metallicity indicator) 32–3, 46
Q_K-parameter see spectrophotometric metallicity indicators
quasars, abundances in 175

$\mathscr{R}$-parameter (metallicity indicator) 33, 184–5, 233–4
rapid burster 322–7
reddenings, galactic globular clusters 5, 27, 29–34, 48
  methods of estimation: stellar photometry 31–2; cosecant law 34; foreground stars 32; integrated spectral types 32; Q-parameter 32; $\mathscr{R}$-parameter 33
reddenings, M31 globular clusters 163
red giant branch (RGB) (see also Omega Cen)
  stellar evolutionary understanding 23–4, 66, 68–70, 73–6, 144–5
  sensitivity to abundances 23, 36, 48–50, 69, 76, 90, 146–7
  abundance anomalies 88, 113–22, 149–57
  studies of 88–100, 113–22, 143–57
  in Dsph galaxies 208
red giant tip 140, 146–7

relaxation times
  data, definitions 53, 55, 57
  abundance gradients 103–4, 109
  dynamics 181–2, 273
resolution of clusters 26, 28, 36–7
richness index 37, 41, 43
rotation of clusters see ellipticities
rotation of cluster system see orbits
rotation, stellar, and abundance anomalies 70, 97–8, 138, 150, 367
RR Lyrae stars
  importance 7, 11, 13, 35, 39, 181–2, 214
  stellar evolutionary understanding 11, 23, 35, 39, 81–2
  pulsation theory 38–9, 79–84, 198
  a, b, c type pulsators 38–9, 51, 81–4, 198
  transition period 39, 81–3, 198
  data compilations 5, 27, 29, 37–8, 84
  and HB morphology 51, 69, 80–2
  Oosterhoff dichotomy 39, 81–3, 198–202
  sensitivity of abundance 35, 37, 39
  in Magellanic Clouds 11
  in Dsph galaxies 198–202, 210

S-parameter (metallicity indicator) 40, 42, 49–50
ΔS-parameter for RR Lyrae stars 47, 88
second-parameter effect
  definition 7, 91–3, 364
  HB morphology 12, 92–4, 364–5
  explanations: age 50, 94, 194, 197, 364, 366; helium 50, 94, 194, 364; CNO 94, 146, 149–50, 156–7, 194, 196–7, 364
  in Dsph galaxies 194–7, 210
soft binaries see binary stars
spectra, integrated 5–6, 40, 42, 44, 46, 185–6
spectrophotometric metallicity indicators
  Δ 27, 29, 45
  ϕ 45
  ψ 45
  Q_K 163–4
spectrophotometry of integrated cluster light 6, 27, 29, 45, 163, 233
s-process elements
  abundance ratios in clusters 91–2, 98–9, 113–22
  abundance ratios in halo stars 98–9
  primordial enhancements in Omega Cen 121–2
  production in stellar evolution 135–7
star counts (see also structure of globular clusters) 5, 56, 249–51, 253, 256–8, 268
stellar atmospheres, modelling in abundance determinations 7, 96, 118–21

stellar evolution
  summary for Population II stars 65–85, 132–4
  comparison of models 12, 22, 70–2, 74, 76, 80
  pulsation theory 11, 79–85
  CMD morphology 22–30
structure of globular clusters (*see also* star counts) 4–5, 21, 51–8, 249–68
subdwarfs 227
sub-giant branch (SGB) stars 30, 49–50
surface brightness *see* surface photometry
surface photometry
  review 249–68
  observations 5, 52–5, 252–6, 260, 262, 264–8, 275–6
  compared with star counts 256
  central luminosity excesses 262–8

temperatures, effective, for red giants 144–5
thermal flashes 133–4, 137
three-body encounters 281–6, 290–1, 297
tidal disruption (*see also* disruption) 193
tidal radius *see* limiting radius
total cluster populations
  in the Galaxy 14, 26, 28, 59, 61, 240
  other galaxies 15, 58, 60, 219, 221–5, 230–3, 245, 312
transition period *see* RR Lyrae stars
transition radius, for onset of anisotropy 274, 276–7
triple systems 284, 288
turnoff 24–5, 36, 41, 43, 50, 70, 75

uniformity
  of galactic globular clusters 3, 259–60
  of cluster populations 2, 6, 15, 163–72, 183–6, 213–14, 217, 219, 225, 228, 233, 240, 245, 309, 312, 366
upper horizontal branch (UHB) stars 30

ΔV-parameter (metallicity indicator) 40, 42, 48
variable stars
  in Dsph galaxies 11, 16, 39, 177, 198–206, 210
  in clusters 3, 5, 11, 27, 29, 37–9, 67, 272

velocities
  of clusters in the Galaxy 5, 13–14, 26, 28, 61
  of stars within clusters: anisotropy 8, 257–8, 263–4, 268, 273–9; observations 53, 55, 58, 113–14, 267, 271–3, 275–8, 293, 302, 305
Virgo cluster (*see also* extragalactic globular clusters)
  distance modulus 213, 221, 240–6
white dwarfs in globular clusters 68, 126, 134, 276–7, 362
W Vir stars
  in clusters 37–8
  in the Galaxy 178
  in other galaxies 177–8, 202–3

X-ray burst sources
  review 315–50
  identifications 328–9, 331
  Type I burst sources 316–19, 321, 327–37, 343
  Type II burst sources 321–8, 343
  Rapid Burster 322–7
  optical bursts 340–3, 346
  radio bursts 341
  infrared bursts 346
  models: accretion in binaries 281, 320, 322, 324, 327, 339–43, 346, 354, 367; black holes 265, 320, 323, 326, 335–6, 346, 352, 354; neutron stars 320, 322–3, 326, 335–6, 339, 345–6, 354, 363, 367; thermonuclear flash 322, 330, 333, 336, 344–5
X-rays from globular clusters – steady sources
  identifications 351, 353, 355
  masses 347, 356–7, 359
  identification with cluster cores 355, 363

Y *see* helium

zero-age horizontal branch (*see also* horizontal branch) 24, 69
zero-age main sequence (*see also* main sequence) 25
Zinn effect (*see also* abundance anomalies) 7, 47, 95–8, 366

RAYMOND H. FOGLER LIBRARY
DATE DUE

BOOKS ARE SUBJECT TO RECALL AFTER TWO WEEKS

AUG 20 1982

MAY 1 1 1984